MODERN METHODS OF FOOD ANALYSIS

ift Basic Symposium Series

Edited by
INSTITUTE OF FOOD TECHNOLOGISTS
221 N. LaSalle St.
Chicago, Illinois

Other Books in This Series

FOOD PROTEINS
Whitaker and Tannenbaum
POSTHARVEST BIOLOGY AND BIOTECHNOLOGY
Hultin and Milner
IMPACT OF TOXICOLOGY ON FOOD PROCESSING
Ayres and Kirschman
FOOD CARBOHYDRATES
Lineback and Inglett
PHYSICAL PROPERTIES OF FOODS
Peleg and Bagley

MODERN METHODS OF FOOD ANALYSIS

Edited by

Kent K. Stewart

Department of Food Science
and Technology
Virginia Polytechnic Institute
and State University
Blacksburg, Virginia

John R. Whitaker

Department of Food Science
and Technology
University of California
Davis, California

AVI PUBLISHING COMPANY, INC.
Westport, Connecticut

Library of Congress Cataloging in Publication Data

Main entry under title:

Modern methods of food analysis.

Papers presented at Symposium on Modern Methods of Food Analysis held June 17 and 18, 1983 in New Orleans; sponsored jointly by the Institute of Food Technologists and the International Union of Food Science and Technology.
Bibliography: p.
Includes index.
1. Food—Analysis—Addresses, essays, lectures.
I. Stewart, Kent K. II. Whitaker, John R. III. Symposium on Modern Methods of Food Analysis (1983 : New Orleans, La.) IV. Institute Food Technologists. V. International Union of Food Science and Technology.
TX541.M56 1984 664'.07 84–14584

DOI 10.1007/978-94-011-7379-7
www.springer.com/mycopy

Contents

Contributors

ALVAREZ, R. (81).[1] Office of Standard Reference Materials, National Bureau of Standards, Gaithersburg, MD 20899

BEECHER, G.R. (29). Nutrient Composition Laboratory, Beltsville Human Nutrition Research Center, U.S. Department of Agriculture, Beltsville, MD 20705

DESSY, R.E. (57). Chemistry Department, Virginia Polytechnic Institute and State University, Blacksburg, VA 24061

HARNLY, J.M. (101). Nutrient Composition Laboratory, Beltsville Human Nutrition Research Center, U.S. Department of Agriculture, Building 161, BARC East, Beltsville, MD 20705

IHNAT, M. (129). Chemistry and Biology Research Institute, Agriculture Canada, Ottawa, Ontario, Canada K1A 0C6

JAY, J.M. (227). Department of Biological Sciences, Wayne State University, Detroit, MI 48202

JENNINGS, W. (319). Department of Food Science and Technology, University of California, Davis, CA 95616

KIRK, J.R. (381). Research and Development, Campbell Soup Company, Camden, NJ 08101

LENTO, H.G. (71). Corporate Analytical Laboratory, Campbell Institute for Research and Technology, Campbell Place, Camden, NJ 08101

MILLER-IHLI, N.J. (101). Nutrient Composition Laboratory, Beltsville Human Nutrition Research Center, U.S. Department of Agriculture, Building 161, BARC East, Beltsville, MD 20705

PANGBORN, R.M. (265). Department of Food Science and Technology, University of California, Davis, CA 95616

PARDUE, H.L. (1). Department of Chemistry, Purdue University, West Lafayette, IN 47907

REINECCIUS, G.A. (293). Department of Food Science and Nutrition, University of Minnesota, St. Paul, MN 55108

SCHWEIGERT, B.S. (xvii). Department of Food Science and Technology, University of California, Davis, CA 95616

STEWART, K.K. (369). Department of Food Science and Technology, Virginia Polytechnic Institute and State University, Blacksburg, Virginia 24061

VANDERSLICE, J.T. (29). Nutrient Composition Laboratory, Beltsville Human Nutrition Research Center, U.S. Department of Agriculture, Washington, DC 20705

[1] Numeral in parentheses indicates the page on which the author's contribution begins.

WHITAKER, J.R. (187). Department of Food Science and Technology, University of California, Davis, CA 95616

WOLF, W.R. (101). Nutrient Composition Laboratory, Beltsville Human Nutrition Research Center, U.S. Department of Agriculture, Building 161, BARC East, Beltsville, MD 20705

ZWEIG, G.[1] (339). Environmental Protection Agency, Office of Pesticide Programs, Washington, DC 20460

[1] Present address: School of Public Health, 322 Warren Hall, University of California, Berkeley, CA 94720

Preface

This Symposium on Modern Methods of Food Analysis was the seventh in a series of basic symposia, begun in 1976, on topics of major importance to food scientists and food technologists. The Symposium, sponsored jointly by the Institute of Food Technologists and the International Union of Food Science and Technology, was held June 17 and 18, 1983, in New Orleans immediately prior to the 43rd annual IFT meeting. Like the other six basic symposia, the program brought together outstanding speakers, from biochemistry, chemistry, food science, microbiology and nutrition, who are at the cutting edge of their specialty, and provided a setting where they could interact with each other and with the participants.

The Symposium and this book are dedicated to the memory of George F. Stewart (1908–1982) who made so many important contributions to the field of food science, including that of food analysis. Bernard S. Schweigert has documented George F. Stewart's contributions in the Dedication of this book.

The field of food analysis touches all of us, whether teacher, scientist, regulator, politician, secretary or consumer. Any time a question of how much of an ingredient or the presence of a compound in food is raised, the answer must always be based upon analysis. For some of us, analysis is our life's work; for others it is only a tool to be used as necessary; for others of us it provides protection and quality assurance. Whatever our role in relation to food analysis, it is important that we all communicate with each other in maximizing the advantages of food analysis.

Selection of Symposia topics and of the Co-Chairs and assistance in planning and executing the program is the responsibility of the Basic Symposium Committee, which includes members from industry, government and academia. The 1983 Basic Symposium Committee members were Darrel E. Goll, chair, Ernest J. Briskey, immediate past chair,

Larry R. Beuchat, John P. Cherry, Richard V. Lechowich, Louis B. Rockland, Richard A. Scanlan and Henry G. Schwartzberg.

The success of the seventh basic symposium was also the result of the expert assistance of Owen Fennema, 1983 President of IFT, Calvert L. Willey, Executive Director of IFT, John B. Klis, Director of Publications, and the IFT staff who provided moral support and publicity and coordinated physical planning including registration, meeting rooms, hotel reservations and the numerous other details of such a two-day symposium.

John Klis coordinated all details of interface with the publisher and Anna May Schenck, JFS Assistant Scientific Editor, served for the seventh time as copy editor for the proceedings. Their capabilities, patience and professionalism in the face of pending deadlines were of immense value.

It is to the authors of the chapters of this book that we owe our deepest gratitude. They heeded the call to teach others—not only at the basic symposium but for many years to come through the written word—the importance of the field of modern food analysis. Their unselfish devotion to knowledge and to the education of others should be an example to all of us.

It is with great humility yet with a strong sense of purpose and pride that we, one the son and the other a junior colleague who learned much through his personal encouragement, join in the dedication of this book to the memory of George F. Stewart.

KENT K. STEWART
JOHN R. WHITAKER

George F. Stewart

Dedication: GEORGE F. STEWART The Man and the Scientist

B.S. Schweigert[1]

It is appropriate to introduce this book with comment on a distinguished colleague Professor George F. Stewart who was keenly interested in methods of analysis as well as in the development of new and modern methods and their applications to food systems.

For perspective, a quote from a resolution adopted by the Executive Committee of the Institute of Food Technologists on March 25, 1982, just a week after his death at age seventy-four, follows:

> George F. Stewart was a man who touched the lives of many in the Institute of Food Technologists, who initiated many of the activities and projects we take for granted. In his roles as a charter member of the society, as executive editor of the IFT journals, as winner of the prestigious International and Appert Awards, as Fellow, and finally as president of the Society, there is hardly an area of IFT which has not felt his guidance and direction.
>
> His concerns were wide-ranging: He was as interested in forming the local IFT section at Ames, Iowa, as he was the International Union of Food Science & Technology; in being an IFT Scientific Lecturer as in helping bring about the First International Congress of Food Science and Technology. He was not narrow in his organizational outlook: He was as at home as a consultant to industry as he was as advisor to many government agencies and committees. He found it as rewarding to practice as to supervise, and was as productive in basic research as in the Experiment Station. He made time for his family, and for an active outdoors life, and truly enjoyed his fellow man.

[1] Department of Food Science and Technology, University of California, Davis, California 95616

Dr. Stewart was recognized as a Fellow by three scientific societies and served as President of the Society of Nutrition Education in its early formative years. See the accompanying tabulation.

GEORGE F. STEWART—AWARDS AND HONORS

Awards	
International, IFT	1968
Nicholas Appert, IFT	1974
Fellow	
American Association for The Advancement of Science	1963
Institute of Food Technologists (IFT)	1971
Poultry Science Association	1949
President	
Institute of Food Technologists	1968
International Union of Food Science & Technology	1970–1974
Society of Nutrition Education	1973

It is also appropriate to provide a few additional highlights on his contributions to teaching, research, and public service. Professor Stewart was active in teaching the introductory course in food science (FS&T 1, Introduction to Food Science), the food packaging course (FS&T 131, Packaging Processed Foods), and after full retirement in 1975, he volunteered to teach FS&T 109, Principles of Quality Assurance in Food Processing that filled a critical teaching need in the Department that of Food Science and Technology, University of California, Davis. This author had the opportunity to work closely with him in the teaching of FS&T 1, and he and another distinguished colleague Professor Maynard Amerine wrote the text published by Academic Press entitled *Introduction to Food Science and Technology*. It is significant to note that the final edited second edition was completed just before Professor Stewart's illness and death.

The following quote from the preface to the second edition illustrates the thinking that he and Professor Amerine provided to readers of the second edition.

> Academic training for technical careers in food science and technology requires a broad, in-depth education both in certain sciences and in selected engineering specialties. It is precisely because of the complex nature of food and its processing and the requirement for a rigorous scientific/technical training that food science and technology offers an exceptional opportunity and a real challenge to the bright applications-oriented science student seeking a rewarding career.

Dr. Stewart's contribution to teaching also included the guidance of graduate students, particularly when he served on the faculty at Iowa State University. A member of the Cooperative Extension Faculty in

Food Science and Technology at the University of California—Davis, Dr. A.W. Brant was one of his graduate students at Iowa State University. Three other faculty members have contributed key treatises to this volume—Professors Walter Jennings, Rose Marie Pangborn, and John Whitaker.

In the area of research and research needs, Professor Stewart was a leader in emphasizing trends occurring in the food and allied industries. This included two relatively new aspects of the interdisciplinary field of food science and technology, namely, food engineering and sensory science. He also highlighted nutrition and food analysis in the address he presented when he was President of the International Union of Food Science & Technology during the Fourth International Congress of Food Science and Technology in 1974. The following is a quote from the paper he developed entitled "Tomorrow's Foods—Obligations and Opportunities for the Food Scientist. Chemical Composition of Processed Foods, Especially Their Nutrient Content."

> We are woefully lacking in reliable data about the nutrient composition of our foods. Equally serious is a lack of sensitive, accurate, and reproducible methods of analysis for nutrients. While many scientists will not find analytical studies very challenging or exciting, it is essential that we obtain reliable information about the nutritional value of tomorrow's food. Someone must address himself to this neglected area of research. Perhaps some of you can be induced to do so.

This leadership is clearly exemplified by this symposium, including the contributions made by his son, Dr. Kent Stewart, the cochairman of this symposium, and head of the Department of Food Science and Technology, Virginia Polytechnic Institute and State University, Blacksburg, Virginia.

Dr. Stewart's leadership and public service have already been referred to with respect to participation and leadership in scientific societies. He also served as Executive Editor of the IFT journals, *Food Technology* and *Journal of Food Science* during the period 1960–1965 and as a coeditor with Academic Press, Inc., for two major publication series: (1) *Advances in Food Research,* and (2) *Monograph Series in Food Science and Technology.* An achievement of major importance was his leadership as a cofounder of *Food Science and Technology Abstracts* in 1969.

Dr. Stewart also led other important developments in public service including working closely with Howard Mattson, Director of Public Information of the Institute of Food Technologists, and in developing food advertising guidelines in a paper entitled "Food Advertising and Promotion—A Plea for Change." This paper is highly recommended.

Dr. Stewart's interest in public service extended beyond the immediate area of professional interest in food science, and he was very active in working with various groups on environmental issues, particularly preservation of wild streams in California and Montana and the protection of habitat for fish and wild birds and other animals. His expertise in this area was increased by his keen interest in fly fishing!

In summary, it is most fitting that this book be dedicated to Professor George F. Stewart in view of his key leadership in emphasizing the basic sciences associated with the interdisciplinary field of food science and technology including food chemistry and even more specifically modern methods of food analysis. His qualities as a person and his perception of the important scientific issues provide the basis for noting further advances in this important field such as those presented in the chapters that follow.

BIBLIOGRAPHY

STEWART, G.F. 1974. Tomorrow's foods—obligations and opportunities for the food scientist, pp. 127–133. *In* Proc. IV Int. Congr. Food Sci. Technol., Vol. 6. Instituto de Agroquimica y Tecnologia de Alimentos, Valencia, Spain.

STEWART, G.F., and AMERINE, M.A. 1982. Introduction to Food Science and Technology, 2nd Edition. Academic Press, New York.

STEWART, G.F., and MATTSON, H. 1978. Food advertising and promotion—A plea for change. Food Tech. *32*(11) 30–33.

1

Systems Approach to Food Analysis[1]

Harry L. Pardue[2]

This chapter addresses two main issues, namely, a problem-oriented role of food analysis and a systematic approach to the analytical process. The chapter is based on two major hypotheses. The first is that the principal role of an analysis is to aid in the resolution of some type of problem. The second is that the total process a food analyst uses to perform this function follows a systematic pattern of thought processes and experimental approaches. Although most food analysts practice their profession this way, they do not incorporate these broader views of analysis into their textbooks, courses, oral presentations, or scientific literature. It is judged to be in the best interests of this profession and its students that a broader, more professional and less technological view of chemical analysis be made an integral part of the educational materials and scientific literature as rapidly as possible. This chapter represents a brief overview of some aspects of this holistic approach to a problem-oriented role of chemical analysis.

RATIONALE FOR PROBLEM-ORIENTED ROLE OF ANALYSIS

The traditional view of food analysis is a process in which a sample is treated in a series of steps to obtain one or more results related to

[1]This work was supported in part by grant No. GM 13326-15 from the NIH, USPHS.
[2]Department of Chemistry, Purdue University, West Lafayette, IN 47907

the composition of the sample. Although this is a very important process, it is only one of several critical steps in the performance of a useful analysis.

If an analyst collects a sample and "analyzes" it without giving any consideration to how the results will be used, then there will be a very low probability that useful information will be obtained. However, if the analyst first considers the problem to be solved, decides what questions need to be answered, identifies the kinds of information needed to answer the questions, selects laboratory methods and procedures that will yield the desired information with needed reliability, and assigns the task of obtaining the desired information to qualified personnel, then it is highly probable that useful information will be obtained. Because the execution of a useful analysis requires careful consideration of the problem to be solved, it is important that planning and data-interpretation steps be considered as much a part of analysis as laboratory experimentation.

Although this is an unconventional idea, it is not without precedence or formal justification. In every other aspect of our lives, when we decide to "analyze a situation," we define the situation in our minds; we decide what information is needed and how it can be obtained; we collect needed information; and we interpret it. This problem-oriented role is inherent in the formal definition of analysis. (Webster's Third New International Dictionary 1961). The association of this problem-oriented role with chemical analysis is not new. Twenty years ago, Hume (1963) stated: ". . . the . . . analytical chemist is one who is a specialist in the methodology of solving problems having to do with properties of chemical systems." More recently, Siggia (1968) stated: "Thus it can be seen that chemical analysis is not concerned with merely detecting or determining a specific component or the general composition of a sample. *It is the resolution or elucidation of a situation.*" Still more recently, Laitinen (1979) stated: "Analysis can now be more accurately described as being applied to a problem rather than to a sample." Lucchesi (1980) has summarized other authors' thoughts on this issue.

Few professional analysts wish to accept samples for processing without also knowing why the results are needed and how they are to be interpreted because that information is required in order to decide how the samples should be processed. Less difficulty would be encountered if the clients also understood this requirement. One way to help future analysts in this regard is to begin now to teach them and their clients that an analysis includes the problem-related thought processes that precede and follow laboratory work. Although this requires a change in some long-established habits, it is in the best interests of the profession and its clients.

RATIONALE FOR SYSTEMS APPROACH TO ANALYSIS

Professional analysts all recognize that a successful analysis, even in its traditional sense, requires the careful integration of a variety of different operations. However, much too often, their textbooks, their oral presentations, and their scientific literature do not reflect such an integrated view of the analytical process; rather, the different steps in an analysis are presented as separate parts that appear to the novice to be selected and used quite independently of one another. This manner of presenting the subject to students and clients fails to give an accurate view of the scientific approach that characterizes all successful analyses. Siggia (1968) has suggested that there is a systematic pattern to handling analytical problems. It is in the best interests of the students, the clients, and the profession that current piecemeal representations of the analytical process be replaced with an integrated approach that more accurately reflects the way the profession is practiced.

Every successful analysis consists of a hierarchical series of functional processes and operational choices that are designed to provide information that will aid in the resolution of a problem. The first level in the hierarchy of functional processes is related primarily to the problem to be resolved. It includes issues such as what hypotheses will be tested, what information is needed to test the hypotheses, what (general) approaches will be used to obtain the information [e.g., library search vs. laboratory work (Siggia 1968)] and how the information will be interpreted. The second level in the hierarchy of functional processes relates to how each function in the first level is implemented. For example, if it is decided that quantitative determinations are to be performed, then the second level of functional processes could include a sampling step, a sample-processing approach (e.g., continuous flow vs. discrete sample), chemical reactions, separations, measurements, data processing, etc. When one has decided which of these functions is needed, then the third and subsequent hierarchical levels should relate to how each function can be implemented (e.g., separation via liquid chromatography or solvent extraction or kinetic vs. equilibrium measurements).

All these ideas will be made more apparent with a specific example.

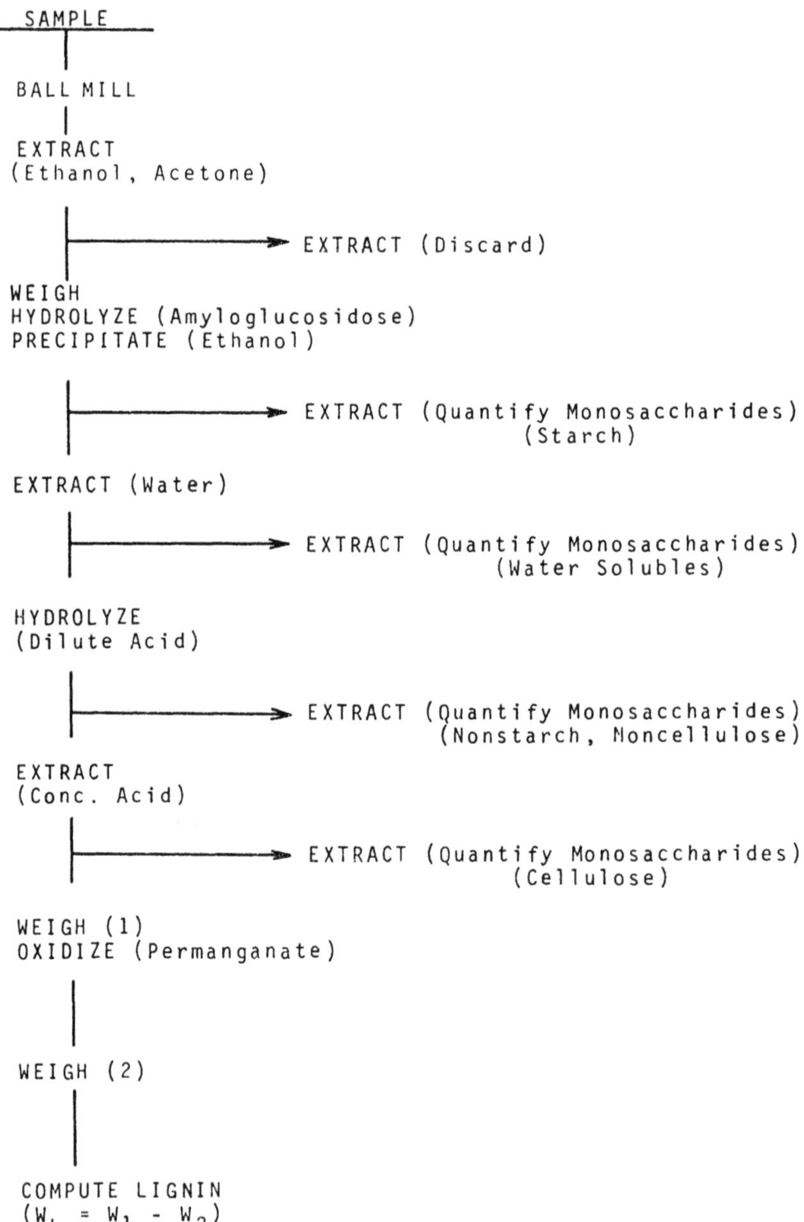

FIG. 1.1. Simplified representation of Southgate procedure used to quantify dietary fiber.

ILLUSTRATIVE EXAMPLE

A recent monograph described a portion of the proceedings of a conference organized to explore possible relationships between cancer and nutrition (James and Theander 1981). The specific example discussed in the monograph involved a cooperative effort among several food scientists to evaluate procedures for the quantitation of dietary fiber in foods. Each participant in the study was given several identical samples (wheat bran, rye flour, rye biscuit, apple pulp, etc.) and a defined protocol based on the Southgate (1969) procedure for dietary fiber. The main features of the protocol are summarized in Fig. 1.1.

The Southgate procedure is based on a definition of dietary fiber as the sum of lignin plus polysaccharides that are not hydrolyzed in the digestive tract. In the prescribed protocol (James and Theander 1981) each participant was instructed to submit the total aliquot of each sample to a ball-milling procedure (Fig. 1.1) and to extract the milled sample with ethanol and acetone. The extract was to be discarded and the residue weighed as total fiber. The remainder of the procedure summarized in Fig. 1.1 is designed to differentiate among the starch, cellulose, noncellulose and nonstarch polysaccharides, and lignin components. At each step in the procedure, monosaccharides in each extract were quantified by colorimetric reactions.

The reader is referred to the original monograph for more detail; the aim here is not to discuss these procedures in detail, but rather to use this study as an example to illustrate the systems approach proposed above. We shall begin with a discussion of the two hierarchical levels of functional processes.

FUNCTIONAL PROCESSES

The first level of functional processes is related to the problem to be solved; the second level is related to sample-oriented operations.

Problem-Oriented Processes (Analysis)

As was mentioned above, the primary problem of interest was the possible relation between nutrition and cancer; the problem addressed in the cited monograph (James and Theander 1981) was the possible relationship between dietary fiber and cancer. Figure 1.2 summarizes the types of thought processes that were necessary before it was useful to implement the protocol for dietary fiber. To un-

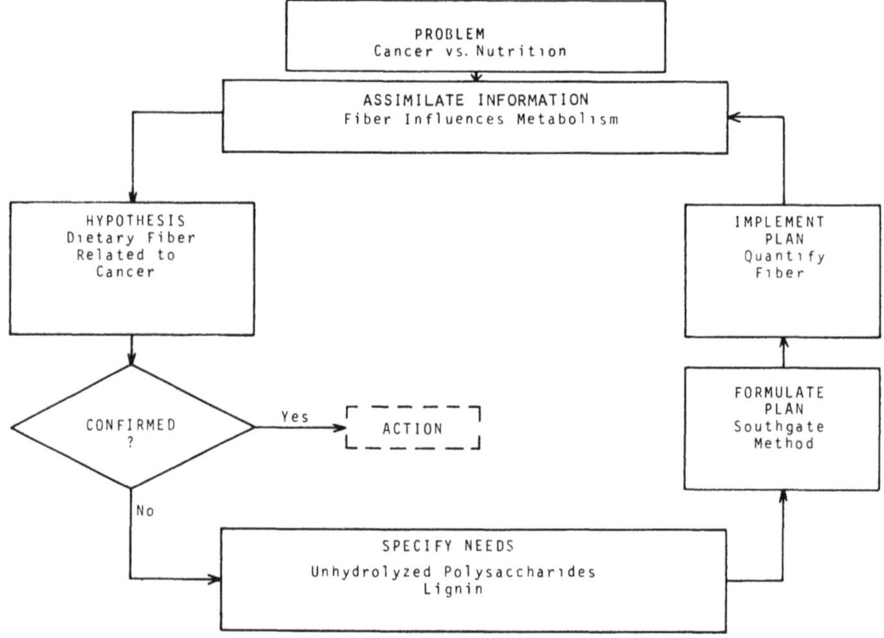

FIG. 1.2. First level of functional processes in a chemical analysis.

derstand Fig. 1.2, one should begin with the block labeled "implement plan" and work clockwise around the cycle.

The block labeled "implement plan" represents the laboratory work that was done to quantify fiber in the food samples. There is no point in beginning such laboratory work without a plan. In this case, the plan was decided upon ("formulate plan") well in advance of the experimental work and involved the Southgate procedure summarized in Fig. 1.1.

Before one can begin to formulate such a plan, one must specify what information is needed. In this case, dietary fiber was defined ("specify needs") as the sum of lignin plus polysaccharides that are not hydrolyzed in the digestive tract. This is the sole basis for the Southgate procedure emphasized in this study.

Before it is useful to try to specify what information is needed, one must establish one or more hypotheses to be tested and decide if additional information is needed to test the hypotheses. In this case, the authors of the monograph justified the study with the "hypothesis" that dietary fiber might be related to cancer.

Before one can formulate a hypothesis, one must assimilate information that is available. In this case, the information used to formulate the hypothesis was quite simply that fiber influences metabolism ("assimilate information") and therefore could be related to cancer.

It should be clear that the value of experimental results obtained in such a study depends upon many factors. It certainly depends upon the care with which experiments are performed; however, it depends on other factors as well. First, if the Southgate procedure (Fig. 1.1) were not sufficiently selective or sensitive to quantify the intended components properly, then regardless of how carefully the experimental work might be performed, it is unlikely that it would yield useful information. Second, even though the Southgate procedure might be designed to quantify exactly what is intended (lignin plus unmetabolized polysaccharides), if it should develop that these were not the most important components of dietary fiber, then again, the laboratory work would be destined to failure before it had begun; regardless of how carefully it might be performed. Third, even though the components quantified in the Southgate procedure might represent accurately the important components of dietary fiber, a poorly formulated hypothesis could negate the value of such a study. For example, suppose such a study is so tightly focused on a particular type of cancer for which there is no connection with dietary fiber that it misses a correlation between fiber and another type of cancer; in such a case, very carefully performed laboratory work could yield potentially misleading conclusions.

This discussion should not be construed in any way as a criticism of the studies to which it relates (James and Theander 1981; Southgate 1969); rather, it is intended to reflect the careful planning that must be done to ensure that such a study will yield useful information. The principal point is that the laboratory work represented by the chart in Fig. 1.1 is just one of several important steps in the analytical process, whatever the situation might be. One might need to quantify the composition of a food to satisfy FDA regulations, to detect a potentially toxic substance, to compare products from different regions or suppliers, to control quality, or for a variety of other reasons. Whatever the reason, laboratory work performed without benefit of the other steps in Fig. 1.2 is unlikely to yield useful information.

Clearly, the functional steps in Fig. 1.2 are all integral and inseparable parts of one larger process, the success of which can be no better than limitations imposed by the weakest link. This author views the steps in Fig. 1.2 as the first level of functional processes in chemical analysis. However, whatever we may choose to call this process, it is important that our students and clients for analytical services recognize that all the steps in Fig. 1.2 are critical and inseparable parts of the analytical process.

Each step in Fig. 1.2 can be subdivided into at least one additional level of functional processes. An example will illustrate the point.

Siggia (1968) has suggested that a chemical analysis may not involve any laboratory work at all; rather the information needed may be more conveniently found in the library. Consider, for example, a situation in which one manufacturer may wish some specific information related to a competitor's product. If the desired information were detailed in a patent for the product, then it could be much faster and less expensive to consult the patent than to set up and perform laboratory experiments. Thus, two functional sublevels of the planning step in Fig. 1.2 (block labeled "formulate plan") would be library search vs. laboratory experimentation. The other steps also include different functional options. The implementation step ("implement plan" in Fig. 1.2) is discussed in detail in the next subsection as an illustrative example.

Sample-Oriented Step (Determinations)

As noted above, the plan to be implemented may involve either library or laboratory work, or a combination of both. Although both these options could be treated in a systematic fashion, it probably is more interesting to address the situation in which one or more components are to be quantified in the food of interest. The procedures summarized in Fig. 1.1 for dietary fiber in food will be used as an illustrative example.

Each participant in the study was supplied with a carefully prepared sample of each material (wheat bran, rye flour, rye biscuit, apple pulp, etc.) and instructed to use the entire sample by initially subjecting it to a ball-milling process and extracting the milled sample with ethanol followed by acetone. The extract was discarded and the residue was dried and treated as the total dietary fiber. The dried residue was to be weighed and hydrolyzed with amyloglucosidase. The resulting mixture was treated with ethanol; the supernatant solution was extracted and saved; and the residue was treated in a continuing series of steps as indicated in Fig. 1.1.

At the end of this series of steps, there were four extracts and a residue. Each extract was subjected to chemical reactions designed to generate colored products representative of monosaccharides that were present, and the colored products were quantified by their absorbances at appropriate wavelengths and related to the amounts of monosaccharides in the extracts. The amounts of monosaccharides were in turn related to the amounts of starch, unspecified water-soluble materials from the first hydrolysis step, nonstarch and noncellulose polysaccharides, and cellulose. The residue was weighed ("weigh 1") and treated with permanganate to oxidize and solubilize lignin, and the mass of the final residue ("weigh 2") was subtracted to quantify the amount of lignin by difference ($W_L = W_1 - W_2$ in Fig. 1.1). More complete details of the process are given in the monograph (James and Theander 1981); the objective here is to use the specific steps in this example to develop a *general pattern* that is applicable to this and other determinations.

Figure 1.3 summarizes the functional operations that were involved in this procedure. The first step was to provide each participant with a *representative sample*. Each participant was also given careful instructions on *sample processing* procedures, which in this case were to be performed manually. The ball-milling procedure produced a *physical change* in the sample, and the initial and subsequent extractions represented *separation* steps. The hydrolysis steps, the steps in which colored products were formed, and the permanganate oxidation all represented *chemical reactions*. The weighing and colorimetric procedures all represent *measurement* steps. Computations of concentration from absorbances and of the mass of lignin as the difference $W_L = W_1 - W_2$ represent *data processing* steps. *Statistical treatments* applied to data obtained by different participants in the study were discussed in the monograph.

The functional steps summarized in Fig. 1.3 are general, and two or more of them are involved in all determinations. For example, all determinations include some type of sampling step and some type of measurement step; these steps may be combined in some specialized situations, but both are present in all determinations. Most determinations will require some type of data processing step to convert the measured signal to the quantity of interest (e.g., mass, concentration, catalytic activity). The other processes (chemical reactions, physical changes, and separations) may or may not be required.

The sampling step merits separate discussion, because it is too often ignored. The primary purpose of the dietary fiber study was to evaluate the performance of the Southgate procedure by comparing results obtained by different investigators for the same food. Clearly, if samples with different compositions had been distributed to different investigators, any attempts to compare results would have been futile, or worse, would have led to improper conclusions regarding the

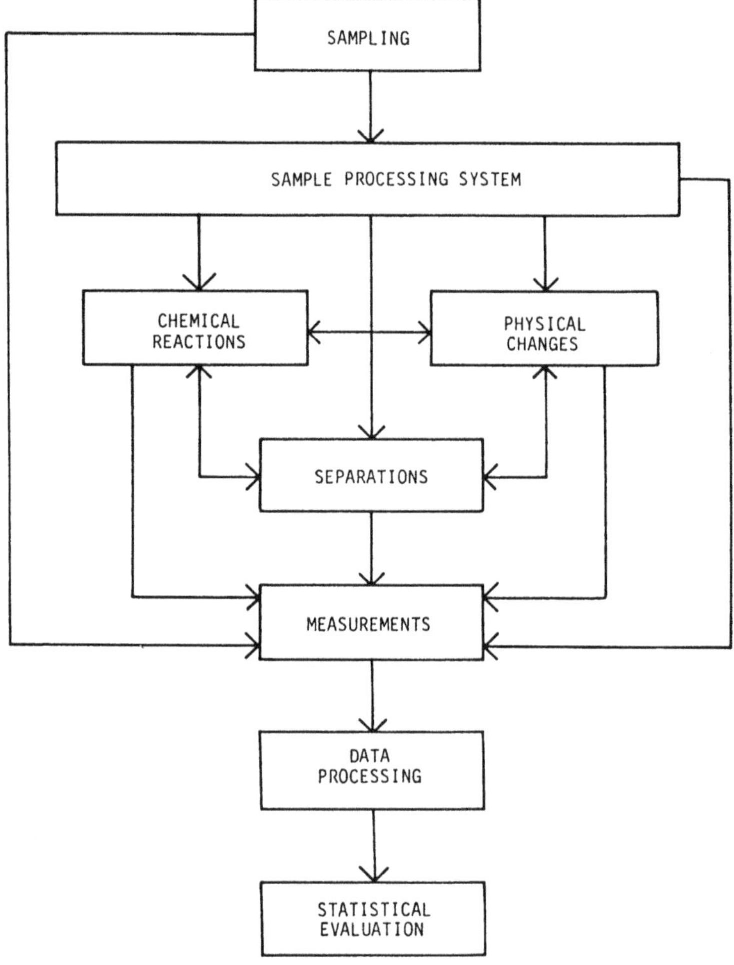

FIG. 1.3. Functional processes in a quantitative chemical determination.

Southgate procedure. Similarly, if a manufacturer is evaluating the nutritional content of a raw material or a food product, it is essential that a *representative sample* be obtained. *The sampling step is a critical part of all determinations, and we should eliminate all habitual usages that imply that it is not.* When we suggest that we "analyze a sample" or "determine a component in a sample," we are implying that the analysis or determination begins *after* the sample is collected, which is not correct. *Determinations should relate to the system that is represented by a sample rather than to the sample itself.* This and other points are discussed in more detail later.

Attention is now turned to operational choices that analysts can make in setting up procedures.

OPERATIONAL CHOICES

When an analyst has decided *what* functions (Fig. 1.3) are needed in a determination, he/she must then decide *how* those functions will be implemented. For each functional step in Fig. 1.3, there are many different operational choices. One can approach these as a random collection of choices, or one can attempt to organize them into a systematic pattern. For example, before designing a sampling procedure, one must decide what type of information is needed. In the dietary fiber example cited above, since the purpose was to compare results on as nearly identical samples as possible, it was very important that each participant receive a sample that was representative of the total batch of food. A similar sampling procedure would be used to quantify the nutritional content of batches of food products manufactured for public consumption. On the other hand, if one wanted to compare the dietary fiber contents of different parts of wheat kernels, then one would want samples taken from carefully selected parts of the grain rather than a sample that would be representative of the total kernel. We could identify these two general sampling approaches as related to *bulk properties* and *localized properties*. Sampling procedures would be very different for these different goals. Other considerations that can influence the sampling approach are invasive vs. noninvasive and static vs. time-dependent sampling. With some effort, the operational choices for all the functions of Fig. 1.3 can be organized into systematic patterns analogous to that implied here for the sampling step, and two of the functions—sample-processing systems and the measurement step—are discussed in some detail below.

Sample-Processing System

The function of a sample-processing system is to do whatever is necessary to or with a sample to obtain the desired information. As indicated in Fig. 1.3, the sample-processing system may implement any or all of the chemical reaction, physical change, separation, and measurement steps to obtain data that can be processed to yield the desired information.

Figure 1.4 summarizes some of the considerations that are involved in selecting a sample-processing strategy. Samples of interest may be either solid, liquid, or gas (e.g., flavor components), and the phase will definitely influence the early stages of the sample-processing strategy. Whatever the phase, one must decide if samples are to be processed manually, semiautomatically, or with fully automated instrumentation. In the dietary fiber example, all procedures were implemented manually. However, there has been significant progress in recent years on semiautomated instrumentation, but, despite claims to the contrary, there are very few fully automated systems that do not require some human intervention at some point. Samples can be transported through the system as discrete aliquots or in fluid flow streams, and they can be processed in discrete steps (e.g., stopped-flow) or with continuous movement. Systems that operate in discrete steps offer greater flexibility in the control of reaction times but continuous-movement systems are simpler to implement. Discrete-batch sample processors involve fully segmented (segregated) sample aliquots; however, fluid-flow systems may involve either segmented or unsegmented flow streams. As examples, the flow systems marketed by Technicon use air bubbles to segment sample aliquots (Snyder 1980), but flow-injection sample processors use unsegmented streams (Ruzicka and Hansen 1980). Similarly, solvent-extraction methods [e.g., Craig machine (Peters *et al.* 1974)] use segmented aliquots, but gas-chromatographic and liquid-chromatographic separations involve unsegmented streams.

This distinction between segmented and unsegmented streams, although simple in concept, has some very important implications. In a segmented system (discrete batch or flow stream), all components in each segregated aliquot can be permitted to reach both chemical and physical (e.g., mixing) equilibrium prior to the measurement step if it is desirable to do so; alternatively, measurements can be made during the kinetic phase of either chemical or physical processes if it is desirable. However, with unsegmented systems (e.g., flow-injection sample processors, liquid-chromatographic separations) one has no option. Measurements must be made during the kinetic phase of the distribution (dispersion) process because at equilibrium, all components in all sample aliquots would be spread uniformly throughout the entire fluid stream.

Returning to Fig. 1.4, samples can be processed sequentially as in the Technicon and flow-injections sample processors (Snyder 1980; Ruzicka and Hansen 1980) or simultaneously (parallel) as in the centrifugal sample processors (Scott and Burtis 1973). Similarly, differ-

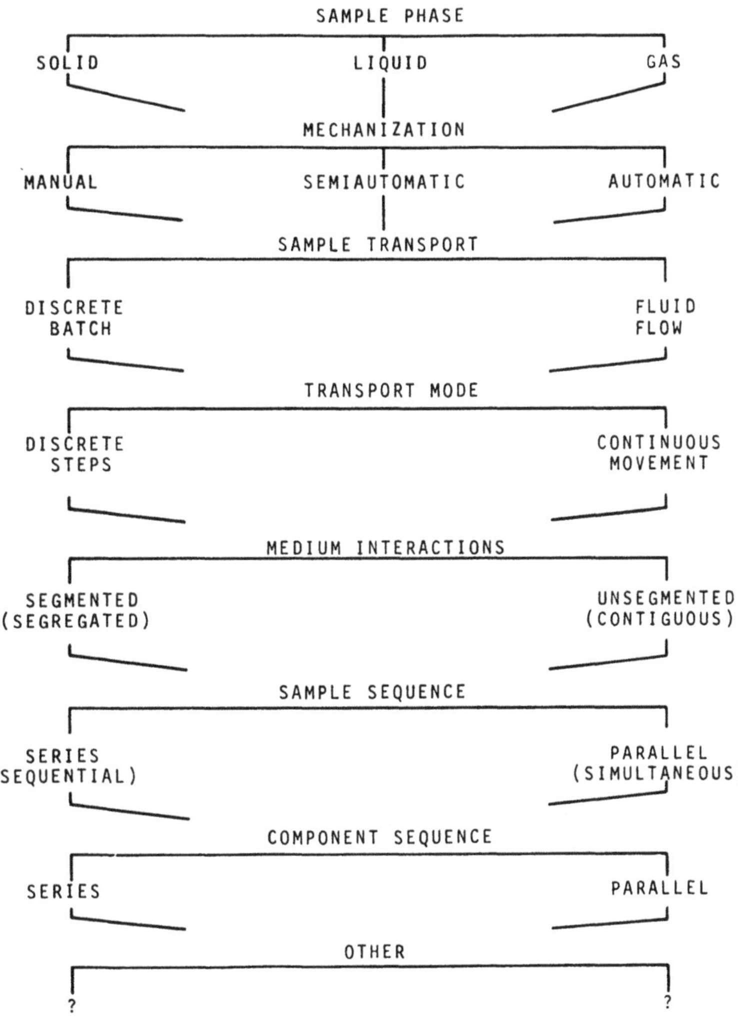

FIG. 1.4. Operational choices for the sample-processing step in a chemical determination.

ent components in samples can be quantified sequentially or simultaneously.

There are other features that can be used to differentiate among sample-processing systems; however, the objective here is not an exhaustive treatment but rather a demonstration that there is a systematic pattern in the design or selection of sample processors. Careful use of such systematic classifications helps to avoid a dichotomy such as that represented by current usage that identifies one group of flow processors as "continuous-flow" and another as "flow-injection" when in fact both are continuous-flow systems and the real difference is that one (Technicon) involves segmented streams and the other (flow injection) involves unsegmented streams. Some of the features mentioned above will be discussed in more detail in a later section; attention is now turned to the measurement step.

Measurement Step

In virtually all qualitative and quantitative determinations, some physical property such as electrode potential, electrolysis current, light absorption, or light emission is measured and related to the chemical property of interest. As indicated in Fig. 1.5, some of the earliest decisions one must make are what properties to measure and what detectors to use. One must also consider the nature of the process to be monitored, whether it is purely chemical, physicochemical (e.g., reaction in a flow stream), or purely physical (e.g., viscosity).

Another important consideration is the time at which measurements are made. As illustrated in Fig. 1.6, any chemical or physical process has two general regions, namely, a *kinetic region* during which the system changes with time and an *equilibrium region* during which there is no change with time. Accordingly, measurement approaches can be grouped into the two general categories of *kinetic methods* and *equilibrium methods* depending on the point in time at which measurements are made relative to the point at which *all* processes that influence the measurement have reached equilibrium. Either chemical or physical processes (e.g., dispersion in flow streams) can impose kinetic character on the measurement step. These points are discussed in more detail in a later section.

Another important consideration in selecting a measurement strategy is the *measurement objective* (Somogyi 1938). The measurement objective is the quantity that is measured and related to sample concentration or another property of interest. The measurement objective may be detector signal (e.g., electrode potential or phototube current), or it may be some quantity other than detector signal such as the volume of titrant or reaction time. In the first case, detector signal is used directly to compute concentration or other desired information; in the second case, detector signal is used only as an indicator that some desired event has taken place (e.g., end point in a titration) and some

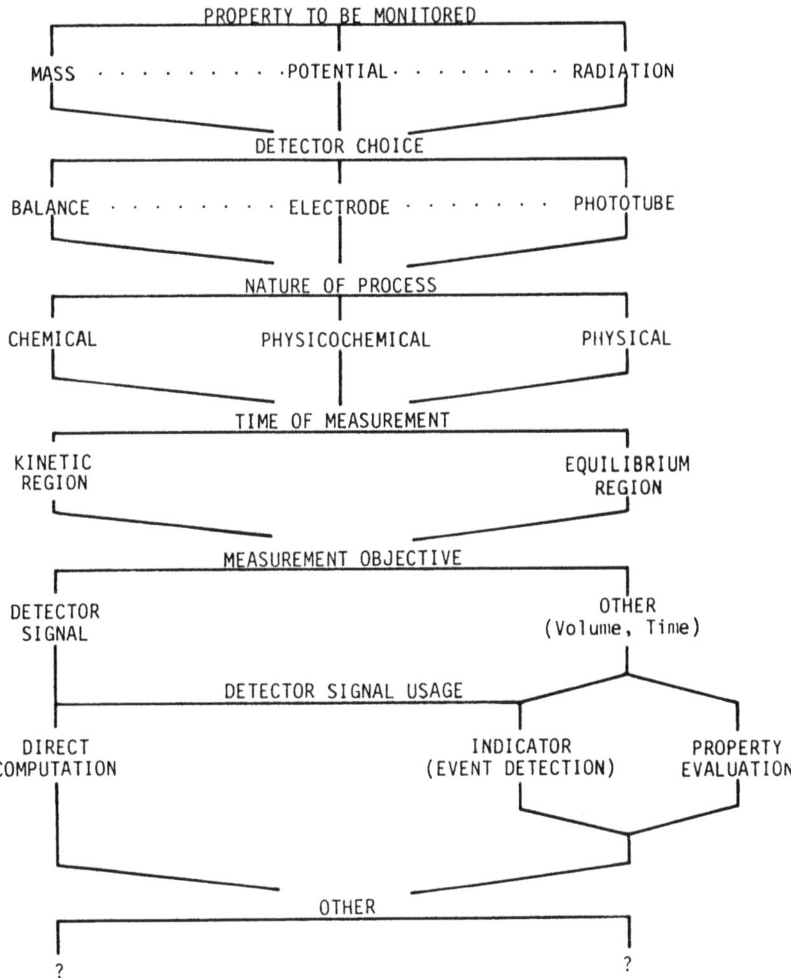

FIG. 1.5. Operational choices for the measurement step in a chemical determination.

other quantity (e.g., titration volume) is measured and used to compute whatever property (e.g., concentration) is desired.

There are other features such as single-dimensional vs. multidimensional data that could be included here; however, the features described above will suffice to illustrate that the measurement step can be represented by a systematic pattern of options.

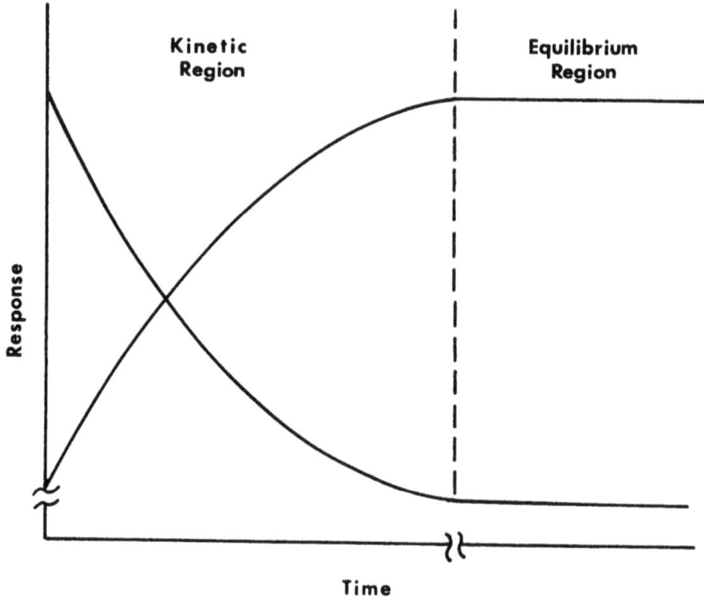

FIG. 1.6. Illustration of kinetic and equilibrium regions in chemical or physical processes.

In the following, selected examples are used to show how these principles can be used to correlate seemingly different methods.

Examples

Colorimetry is used as a basis for this discussion because this is one of the more commonly used and better understood measurement approaches. First two classical applications of colorimetric measurements are given and then other more recently developed approaches are presented.

One common approach is to measure the absorbance A of a colored solution and to compute the concentration of the colored species with the Beer-Lambert law ($C = A/ab$); a second approach is to use a color change to detect the end point in a titration and to use the volume of titrant to compute concentration ($C_s = V_t C_t / V_s$, where subscripts s and t refer to sample and titrant). In the first case, the measured signal A is used directly in the computation of concentration (see Fig. 1.5); in the second case, the measured signal A is used only as an indicator that the end point has been reached and titrant volume (V_t) is used to compute concentration. This is an important distinction. In the first case, it is necessary to calibrate the measured signal in concentration units while in the second case this calibration is not necessary; rather all that is needed is a detectable change. This is the reason that the human eye is a very good detector for titrations but the human eye is a very poor detector for procedures in which concentrations are computed from the actual absorbance value. The eye can detect color changes easily but it can not quantify color intensity accurately.

Both of the above measurement approaches are usually performed under equilibrium conditions. There are analogous options for kinetic measurements.

Blaedel and Hicks identified two groups of kinetic methods that they called *fixed-time* and variable-time methods (Pardue 1977). Principles involved in the two approaches are illustrated in Fig. 1.7. In the fixed-time method (Fig. 1.7A), one measures the signal change ΔA during a fixed period of time and uses that signal change to compute concentrations (usually $C \propto \Delta A$). In the variable-time method (Fig. 1.7B), one measures the time interval Δt required for the signal to change by a fixed amount, and computes concentration from the measured time interval (usually $C \propto 1/\Delta t$). Notice the analogy with the classical colorimetric and titration methods described above. In the fixed-time kinetic method and the classical colorimetric method, the signal A or ΔA is measured and used to compute concentration; in the variable-time kinetic method and the titration method, the signal is used *only as an indicator* (see Fig. 1.5) that a change has occurred, and another quantity (V_t or t) is measured and used to compute concentration. In the first two cases, the detector signal must be calibrated in concentration units; in the second two methods, the detector signal need not be calibrated because concentration is not computed with the signal. The human eye could be (and has been) used as a satisfactory detector for variable-time kinetic methods but would be a poor detector for fixed-time kinetic methods. Thus, this systematic approach helps to correlate characteristics of seemingly unrelated measurement approaches.

In each of the above cases, it was assumed implicitly that the signal change occurred from chemical reactions. The same principles and correlations apply for purely physical and combined physicochemical processes, and two examples are used to illustrate the point.

Viscosity must be determined for some foods. One way to determine viscosity involves the rate at which a spherical ball falls through a fluid (Siggia 1968). One approach is to measure the distance the ball

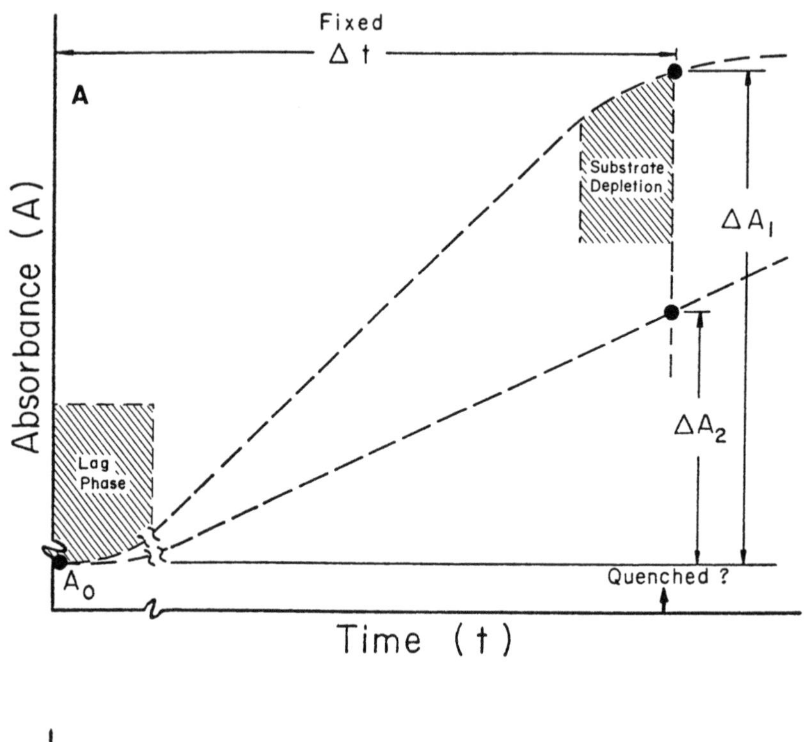

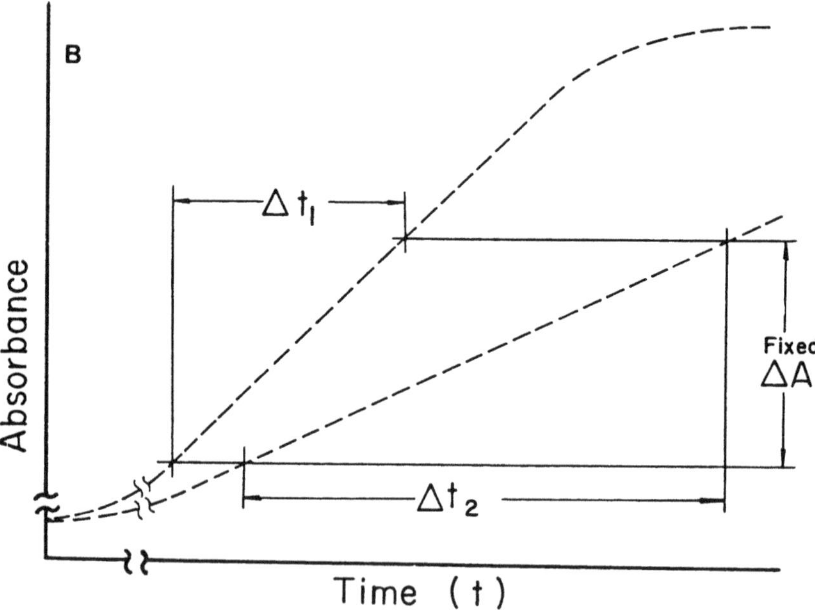

FIG. 1.7. Two-point kinetic measurement approaches: (A) Fixed-time method; (B) Variable-time method.

falls during a fixed-time interval and another is to measure the time required for the ball to fall a fixed distance. The latter approach is usually used because it is easier to measure time precisely than to measure the distance the ball falls in a fixed time. Clearly, these are examples of fixed-time and variable-time kinetic approaches applied to a purely physical process. The same principles discussed above for chemical processes also apply for purely physical processes.

A second and perhaps more interesting example involves flow-injection sample processors (Ruzicka and Hansen 1980). In the most common approach to flow-injection sample processing (FISP), an aliquot of fluid sample is introduced into a flowing stream where it mixes and reacts with reagents in the stream as it moves toward a detector. When the sample aliquot reaches the detector, a signal (e.g., absorbance) is measured, and the amplitude of the signal is used to compute concentration.

Figure 1.8A shows how the signal for a colored species changes as it moves along the flow stream. Figure 1.8B shows the same signal profiles with a smooth curve drawn to connect the peak heights. The reader should note that this curve decreases continually as the sample progresses along the flow stream. Clearly, because the signal changes with time, this is a kinetic process. What is usually done is to measure the signal at a carefully controlled, fixed time after the

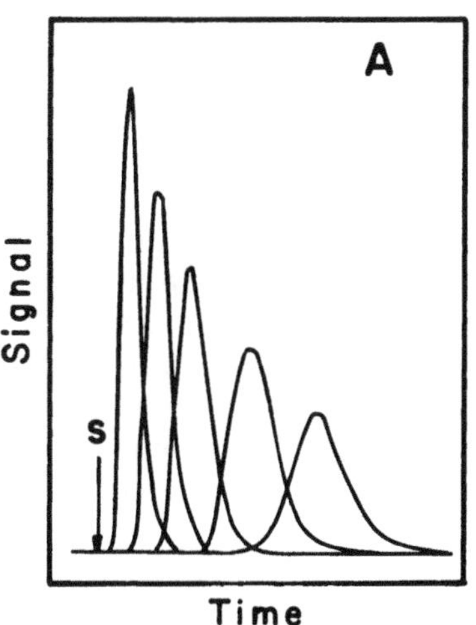

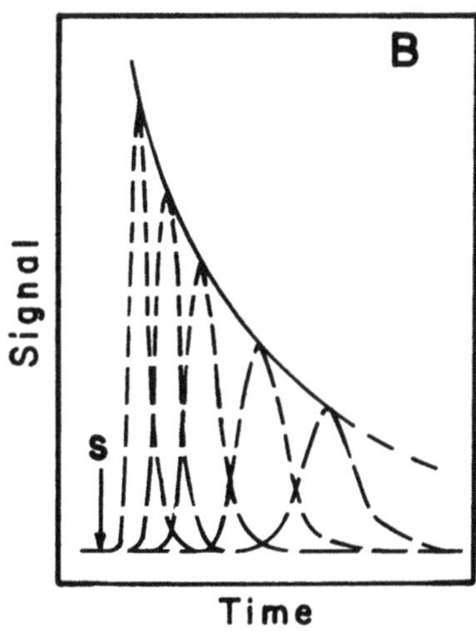

FIG. 1.8. Effects of time or length of flow stream on dispersion in a flow-injection sample processor: (A) Simulated peaks; (B) Time-dependent character of peak heights.

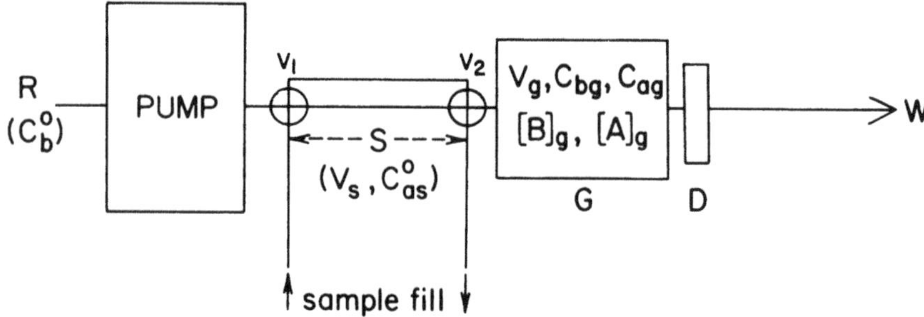

FIG. 1.9. Schematic diagram of flow system with gradient chamber.
From Pardue and Fields (1981).

sample is introduced into the flow stream. Therefore, *the most common approach to flow-injection sample processing is properly classified as a fixed-time kinetic method.* The same is true of chromatographic methods in which the signal is measured after a fixed time; these are fixed-time kinetic methods.

Figure 1.9 illustrates another approach to FISP (Ruzicka and Hansen 1980). In this approach, the sample aliquot moves quickly into a gradient chamber G where it is diluted and reacts with a reagent of concentration, C_b^o, as it flows into the gradient chamber. This approach was first implemented with acid–base reactions (Ruzicka and Hansen 1975). In the initial (Ruzicka and Hansen 1975) and subsequent studies (Pardue and Fields 1981) of this approach to FISP, samples of hydrochloric acid were injected into a flow stream that contained a fixed concentration C_b^o of sodium hydroxide, and a colored indicator was used to monitor the reaction. The sample aliquot was injected into the flow stream so that it moved quickly into the gradient chamber where the HCl reacted with NaOH in the flow stream as it passed into the gradient chamber. The acid–base indicator exhibited two abrupt changes during the experiment: (1) when the HCl sample first entered the gradient chamber to make the solution acidic and (2) when all the HCl disappeared from the chamber and the solution became basic due to excess NaOH flowing into the chamber.

Figure 1.10A illustrates a response curve (signal vs. time) similar to those reported in the original work (Ruzicka and Hansen 1980). The data in Fig. 1.10A were obtained with an oxidation-reduction reaction between triiodide (I_3^-) and thiosulfate ($S_2O_3^{2-}$). In this experiment, a triiodide sample was injected into a flow stream that contained thiosulfate. The aliquot of triiodide solution moved quickly into the gra-

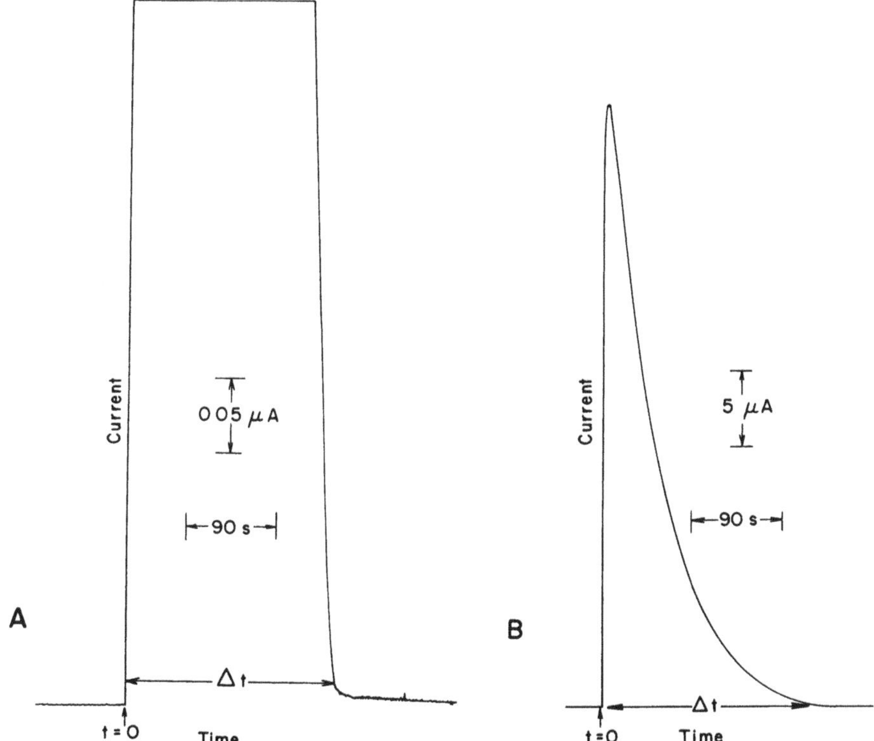

FIG. 1.10. Effects of recorder sensitivity on shapes of response curves for thisulfate/iodine reaction in a flow system with gradient chamber: (A) 0.5 μA full scale; (B) 50 μA full scale. $C^0_{I_3^-} = 2$ m*M*, $C^0_{S_2O_3^{2-}} = 10$ μ*M*, flow rate = 33.3 μL/s, chart speed = 0.33 mm/s.

dient chamber where it reacted with thiosulfate in the flow stream as follows

$$I_3^- + 2S_2O_3^{2-} \rightarrow 3I^- + S_4O_6^{2-}$$

The process was monitored by the electrolysis current for the electrochemical reduction of triiodide ($I_3^- + 2e \rightarrow 3I^-$) at a platinum electrode in the flow stream immediately following the gradient chamber.

For the data in Fig. 1.10A, the recorder sensitivity was set very high (0.5 μA full scale) so that the first small increments of I_3^- to enter the gradient chamber caused the pen to move off scale rapidly and to remain off scale until the last traces of I_3^- had disappeared from the gradient chamber, at which point the pen returned to the baseline rapidly. This behavior is completely analogous to the abrupt change in color of an acid–base indicator as used in the initial and subsequent studies of this method (Ruzicka and Hansen 1975; Pardue and Fields 1981).

Because there is some correlation between sharp changes in signal amplitude shown in Fig. 1.10A and the sharp changes observed in a titration, and because the reagent in the flow stream reacts with the component of interest in the sample, this procedure was originally identified as a titration. However, there is a major problem with this classification. The problem involves the equivalency relationship that is involved in titrations. By definition (Irving *et al.* 1978), the equivalence point in a titration corresponds to the point at which all the titrant added to a sample aliquot has reacted with all the titrant in that aliquot. The experiment described above does not satisfy this critical feature of a titration.

Because the gradient chamber has a fixed volume, as reagent flows into the chamber some of the solution must flow out of the chamber to make room for it. That portion of the sample that flows out of the chamber cannot react with reagent in the flow stream, and, therefore, the method cannot qualify as a titration because the amount of reagent that reacts is not equivalent to the amount of species of interest in the sample aliquot.

If the method is not properly classified as a titration, then how is it best classified? Stewart (1983) has noted that the method does not satisfy the equivalency criterion of titrations and accordingly has identified it as a *pseudo-titration*. However, there is a more exact classification that is readily illustrated with the aid of Fig. 1.10B. Results in Fig. 1.10B were obtained for exactly the same conditions as those in Fig. 1.10A except that the recorder sensitivity was changed (50 μA full scale) so that time-dependent changes that occur during the process are made more apparent than in Fig. 1.10A and in an earlier work (Ruzicka and Hansen 1975). It is quite clear from the response curve in Fig. 1.10B that this is a kinetic process because the signal changes gradually with time. Exactly the same kinetic processes were occurring in all other experiments of this type; the only difference is that the detector systems in Fig. 1.10A and earlier reports (Ruzicka and Hansen 1980, 1975) were not designed to detect the time-dependent character of the processes. In other words, the mode of detection tended to mask the true kinetic character of the process.

The quantity that was measured and related to analyte concentration was the time interval, Δt (see Figs. 1.10A and B), required for the detector signal to rise above and fall below a predetermined point. That time interval Δt varies with concentration, being larger for high concentrations and smaller for low concentrations. This is completely analogous to the purely chemical processes illustrated in Fig. 1.7B. That procedure was identified as a variable-time kinetic method and the method being discussed here is also properly classified as a *variable-time kinetic method* rather than as a titration as was suggested initially.

At this point the reader is probably saying "so what?"; what possible difference can this make? It will be shown that it makes a lot of difference.

In the original study (Ruzicka and Hansen 1975), a titration model was used to derive an equation of the form

$$\Delta t = \frac{V_g}{f} \ln C^0_{as} - \frac{V_g}{f} \ln C^0_b + \ln \frac{V_s}{V_g} \tag{1}$$

to relate the measured time interval Δt to initial analyte concentrations C^0_{as}. In this equation, V_g is the volume of the gradient chamber, f is the flow rate, C^0_b is the initial concentration of reactant in the flow stream, and V_s is the volume of sample injected. In a later study (Pardue and Fields 1981), the variable-time kinetic model was used to develop an equation of the form

$$\Delta t_{ep} = (V_g/f)\ln\left\{C^0_{as}\,\frac{C^0_{as}[\exp(V_s/V_g)](zC^0_b+1)-C^0_{as}-C^0_{bg}}{(C^0_{as}+zC^0_{bg})}\right\} \tag{2}$$
$$-(V_g/f)\ln\{[A]^{ep}_g+zC^0_b\}$$

in which C^0_{bg} is the initial concentration of reactant in the flow chamber, $[A]^{ep}_g$ is the concentration of the species being detected at the end point, and other symbols are as defined above.

Comparing Eqs. (1) and (2), it is clear that the titration model and variable-time kinetic model lead to very different mathematical descriptions of the process, emphasizing the importance of selecting the correct model. Data and calculated curves summarized in Fig. 1.11 are aids for deciding which is the better model. In the figure, the circles and squares represent experimental data obtained for different conditions; the bold lines represent results predicted for those conditions with Eq. (1) (titration model); and the dashed curves represent results predicted with Eq. (2) (variable-time kinetic model). The variable-time kinetic model satisfied all conditions examined (Pardue and Field 1981). The titration model did not satisfy any condition examined, and errors were very large in some cases (e.g., curves 2, 3, and 6).

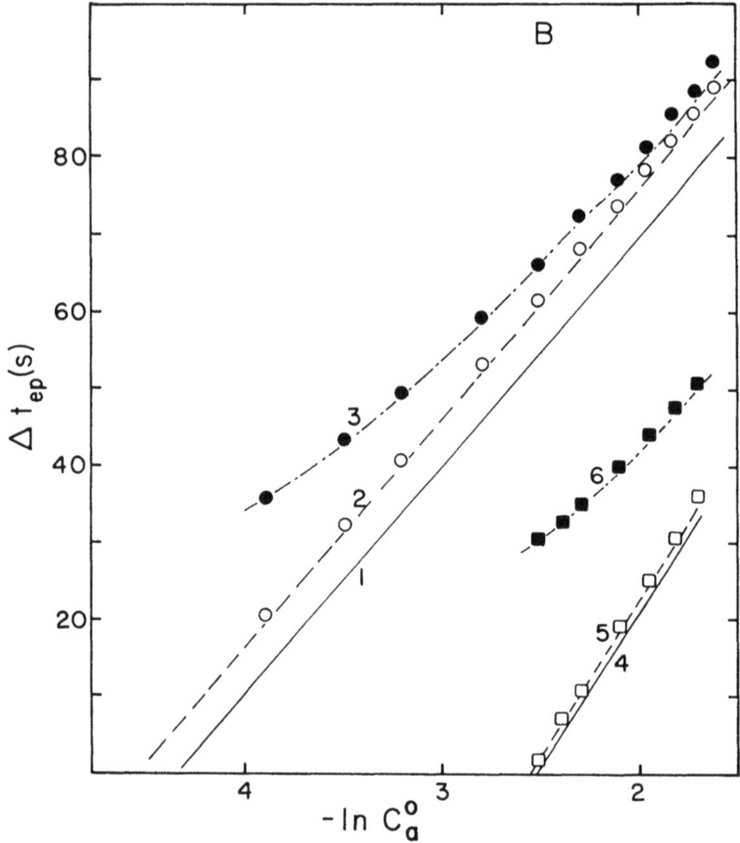

FIG. 1.11. Calibration plots for flow-injection experiment with gradient chamber. Experimental: ○, ●, □, ■; computed with equation from titration model: —; computed with equation from kinetic model: – – and – - –. For more detail see Pardue and Fields (1981).

This example emphasized the importance of a systematic classification scheme that can help to avoid both conceptual and quantitative errors such as those illustrated above. Other similar situations in which misclassifications of quantitative procedures led to less than optimum performances have been discussed (Pardue and Fields 1981).

Returning now to the broader picture, it is apparent that flow-injection sample-processors can be classified as either fixed-time or variable-time kinetic methods. In the fixed-time FISP approach described first, detector signal is measured and used to compute concentration;

in the variable-time method (Figs. 1.9 and 1.10), detector signal is used only as an indicator that predetermined events have occurred, and another quantity (Δt) is measured and related to concentration. In the first case, it is necessary to measure the detector signal accurately and to have it calibrated in concentration units; in the second case, it is only necessary to detect a signal change and it is not necessary to have the signal calibrated in concentration units.

Thus, there is a common pattern that relates purely chemical (Fig. 1.7), purely physical (viscosity), and combined physicochemical (Figs. 1.8–1.10) processes; and that pattern applies to both equilibrium and kinetic measurement approaches.

The detailed discussion focused here on selected aspects of the sample-processing measurement steps could be extended to other aspects of these steps and to other steps in the quantitative process (Fig. 1.5). However, the objective here is not an exhaustive treatment but rather an illustration of how one can proceed to use a systematic approach to correlate similar features of seemingly different approaches. For example, it is judged that titrations and variable-time kinetic methods or flow-injection methods and fixed-time kinetic methods are not generally recognized to have the kindred relationships illustrated above. Although many other unexpected correlations could be made, those are beyond the scope of this paper, and attention is now returned to the general significance of these ideas.

DISCUSSION

Although the problem-oriented role of chemical analysis proposed herein differs from the traditional view of analysis, this is not a new idea. Similar views of other well-known analytical chemists are summarized above. However, the central issue here is not what others have said but why this view of analysis is important to analysts and their clients alike. The traditional sample-oriented view of analysis suggests to clients that they can submit samples with a request that one or more components be identified or quantified, and expect to get useful information with little or no indication of how the information is to be used. This is an outmoded and dangerous idea. As illustrated in Fig. 1.12, chemical analysis is a cooperative process, with clients, analysts, and technicians all contributing. In many cases, the client will be most familiar with the problem to be solved; the analyst will be most familiar with what kinds of information can and cannot be provided; and technicians will be most proficient in executing intricate

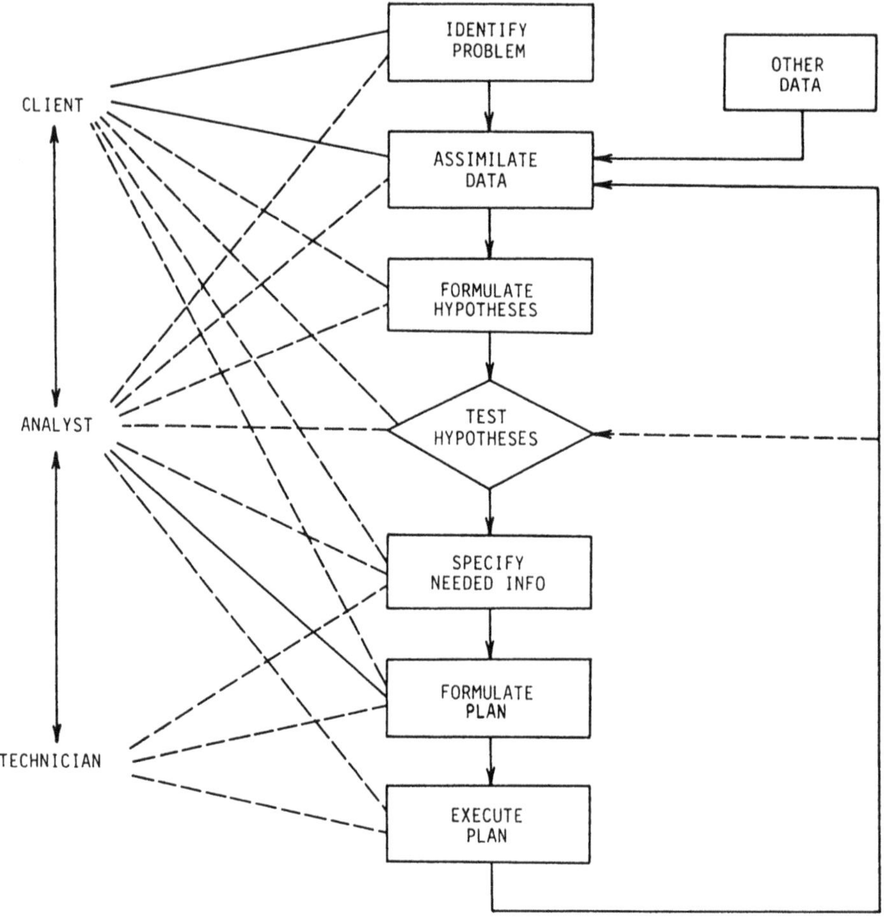

FIG. 1.12. Illustration of cooperative effort involved in a chemical analysis.

procedures. A critical key to a successful analysis is effective communication and planning, sometimes well in advance of when the problem must be solved. For example, if a problem is to require results from a very complex and/or expensive piece of equipment, it will usually not be satisfactory to wait until the problem arises to acquire the instrument and train personnel to use it. This need for effective communication among clients, analysts, and technicians is the primary reason why the outdated sample-oriented view of analysis should be replaced by a problem-oriented view of analysis.

It is suggested that analysts should continue to accept and promote the conventional sample-oriented role of analysis *only* if they are satisfied that their *only* role should be to accept samples and report results. However, if analysts believe that they have a need and responsibility to help make decisions on what components can and should be determined, what trade-offs should be exercised among time, simplicity, cost, and reliability, what and how samples should be collected, how samples should be handled, how components should be quantified, and what interpretations of data are supported by the reliability of methods used, then they should make every effort to dispel the outdated, sample-oriented view of analysis and support a more realistic, problem-oriented view of the process.

Another issue that merits explicit thought is the technique-oriented view of analysis that so many analysts are prone to promote. The technique-oriented view of analysis is manifested in the tendency they have to focus on their favorite part of the quantitative process with terms such as continuous-flow analysis, flow-injection analysis, enzymatic analysis, liquid chromatographic analysis, kinetic analysis (the author's pet), colorimetric analysis, etc. As shown in Fig. 1.3, these and other similar usages relate to just one part of a total determination, which is itself just one part of the problem-oriented role of analysis (Fig. 1.2). These usages tend to associate the role of an analyst with small parts of the total analytical process. They tend to confuse the role of the analyst who plans the procedures with the role of the technician who executes the plan. Certainly, if one were to participate in a study such as that involving dietary fiber (James and Theander 1981), it would not be necessary to perform the operations summarized in Fig. 1.1. Professional food analysts would handle the planning phases in Figs. 1.2 and 1.3 and then delegate the execution of the plan to a technician. Analysts would serve themselves, their clients, and the discipline much better if they would replace usages cited above with more accurate terms, such as continuous-flow or flow-injection sample processing, enzymatic reaction, chromatographic separation and kinetic measurement, and reserve the concept of chemical analysis for the thought processes involved in solving problems as is done in every other aspect of their lives.

EPILOGUE

Some are certain to regard all of the above as little more than a matter of semantics. However, it is important to remember that the words that are used convey impressions to others who make decisions that are critical to the success or failure of this scientific discipline. Those decisions involve making funds available for research, for space and other resources used for education, and for contributions to the resolution of problems that society faces. Whatever the reason, it is fact that funding levels for research in chemical analysis lag behind other areas; support for analytical curricula has suffered in many ways; and appointments of analysts on scientific commissions have not kept pace with other areas. It can do no harm to formulate new usages that convey the scientific nature of an analysis rather than the technique-oriented view that has characterized the field for well over a century; it could do a lot of good. It will not be easy, but it could be rewarding. The problem-oriented view of food analysis coupled with a systematic approach could help the discipline approach the stature that has eluded it for decades.

BIBLIOGRAPHY

HUME, D.N. 1963. The analysis of a profession. Anal. Chem. *35*(13), 29A.

IRVING, M.N.H., FREISER, H., and WEST, T.S. 1978. IUPAC Compendium of Analytical Nomenclature, Definitive Rules 1977, p. 41. Pergamon, Oxford.

JAMES, W.P.T., and THEANDER, O. 1969. The Analysis of Dietary Fiber in Food. Marcel Dekker, New York.

LAITINEN, H.A. 1979. The essence of modern analytical chemistry. Anal. Chem. *51*, 2065.

LUCCHESI, C.A. 1980. The analytical approach. The analytical chemist as problem solver. Am. Lab., Oct., 113.

PARDUE, H.L. 1977. A comprehensive classification of kinetic methods of analysis used in clinical chemistry. Clin. Chem. *23*, 2189.

PARDUE, H.L., and FIELDS, B. 1981. Kinetic treatment of unsegmented flow systems. I. Subjective and semi-quantitative evaluations of three methods. Anal. Chim. Acta *124*, 39, 65.

PETERS, D.G., HAYES, J.M., and HIEFTJE, G.M. 1974. Chemical separations and measurements, pp. 505–507. W.B. Saunders, Philadelphia, PA.

RUZICKA, J., and HANSEN, E.H. 1975. Flow injection analysis. Part I. A new concept of fast continuous flow analysis. Anal. Chim. Acta *78*, 145.

RUZICKA, J., and HANSEN, E.H. 1980. Flow injection analysis. Principles, applications and trends. Anal. Chim. Acta *114*, 19.

SCOTT, C.D., and BURTIS, C.A. 1973. A miniature fast analyzer system. Anal. Chem. *45*, 327A.

SIGGIA, S. 1968. Survey of Analytical Chemistry, pp. 1–8. McGraw-Hill, New York.

SNYDER, L.R. 1980. Continuous-flow analysis: Present & future. Anal. Chim. Acta *114*, 3.

SOUTHGATE, D.A.T. 1969. Determination of carbohydrates in food. Part I. Available carbohydrate; Part II. Unavailable carbohydrates. J. Sci. Food Agric. *20*, 331.

STEWART, K.K. 1983. Flow injection analysis. New tool for old assays, new approach to analytical measurements. Anal. Chem. *55*, 931A.

WEBSTERS THIRD NEW INTERNATIONAL DICTIONARY OF THE ENGLISH LANGUAGE, Unabridged. 1961. G. & C. Merriam Co., Springfield, MA.

2

Determination of Nutrients in Foods: Factors That Must Be Considered

G. R. Beecher and J. T. Vanderslice[1]

INTRODUCTION

The importance and awareness of nutrition in public health issues has resulted in increased demands for knowledge of the nutrient content of foods. This knowledge is required by scientists conducting research in such areas as nutrition, food science, clinical chemistry, and epidemiology. Dieticians and other professionals responsible for formulating diets are requesting detailed information about the nutrient content of foods. Food industry personnel and government officials concerned with nutrient labeling also have an increased need for detailed nutrient composition data.

Quantification of all of the nutrients important to human health in all of the available foods is an overwhelming, if not impossible, task. A conservative estimate of about one million analyses would be required to quantify a reasonable number of nutrients in a few representative samples of each of the generic food items available in the United States (Table 2.1). A substantial base of nutrient composition information currently exists; nonetheless, data on many nutrients in many foods are unavailable. Budgetary and manpower constraints require that priorities be established for nutrients to be analyzed and for the foods selected for nutrient analysis.

[1] Nutrient Composition Laboratory, Beltsville Human Nutrition Research Center, U.S. Department of Agriculture, Beltsville, MD 20705

TABLE 2.1. NUTRIENT ANALYSIS IN U.S. FOODS

Total generic food items	
Baked products, soups, etc.	~4000[a]
Nutrients	
Recommended dietary allowances and recommended ranges[b]	30
Other (caffeine, carotenes, fiber, etc.)	~10
Total	~40
Food sample replication	5–100
Total analyses	>0.8 million

[a] Adapted from Stewart (1981A).
[b] National Research Council, Food and Nutrition Board (1980).

The purpose of this chapter is to discuss (1) several factors that must be considered in conducting the analysis of nutrients in foods, and (2) factors that are important in selecting foods for nutrient composition analysis. It is beyond the scope of this chapter to provide detailed analytical schemes for each nutrient and detailed sampling schemes for each food or food item.

SELECTION OF NUTRIENTS FOR ANALYSIS

There are several reasons for quantifying nutrients in foods. These include research in plant and animal genetics, modifications in postharvest technology, research in food science, food product development, and nutrient composition analysis per se. In the case of those research areas primarily concerned with nutrient analysis, there must be a rational approach to the selection of the nutrient(s) to be analyzed. Stewart (1981A) recently outlined such an approach; it is shown schematically in Fig. 2.1. The rectangle represents the large domain of all nutrients. Within the large domain, there are three smaller domains, represented by circles, that correspond to (1) nutrients associated with public health problems, (2) nutrients for which data are inadequate, and (3) nutrients for which analytical methods are good. The overlap of all three smaller domains, represented by region 1 (Fig. 2.1), corresponds to those nutrients for which it is appropriate to generate nutrient composition information, i.e., those nutrients related to public health problems for which there are adequate methods but inadequate data. Region 2 in Fig. 2.1 represents those nutrients of public health concern for which there are inadequate data and inadequate methods. The nutrients contained in this region are those nutrients that should receive high priority research on analytical

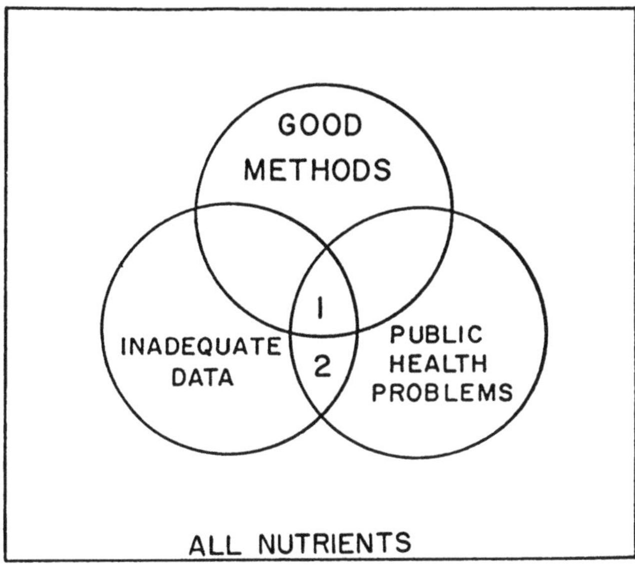

FIG. 2.1. Schematic representation of factors used in the selection of nutrients for analysis. The square represents the domain of all nutrients. Each circle represents a smaller domain of nutrients associated with a specific factor. Region 1 represents those nutrients for which it is appropriate to generate composition data. Those nutrients represented by region 2 should receive high priority research on analytical methodology. Adapted from Stewart (1981A).

methodology. The nutrients represented by the other regions have a lower priority for nutrient analysis. A detailed discussion of each domain and region follows.

Nutrients and Public Health

Standards of intakes for individual nutrients that are intended to protect the population against deficiencies have been developed in many countries. In the United States, these standards are published as Recommended Dietary Allowances (RDA); the most recent edition appeared in 1980 (Nutrition Research Council, Food and Nutrition Board 1980). Despite the RDA guidelines, several diseases and disease states may be caused, in part, by an inadequate or excessive intake of a nutrient or nutrients. The contribution of nutrient intake to U.S. public health problems is tabulated in Table 2.2. The assessment of the con-

TABLE 2.2. CONTRIBUTION OF INADEQUATE OR EXCESSIVE NUTRIENT INTAKES TO PUBLIC HEALTH PROBLEMS IN THE UNITED STATES[a]

	Contribution to U.S. public health problems		
Nutrient category	None known	Suspected	Accepted
Carbohydrates, fiber and sugars	Starch Maltose	Fructose	Lactose Fiber Sucrose
Energy			Food energy
Lipids	—	Fatty acids Other sterols Trans-fatty acids	Cholesterol Total fat
Minerals/inorganic nutrients	Cobalt Nickel Vanadium Tin	Arsenic Chromium Copper Magnesium Manganese Molybdenum Selenium Silicon	Calcium Fluorine Iodine Iron Phosphorus Sodium Zinc
Proteins and amino acids	—	Amino acids[b] Total protein[b]	—
Vitamins	Biotin Choline Pantothenic acid	Carotenes Niacin Vitamin E Vitamin K	Folacin Riboflavin Thiamin Vitamin A Vitamin B_6 Vitamin B_{12} Vitamin C Vitamin D

[a] Adapted from Stewart (1981A).
[b] It is unlikely that increased information on the nutrient composition of foods for these nutrients will significantly help in combating the public health problems associated with these nutrients.

tributions is general in nature and intended to reflect the concerns of the nutrition research community for the free-living population. Inclusion of a nutrient in the category "none known" implies that an association between these nutrients and a public health problem has not been raised (Table 2.2). It does not imply that the nutrient is not required or that it has no biological function. Inclusion of a nutrient in the category "suspected" indicates that a respected research nutritionist has voiced concern about a public health problem associated with that nutrient, whereas, inclusion in the "accepted" category means that many professionals, including physicians and nutritionists, generally accept a relationship between a nutrient and a public health problem. It does not imply that a relationship between the nutrient and a public health problem has been proven.

Nutrition plays an important role in determining the capacity of the body to adapt to stress and ward off disease (Krehl 1956). Knowledge of the nutrient composition of foods is only one of many factors involved in attaining optimum human nutrition (Stewart 1981A). Nonetheless, improvements in human health resulting from improved nutrition cannot be affected until an accurate and precise data base containing the nutrient content of all foods is available.

Knowledge of Nutrient Composition of Food

The U.S. Department of Agriculture has collected and disseminated data on the nutrient content of foods for many years (Watt and Merrill 1963; Exler 1982; Consumer and Food Economics Institute 1976, 1977, 1978, 1979A, 1979B, 1980; Consumer Nutrition Division 1980, 1982A, 1982B, 1983, 1984). This activity is currently vested in the Nutrient Data Research Branch, Consumer Nutrition Division of the Human Nutrition Information Service. Scientists in this branch have recently reevaluated the state of knowledge on the nutrient content of foods; the results of their most recent evaluation are shown in Figs. 2.2–2.9.

Careful examination of Figs. 2.2–2.9 reveals that a considerable amount of nutrient content data is available for commodities. These data have resulted, in part, from the research effort in such disciplines as plant and animal genetics, food science and technology, postharvest physiology, and agricultural engineering during the past two decades. Analytical information gathered for nutrient labeling has also contributed to these data. During the past few years, a considerable volume of new nutrient content information has become available; see Stewart (1980) for the state of knowledge of nutrient composition in early 1980. This phenomenon can be primarily attributed to the public demand for these data and the response to that demand as a result of the awareness of nutrition in public health problems.

Nonetheless, nutrient composition information is lacking for several food groups and for several categories of nutrients. A dearth of information exists for those food groups consisting primarily of highly processed or manufactured foods, i.e., baked products, snack foods, etc. Changes in the lifestyle of Americans during the past quarter century have resulted in the proliferation of convenience foods (fast foods, frozen dinners, restaurant food, etc.) for which limited nutrient composition data are available. In the case of entire categories of nutrients, biological effects and requirements have only recently been described. For some nutrients, lack of knowledge of their level in foods must be attributed to the lack of accurate, precise, and inexpensive analytical methods. The complexity of foods, the multiplicity of the chemical forms of the nutrients, and the sensitivity required for quantitation place special demands upon analytical methodology and the analyst.

	INDIVIDUAL SUGARS	STARCH	NUTRIENT FIBER
BABY FOODS			□
BAKED PRODUCTS, BREAD		□	□
SWEET GOODS		□	□
COOKIES, CRACKERS		□	□
BEVERAGES	□	☆	☆
BREAKFAST CEREALS	●	□	●
CANDIES	□		
CEREAL GRAINS, WHOLE	□	□	□
FLOUR	□	□	□
PASTA	□	□	□
DAIRY PRODUCTS	●	☆	☆
EGGS & EGG PRODUCTS	☆	☆	☆
FAST FOODS			□
FATS AND OILS	☆	☆	☆
FISH & SHELLFISH, RAW	☆	☆	☆
COOKED	☆	☆	☆
FROZEN DINNERS			□
FRUITS, RAW	●		□
COOKED	□		□
FROZEN OR CANNED	□		□
INFANT FORMULA			
INSTITUTIONAL FOOD	□	□	□
LEGUMES, RAW	□	□	□
COOKED	□	□	□
PROCESSED	□	□	□
MEAT, BEEF	☆	☆	☆
LAMB	☆	☆	☆
PORK	☆	☆	☆
SAUSAGE	☆	☆	☆
VEAL	☆	☆	☆
MIXED DISHES, COMMERCIAL			
HOME PREPARED			
NUTS & SEEDS	□	□	□
POULTRY	☆	☆	☆
RESTAURANT FOOD	□	□	□
SNACK FOODS			
SOUPS			
VEGETABLES, RAW	□	●	□
COOKED	□	□	□
FROZEN	□	□	□
CANNED	□	□	□

(blank) LITTLE OR NO DATA   □ INADEQUATE DATA

● SUBSTANTIAL DATA   ☆ NOT APPLICABLE

NCL
1983

FIG. 2.2 State of knowledge of the carbohydrate composition of foods.

	TOTAL FAT	FATTY ACIDS	CHOLESTEROL	OTHER STEROLS	TRANS-FATTY ACIDS
BABY FOODS	●	□			
BAKED PRODUCTS, BREAD	□	□			□
SWEET GOODS	□	□			□
COOKIES, CRACKERS	□	□			□
BEVERAGES			☆	☆	☆
BREAKFAST CEREALS	●	□			
CANDIES	□	□	□	□	□
CEREAL GRAINS, WHOLE	□	●	☆		☆
FLOUR	□	●	☆		☆
PASTA	□	□	☆		☆
DAIRY PRODUCTS	●	●	●	☆	☆
EGGS & EGG PRODUCTS	●	●	●	☆	☆
FAST FOODS	●	□	□	□	□
FATS AND OILS	●	●	●	□	●
FISH & SHELLFISH, RAW	●	●	●	□	
COOKED	●	●	●		
FROZEN DINNERS	□	□			
FRUITS, RAW	●	□	☆	□	☆
COOKED			☆		☆
FROZEN OR CANNED	●		☆		☆
INFANT FORMULA	□	□			
INSTITUTIONAL FOOD	□	□	□	□	
LEGUMES, RAW	●	□	☆	□	☆
COOKED	●	□	☆		☆
PROCESSED	□	□	☆		☆
MEAT, BEEF	●	●	●	☆	☆
LAMB	●	●	●	☆	☆
PORK	●	●	●	☆	☆
SAUSAGE	●	●	●	☆	☆
VEAL	●	●	●	☆	☆
MIXED DISHES, COMMERCIAL	□	□	□	□	
HOME PREPARED	□	□	□		
NUTS & SEEDS	●	●	☆	□	☆
POULTRY	●	●	●	☆	☆
RESTAURANT FOOD					
SNACK FOODS		□			□
SOUPS	●	□	□	□	
VEGETABLES, RAW	□	□	☆	□	☆
COOKED			☆		☆
FROZEN	●		☆		☆
CANNED	●		☆		☆

(blank)	LITTLE OR NO DATA
□	INADEQUATE DATA
●	SUBSTANTIAL DATA
☆	NOT APPLICABLE

NCL
1983

FIG. 2.3. State of knowledge of the lipid composition of foods.

	CALCIUM	IRON	PHOSPHORUS	SODIUM	MAGNESIUM	POTASSIUM
BABY FOODS	●	●	●	●	□	●
BAKED PRODUCTS, BREAD	●	●	□	□	□	□
SWEET GOODS	□	□	□	□	□	□
COOKIES, CRACKERS	□	□	□	□	□	□
BEVERAGES	□	□	□	□	□	□
BREAKFAST CEREALS	●	□	●	●	●	●
CANDIES	□	□	□	□	□	□
CEREAL GRAINS, WHOLE	□	□	□	□	□	□
FLOUR	□	□	□	□	□	□
PASTA	□	□	□	□	□	□
DAIRY PRODUCTS	●	●	●	●	●	●
EGGS & EGG PRODUCTS	●	●	●	●	□	●
FAST FOODS	□	□	□	□	□	□
FATS AND OILS						
FISH & SHELLFISH, RAW	●	●	●	●	●	●
COOKED						
FROZEN DINNERS	□	□	□	□	□	□
FRUITS, RAW	●	●	●	●	●	●
COOKED						
FROZEN OR CANNED	●	●	●	●	●	●
INFANT FORMULA	□	□	□	□	□	□
INSTITUTIONAL FOOD						
LEGUMES, RAW	●	●	●	●	●	●
COOKED	●	●	●	●	●	●
PROCESSED	□	□	□	□	□	□
MEAT, BEEF	●	●	●	●	●	●
LAMB	●	●	●	●	●	●
PORK	●	●	●	●	●	●
SAUSAGE	●	●	●	●	●	●
VEAL			□		□	
MIXED DISHES, COMMERCIAL	●	●	●	●	●	●
HOME PREPARED	□	□	□	□	□	□
NUTS & SEEDS	□	□	□	□	□	□
POULTRY	●	●	●	●	●	●
RESTAURANT FOOD						
SNACK FOODS						
SOUPS	●	●	●	●	●	●
VEGETABLES, RAW	□	□	□	□	□	□
COOKED						
FROZEN	●	●	●	●	●	●
CANNED	●	●	●	●	●	●

Symbol	Meaning	Symbol	Meaning
(blank)	LITTLE OR NO DATA	□	INADEQUATE DATA
●	SUBSTANTIAL DATA	☆	NOT APPLICABLE

NCL
1983

FIG. 2.4. State of knowledge of the mineral composition of foods.

	ZINC	COPPER	MANGANESE	CHROMIUM FLUORINE IODINE SULFUR SELENIUM	COBALT NICKEL SILICON TIN VANADIUM
BABY FOODS	□	□			
BAKED PRODUCTS, BREAD	□	□			
SWEET GOODS	□	□			
COOKIES, CRACKERS	□	□			
BEVERAGES	□	□			
BREAKFAST CEREALS	●	●	□		
CANDIES	□	□			
CEREAL GRAINS, WHOLE	□	□	□	□	
FLOUR	□	□			
PASTA	□	□			
DAIRY PRODUCTS	●	●	□		
EGGS & EGG PRODUCTS	●	□	□		
FAST FOODS	●	●			
FATS AND OILS	□	□			
FISH & SHELLFISH, RAW	●	●	●		
COOKED					
FROZEN DINNERS					
FRUITS, RAW	□	□	□		
COOKED					
FROZEN OR CANNED	□	□			
INFANT FORMULA					
INSTITUTIONAL FOOD					
LEGUMES, RAW	●	●	●		
COOKED	●	●	●		
PROCESSED					
MEAT, BEEF	●	●	●		
LAMB	●	●	●		
PORK	●	●	●		
SAUSAGE	●	●	●		
VEAL	●	●	●		
MIXED DISHES, COMMERCIAL	□	□			
HOME PREPARED					
NUTS & SEEDS	□	□	□	□	
POULTRY	●	●	●		
RESTAURANT FOOD					
SNACK FOODS					
SOUPS	●	●			
VEGETABLES, RAW	□	□			
COOKED					
FROZEN	●	●			
CANNED	●	●			

(blank) LITTLE OR NO DATA   □ INADEQUATE DATA

● SUBSTANTIAL DATA   ☆ NOT APPLICABLE

NCL
1983

FIG. 2.5. State of knowledge of the mineral composition of foods, continued.

	TOTAL PROTEIN	CYSTINE METHIONINE	TRYPTOPHAN
BABY FOODS	●	□	□
BAKED PRODUCTS, BREAD	●	□	□
SWEET GOODS	●	□	□
COOKIES, CRACKERS	●	□	□
BEVERAGES	☆	☆	☆
BREAKFAST CEREALS	●	□	□
CANDIES	☆	☆	☆
CEREAL GRAINS, WHOLE	●	□	●
FLOUR	●	□	●
PASTA	●	□	●
DAIRY PRODUCTS	●	●	●
EGGS & EGG PRODUCTS	●	●	●
FAST FOODS	●	□	□
FATS AND OILS	☆	☆	☆
FISH & SHELLFISH, RAW	●	●	●
COOKED	□	□	□
FROZEN DINNERS	□		
FRUITS, RAW	●	□	□
COOKED	□	□	□
FROZEN OR CANNED	●	□	□
INFANT FORMULA	●	●	●
INSTITUTIONAL FOOD			
LEGUMES, RAW	●	●	●
COOKED	●	□	□
PROCESSED	●	□	□
MEAT, BEEF	●	●	●
LAMB	●	●	●
PORK	●	●	●
SAUSAGE	●	●	●
VEAL	●	●	●
MIXED DISHES, COMMERCIAL	●		
HOME PREPARED	□		
NUTS & SEEDS	●	□	□
POULTRY	●	●	●
RESTAURANT FOOD			
SNACK FOODS	□		
SOUPS	●	□	□
VEGETABLES, RAW	●	□	□
COOKED	●	□	□
FROZEN	●	□	□
CANNED	●	□	□

(blank) LITTLE OR NO DATA — □ INADEQUATE DATA — ● SUBSTANTIAL DATA — ☆ NOT APPLICABLE

NCL 1983

FIG. 2.6. State of knowledge of the protein and amino acid content of foods.

	ISOLEUCINE LEUCINE VALINE	SERINE THREONINE TYROSINE	ALANINE ARGININE ASPARTATE GLUTAMATE	GLYCINE HISTIDINE LYSINE PROLINE
BABY FOODS	□	□	□	□
BAKED PRODUCTS, BREAD	●	●	●	●
SWEET GOODS	□	□	□	□
COOKIES, CRACKERS	□	□	□	□
BEVERAGES	☆	☆	☆	☆
BREAKFAST CEREALS	□	□	□	□
CANDIES	☆	☆	☆	☆
CEREAL GRAINS, WHOLE	●	●	●	●
FLOUR	●	●	●	●
PASTA	●	●	●	●
DAIRY PRODUCTS	●	●	●	●
EGGS & EGG PRODUCTS	●	●	●	●
FAST FOODS	□	□	□	□
FATS AND OILS	☆	☆	☆	☆
FISH & SHELLFISH, RAW	●	●	●	●
COOKED	□	□	□	□
FROZEN DINNERS				
FRUITS, RAW	□	□	□	□
COOKED	□	□	□	□
FROZEN OR CANNED	□	□	□	□
INFANT FORMULA	●	●	●	●
INSTITUTIONAL FOOD				
LEGUMES, RAW	●	●	●	●
COOKED	□	□	□	□
PROCESSED	□	□	□	□
MEAT, BEEF	●	●	●	●
LAMB	●	●	●	●
PORK	●	●	●	●
SAUSAGE	●	●	●	●
VEAL	●	●	●	●
MIXED DISHES, COMMERCIAL				
HOME PREPARED				
NUTS & SEEDS	□	□	□	□
POULTRY	●	●	●	●
RESTAURANT FOOD				
SNACK FOODS				
SOUPS	□	□	□	□
VEGETABLES, RAW	●	●	●	●
COOKED	□	□	□	□
FROZEN	□	□	□	□
CANNED	□	□	□	□

(blank) LITTLE OR NO DATA   □ INADEQUATE DATA

● SUBSTANTIAL DATA   ☆ NOT APPLICABLE

NCL 1983

FIG. 2.7. State of knowledge of the protein and amino acid content of foods, continued.

	FOLACIN	VITAMIN D	VITAMIN E	BIOTIN	CHOLINE	PANTOTHENIC ACID
BABY FOODS	□	☆	□			
BAKED PRODUCTS, BREAD	□	☆	□			□
SWEET GOODS	□	☆	□			
COOKIES, CRACKERS	□	☆	□			
BEVERAGES	□	☆	☆	☆	☆	□
BREAKFAST CEREALS	□		□			□
CANDIES	□	☆				□
CEREAL GRAINS, WHOLE	□	☆	□		●	□
FLOUR	□	☆	□		□	□
PASTA	□	☆				
DAIRY PRODUCTS	□	●	□		□	●
EGGS & EGG PRODUCTS	□	□	□		●	●
FAST FOODS			□			
FATS AND OILS	☆	□	□	☆	☆	☆
FISH & SHELLFISH, RAW		□	□		□	□
COOKED		□	□			
FROZEN DINNERS		☆				
FRUITS, RAW	□	☆			□	□
COOKED	□	☆				□
FROZEN OR CANNED	□	☆				□
INFANT FORMULA						
INSTITUTIONAL FOOD			□			
LEGUMES, RAW	□	☆	●		□	□
COOKED	□	☆	□			□
PROCESSED		☆	□			
MEAT, BEEF	●	□	□		□	□
LAMB	●	□	□			□
PORK	□	□	□		□	□
SAUSAGE	□	□				□
VEAL	●	□	□			□
MIXED DISHES, COMMERCIAL	□		□			
HOME PREPARED						
NUTS & SEEDS	□	☆	□	□	□	□
POULTRY	□	□	□		□	□
RESTAURANT FOOD						
SNACK FOODS						
SOUPS		☆				
VEGETABLES, RAW	□	☆	□		□	
COOKED	□	☆	□			
FROZEN	□	☆				□
CANNED	□	☆				□

(blank) LITTLE OR NO DATA	□ INADEQUATE DATA
● SUBSTANTIAL DATA	☆ NOT APPLICABLE

NCL
1983

FIG. 2.8. State of knowledge of the vitamin content of foods.

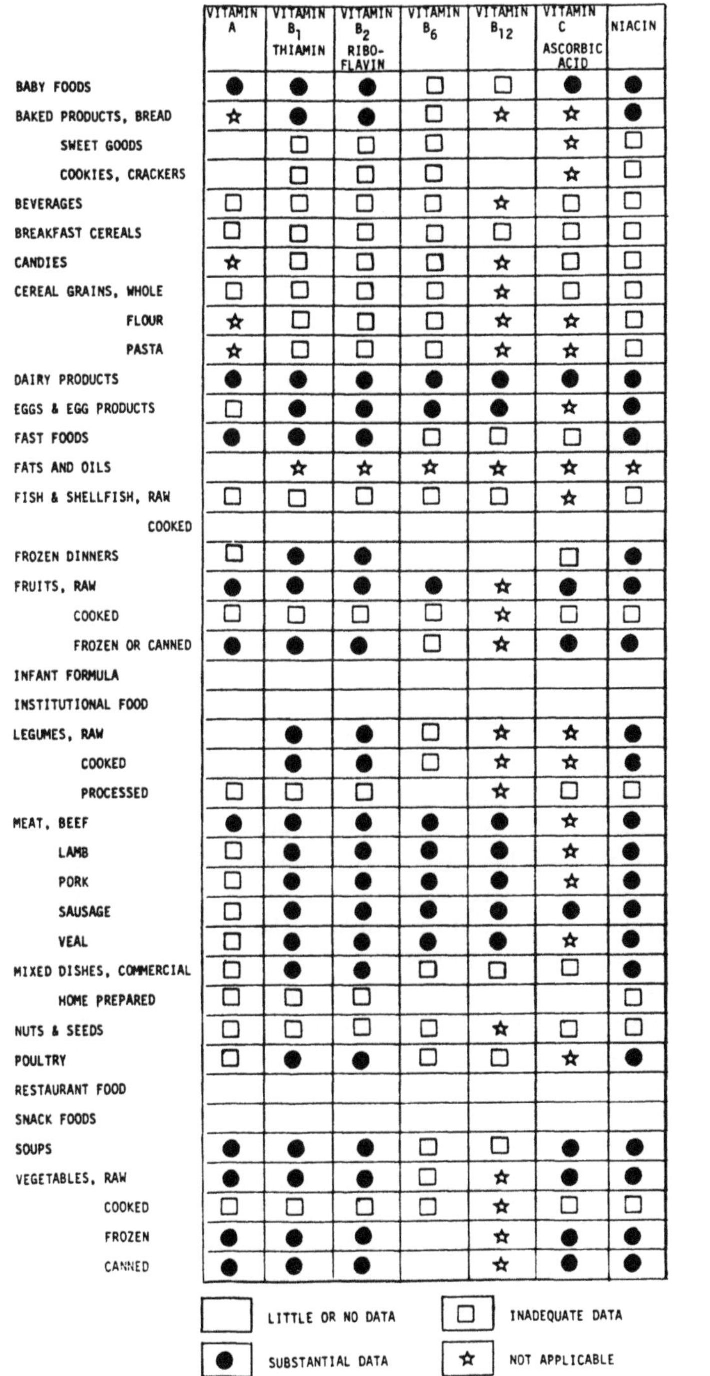

	VITAMIN A	VITAMIN B_1 THIAMIN	VITAMIN B_2 RIBO-FLAVIN	VITAMIN B_6	VITAMIN B_{12}	VITAMIN C ASCORBIC ACID	NIACIN
BABY FOODS	●	●	●	□	□	●	●
BAKED PRODUCTS, BREAD	☆	●	●	□	☆	☆	●
SWEET GOODS		□	□	□		☆	□
COOKIES, CRACKERS		□	□	□		☆	□
BEVERAGES	□	□	□	□	☆	□	□
BREAKFAST CEREALS	□	□	□	□	□	□	□
CANDIES	☆	□	□	□	☆	□	□
CEREAL GRAINS, WHOLE	□	□	□	□	☆	□	□
FLOUR	☆	□	□	□	☆	☆	□
PASTA	☆	□	□	□	☆	☆	□
DAIRY PRODUCTS	●	●	●	●	●	●	●
EGGS & EGG PRODUCTS	□	●	●	●	●	☆	●
FAST FOODS	●	●	●	□	□	□	●
FATS AND OILS		☆	☆	☆	☆	☆	☆
FISH & SHELLFISH, RAW	□	□	□	□	□	☆	□
COOKED							
FROZEN DINNERS	□	●	●			□	●
FRUITS, RAW	●	●	●	●	☆	●	●
COOKED	□	□	□	□	☆	□	□
FROZEN OR CANNED	●	●	●	□	☆	●	●
INFANT FORMULA							
INSTITUTIONAL FOOD							
LEGUMES, RAW		●	●	□	☆	☆	●
COOKED		●	●	□	☆	☆	●
PROCESSED	□	□	□		☆	□	□
MEAT, BEEF	●	●	●	●	●	☆	●
LAMB	□	●	●	●	●	☆	●
PORK	□	●	●	●	●	☆	●
SAUSAGE	□	●	●	●	●	●	●
VEAL	□	●	●	●	●	☆	●
MIXED DISHES, COMMERCIAL	□	●	●	□	□	□	●
HOME PREPARED	□	□	□				□
NUTS & SEEDS	□	□	□	□	☆	□	□
POULTRY	□	●	●	□	□	☆	●
RESTAURANT FOOD							
SNACK FOODS							
SOUPS	●	●	●	□	□	●	●
VEGETABLES, RAW	●	●	●	□	☆	●	●
COOKED	□	□	□	□	☆	□	□
FROZEN	●	●	●		☆	●	●
CANNED	●	●	●		☆	●	●

FIG. 2.9. State of knowledge of the vitamin content of foods, continued.

Analytical Procedures for Nutrient Analysis in Foods

Evaluation of Existing Methods. Valid analytical procedures are essential for reliable nutrient analysis of foods. Only trained analysts using accurate and precise analytical techniques can acquire reliable data on the nutrient content of foods. If analytical techniques are inadequate, then even data collected by trained analysts should be questioned and as a result, resources may be wasted. The state of methodology for the analysis of nutrients in foods is tabulated in Table 2.3. The boundary between acceptable and unacceptable methods lies between substantial and conflicting states of methodology (Stewart 1980, 1981A). Thus, if appropriate methods are used by trained analysts, values will probably be correct for nutritionally significant levels of those nutrients listed as having adequate and substantial methodologies. In the case of most of the other nutrients, it is doubtful that valid results can be obtained during routine analysis. For these nutrients, different methods generally yield different values. Nonetheless, for a few of the nutrients in the conflicting and lacking category, reliable values may be obtained if extreme care is exercised by the analyst. There are also some very promising new methods in these categories and hopefully they will be quickly validated and adopted.

Accuracy and precision of the resulting data were used to classify the state of methods for nutrient analysis of food (Table 2.3). As suggested by Stewart (1981A), the analytical value should be within 10% of the true value when the nutrient of interest is present at nutritionally significant levels (greater than 5% of the RDA per standard serving or daily intake, whichever is greater). Many methods fail this criterion. The usual causes of failure are the presence of interfering compounds or the loss or destruction of the nutrient during extraction and sample preparation. Sometimes samples are contaminated from external sources and the resulting nutrient level is apparently higher than the true level. Some methods lack specificity which is a particular problem when accurate analytical data are required for closely related molecular structures. Lastly, some methods lack the sensitivity to accurately measure nutritionally significant levels of a nutrient.

Precision is also an important consideration in the evaluation of a method. Common sense must be applied during the evaluation of analytical precision in nutrient analysis. The method must be sufficiently precise to yield credible nutrient composition information, and at the same time some imprecision must be tolerated in the interest

TABLE 2.3. STATE OF DEVELOPMENT OF METHODS FOR ANALYSIS OF NUTRIENTS IN FOODS[a]

Nutrient category	State of methodology[b]			
	Adequate	Substantial	Conflicting	Lacking
Carbohydrates, fiber and sugars		Individual sugars	Fiber Starch	
Energy			Food energy	
Lipids		Cholesterol Fat (total) Fatty acids (common)	Sterols Fatty acids (isomeric)	
Minerals/inorganic nutrients	Calcium Copper Magnesium Phosphorus Potassium Sodium Zinc	Iron (total) Selenium	Arsenic Chromium Fluorine Iodine Manganese	Cobalt Heme–iron Molybdenum Nonheme–iron Silicon Tin Vanadium
Proteins and amino acids	Nitrogen (total)	Amino acids (most)	Amino acids (some) Protein (total)	
Vitamins		Niacin Riboflavin Thiamin Vitamin B_6	Vitamin A Carotenes Vitamin B_{12} Vitamin C Vitamin D Vitamin E Folacin Pantothenic acid	Biotin Choline Vitamin K

[a] Adapted from Stewart (1981A).
[b] Description of methodology states the following:

Factors	Adequate	Substantial	Conflicting	Lacking
Accuracy	Excellent	Good	Fair	Poor
Speed of analysis	Fast	Moderate	Slow	Slow
Cost per analysis	Modest (<$100)	Modest to high	High	?
Development needs	—	Method modification Extraction procedure Applications	Method development/modification Extraction procedure Applications	Method development Extraction procedure Applications

of time and economics. With pure standards, the maximum acceptable relative standard deviation (RSD) should range between 10 and 15%; attempts to obtain RSDs of 1% or less are probably impractical uses of resources. In the case of nutrient analysis in foods, acceptable precision should be tempered by the level of the nutrient relative to its nutritional significance. Stewart (1981A) has recently shown that

RSDs in the range of 10 to 15% are adequate for all nutrients except for those that occur at low levels (5–30% RDA per serving) relative to their nutritional significance. In these cases precision criteria should be relaxed by some predetermined standard.

Analysis of identical samples by several laboratories often show unacceptably large variations among laboratories when different methods are used or even when the same method is used. Interlaboratory quality control of nutrient analysis must be improved because most nutrient composition tables and data bases are compiled from data generated by many laboratories. The appropriate use of quality control samples and the development and use of nutrient standards and standard reference materials will greatly improve the quality of nutrient composition data.

Modifications required to improve existing nutrient analysis methods are tabulated in Table 2.3 (footnote b). Changes to move a method from the substantial to adequate category may require only minor modifications of the extraction and quantitation steps and/or application of the method to more food groups. Changes, however, required to recategorize an unacceptable (conflicting or lacking) method as acceptable (adequate or substantial) may require extensive method modification or development, development of extraction and sample preparation procedures, and application of the new method to several different food groups. As an example of extensive changes of a procedure, the following section describes the development and application of an extraction, separation, and quantitation scheme for all known forms of vitamin B_6.

Development of Analytical Procedures. Substantial advancements in analytical instrumentation have been introduced during the past decade. The application of analytical instrumentation, however, constitutes only part of a total analytical procedure. Stewart (1980) has outlined the components of an ideal analytical method for the analysis of nutrients in foods. These components consist of homogenization and subsampling, extraction, separation, detection, and identification; calculation of results and report generation; application of standards, standard reference materials, and control samples; and finally validation of results. The development of a reliable analytical procedure is complex but relatively straightforward if the problem is approached in a systematic manner.

An example of the systematic development of an analytical procedure is the recent work of Vanderslice and co-workers who have described an extraction, separation, and quantitation scheme for all known vitamers of vitamin B_6 (Vanderslice *et al.* 1979, 1980, 1981A, 1981B, 1983; Vanderslice and Maire 1980). Separation of the vitamers takes advantage of their ionic characteristics and is accomplished by elution with an isocratic buffer from a column packed with anion exchange resin. A typical chromatogram of standards is shown in Fig. 2.10. The internal standard, 3-hydroxypyridine (HOP), is also resolved from the B_6 vitamers (Fig. 2.10), has fluorescent characteristics similar to the vitamers, and is completely recovered from food matrices. A single change of eluant permits the resolution and quantification of one of the metabolites of vitamin B_6, pyridoxic acid (4-PA) concomitantly with all six vitamers of B_6 and the internal standard. Such a system is applicable to metabolic studies.

The hardware configuration used to accomplish separation of the B_6 vitamers is shown in Fig. 2.11. The system is a dual-column high performance liquid chromatograph (HPLC) with column switching capabilities (Vanderslice *et al.* 1981A). A fluorescence spectrophotometer capable of preprogrammed changes of excitation and emission wavelengths serves as a detector. Data are recorded and presented on a computing integrator. A typical time program for eluants, columns, and

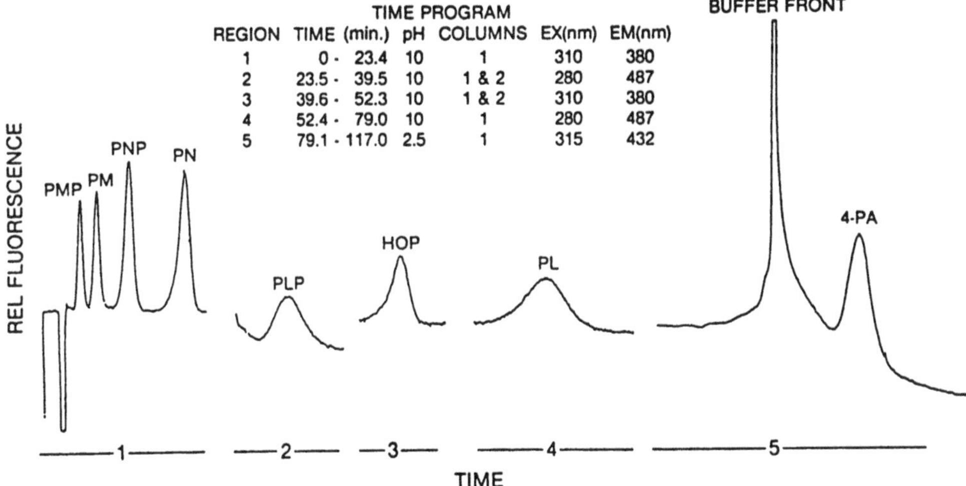

FIG. 2.10. Anion exchange separation of vitamin B_6 vitamers. Abbreviations are as follows: PMP = pyridoxamine phosphate; PM = pyridoxamine; PNP = pyridoxine phosphate; PN = pyridoxine; PLP = pyridoxal phosphate; PL = pyridoxal; HOP = 3-hydroxypyridine (internal standard); and 4-PA = 4 pyridoxic acid. The numbers on the time axis correspond to the regions defined in the time program table. All times are listed in minutes. *Reprinted with permission from Vanderslice et al. (1983).*

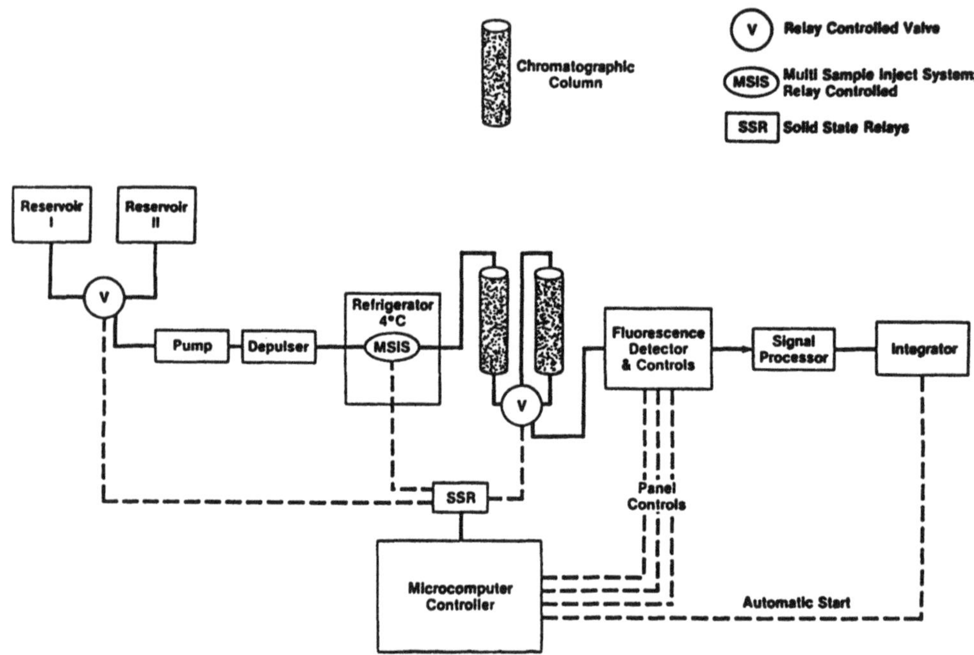

FIG. 2.11. Block diagram of automated anion exchange chromatographic system. Dashed lines correspond to electrical leads from the microcomputer to various components. *Reprinted with permission from Vanderslice et al. (1981A).*

excitation and emission wavelengths is shown in Fig. 2.10. Addition of an automatic sampler, integrator, and microcomputer controller has allowed the system to process a number of samples with a minimum of operator attention (Brown *et al.* 1981).

The successful separation and quantification of nutrient standards on a chromatographic system does not necessarily guarantee quantification of these nutrients in foods. Careful attention must be given to the extraction of the desired nutrient(s) from food, and the subsequent "preparation" of the sample for analysis, i.e., liquid chromatography, etc. Of primary concern is the quantitative extraction of the "naturally" occurring forms of a nutrient from a complex and variable matrix presented by the food or food product. The extracted sample must be subsequently processed to remove components that would interfere during analysis. In addition, the concentration of the desired nutrients must be adjusted to the appropriate range for analysis and the nutrients must appear in a solvent or buffer compatible with the analytical system. It is necessary for all of these manipulations to be accomplished without loss or conversion of the "naturally" occurring form(s) of the nutrient.

A number of sample extraction and preparation procedures have been reviewed (Lawrence 1981). Vanderslice *et al.* (1980) have recently described such a system for the extraction and preparation of extracts of vitamin B_6 from food. A block diagram of the procedure is shown in Fig. 2.12. The sample purification system, referred to in Fig. 2.12, consists of preparative column chromatography using an anion exchange resin to remove sulfosalicylic acid (SSA) from the original extract (Vanderslice *et al.* 1980). Using this extraction scheme and the analytical procedure described above, the vitamin B_6 vitamer content in a number of foods and biological materials has been quantified

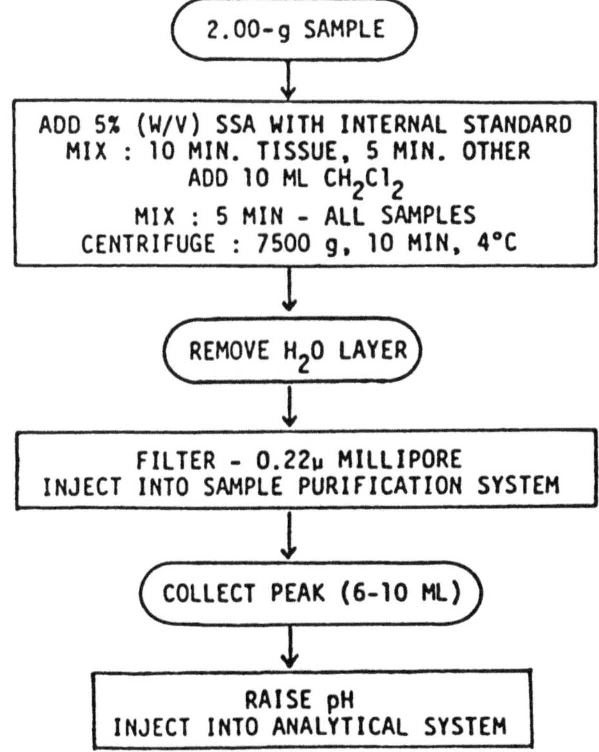

FIG. 2.12. Block diagram of an extraction and sample preparation for several foods.
Reprinted with permission from Vanderslice et al. (1980).

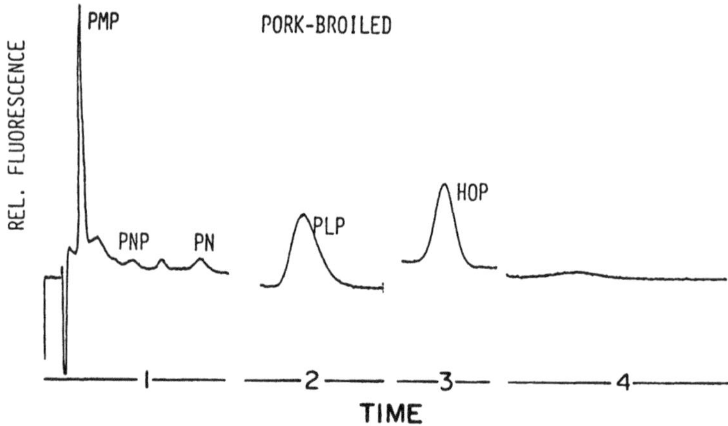

FIG. 2.13. Chromatogram of vitamin B_6 extract from broiled pork. Integrator instructions and data printout have been deleted. See Fig. 2.10 for abbreviations and details of the chromatographic program.
Reprinted with permission from Vanderslice et al. (1980).

(Vanderslice *et al.* 1983). A typical chromatogram of B_6 vitamers extracted from broiled pork is shown in Fig. 2.13. It should be noted that the baseline of this chromatogram is stable and that only a few minor unidentified peaks appear on the chromatogram. These characteristics are essential if precise quantification of components from peak areas is expected.

Prior to accepting a new extraction and analytical procedure for routine analysis, a number of experiments must be conducted to validate the entire procedure for the matrix that will be encountered. First, the recovery of the nutrient(s) from a food matrix using the extraction and analytical procedure of choice should be determined. Data from such an experiment are tabulated in Table 2.4. Recoveries from these

TABLE 2.4. B_6 VITAMER CONTENT OF BROILED PORK AND RESULTS OF STANDARDS ADDITION[a]

Sample	PMP[b]	PNP	PN	PLP
(1) Original analysis, μg/g	3.89	0.09	0.17	2.11
(2) Vitamer added, μg/g	1.47	1.31	1.49	1.57
(3) 1+3, μg/g	5.36	1.40	1.66	3.68
(4) Analysis of (3)	5.12	1.22	1.65	3.62
(5) % recovery	96	87	99	98

[a] Adapted from Vanderslice *et al.* (1981B).
[b] See Fig. 2.10 for explanation of abbreviations.

types of experiments should range between 95 and 105% for those nutrients or forms of a nutrient that are present in significant levels in the food. Deviations from these ranges demand that the investigator examine all steps of the extraction/analysis procedure and make modifications to prevent the loss or conversion of the nutrient(s) (Vanderslice *et al.* 1981B). Appropriate standard reference materials (SRMs), if available, and quality control samples should also be analyzed and the data compared with validated numbers, in the case of SRMs, or with means for the quality control samples. Finally, data obtained from any new procedure should be directly compared with data derived from a "traditional" or "established" procedure. Data from a typical experiment are shown in Table 2.5. The maximum difference between the two methods for any sample was 11%; however, the average difference for the nine samples was less than 1%, which is considered excellent agreement. Large random or uniform differences between a new procedure and an established procedure must be reconciled before routine analysis can proceed.

During the routine analysis of samples with any analytical technique, adequate standards and control samples must also be analyzed to provide a basis for the validation of the results from unknown samples (Stewart 1980). In addition to standards used for instrument calibration, analysis of standard reference materials, quality control samples and unknowns to which known amounts of a nutrient have been added (methods of standard additions) all provide data for the validation of results from unknown samples. Only after careful validation of results can data be published and collated with other data in data bases.

TABLE 2.5. COMPARISON OF HPLC AND MICROBIOLOGICAL ASSAY OF VITAMIN B_6 IN RAW PORK

Sample	HPLC summation	Microbiological	% Difference
	nmol/g		
1	22.4	25.1	−11.3
2	35.4	37.1	− 4.7
3	29.0	29.8	− 2.7
4	37.4	33.7	10.4
5	36.9	36.4	1.4
6	32.1	34.1	− 6.0
7	40.2	38.1	5.4
8	31.9	32.7	− 2.5
9	34.1	31.9	6.7
		Mean	−0.37

Development of Instrumentation. Development of new or improved analytical procedures for nutrient analysis of foods generally takes advantage of existing analytical instrumentation. During the past decade however, certain characteristics of the analysis of nutrients in foods, i.e., complex matrices, and large numbers of nutrients of interest in a large number of samples and economics, have provided the impetus for the development of new analytical instrumentation. Two examples are the development of the concepts and equipment for (1) flow injection analysis, and (2) simultaneous multielement atomic absorption spectrometry.

Flow injection analysis (FIA), recently reviewed by Stewart (1981B; Chapter 15, this volume), was developed simultaneously by a group of scientists in Denmark and by an independent group in the United States. The U.S. group developed FIA to analyze a large number of samples during a relatively short period of time with high precision and accuracy. As the theoretical basis of FIA is evaluated and established (Vanderslice *et al.* 1981C), the applications of this new instrumentation appear to be very broad.

Simultaneous multielement atomic absorption spectrometry (SIMAAC), reviewed by Harnly (1983) and Harnly *et al.* (Chapter 6, this volume), represents a combination of the advantages of atomic absorption spectrometry (AAS) and inductively coupled plasma, atomic emission spectrometry (ICP–AES). The AAS offers the advantages of ease of operation, low detection limits (particularly when a furnace is used), extensive documentation, and modest cost. On the other hand, the ICP–AES offers multielement capability, extended calibration ranges and insensitivity to interferences. The SIMAAC currently operating in the Nutrient Composition Laboratory is capable of quantifying 16 elements simultaneously from either flame or furnace atomization, has extended analytical range capabilities, and is insensitive to interferences because background correction is determined for each element analyzed. The extensive computerization of SIMAAC allows several quality assurance features to be built into the data acquisition phase of the system. Current research with SIMAAC is oriented toward development of furnace atomization as a more reliable atomization source and the development of a microprocessor controlled instrument.

SELECTION OF FOODS FOR NUTRIENT ANALYSIS

The summaries of the present state of knowledge of nutrient composition presented in Figs. 2.2–2.9 show that a considerable amount of information on the nutrient content of many foods is inadequate or lacking. In order to make advances in human nutrition, nutrient composition information for many food items must be generated as rapidly as possible, consistent with acceptable techniques of sampling and analysis. The numbers of foods available for sampling and analysis (Table 2.1) could overwhelm analytical resources unless nutrient composition studies are carefully planned.

An approach to the selection of foods for nutrient analysis, suggested by Stewart (1981A), is shown in Fig. 2.14. The rectangle represents the large domain of all foods. Within the large domain, there are four smaller domains, represented by circles that correspond to: (1) core foods, (2) categories of foods that lack data (inadequate data), (3) foods having high concentrations of nutrient(s), and (4) foods as eaten. The overlap of the four smaller domains, represented by region 1 (Fig. 2.14), corresponds to those foods for which it is appropriate to generate nutrient composition data immediately. Foods represented

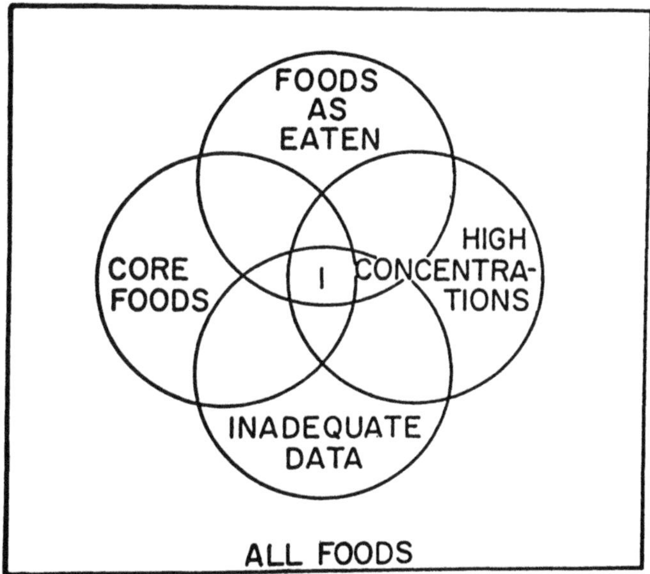

FIG. 2.14. Schematic representation of factors used in the selection of foods for nutrient analysis. The square represents the domain of all foods. Each circle represents a smaller domain of foods associated with each factor indicated. Region 1 represents those foods that should be sampled for nutrient analysis.
Adapted from Stewart (1981A).

by the other regions have a lower priority for nutrient analysis, but it is hoped that all nutrients in all foods would ultimately be quantitated.

Recent studies have demonstrated that a small percentage of the total food items available in the United States make up a large percentage of the total food consumed. About 15% of the available food items account for 90% of the weight of the diet consumed (Pennington 1983), whereas about 4% of the total food items account for 90% of the reported frequency of consumption (Wolf 1981). These foods with high consumption have been termed "core foods" (Stewart 1981A). The data in Table 2.6 compare the 15 highest ranking food items based on frequency of consumption and weight of consumption. There is remarkable similarity in the food items that appear in these two lists. Core food lists identified with specific socioeconomic or cultural groups may vary somewhat from these lists. Nonetheless, a great deal of information on nutrient intake can be provided by using the nutrient composition of core foods.

The extent of the knowledge of the nutrient composition of food is presented in Figs. 2.2–2.9 and has been discussed earlier in this chapter. Analysis of foods containing a high concentration of nutrient(s) can be justified, because foods that fall in this category as

TABLE 2.6. COMPARISON OF CORE FOOD LISTS BASED ON FREQUENCY AND WEIGHT OF CONSUMPTION

Rank	Food item based on frequency of consumption[a]	Food item based on weight of consumption[b]
1	Water	Water
2	Bread, white	Coffee
3	Milk, cow, whole	Milk, cow, whole
4	Coffee	Soda, carbonated, cola type
5	Orange juice, frozen, reconstituted	Tea
6	Tomatoes, fresh	Beer
7	Lettuce, raw	Milk, cow, 2% fat
8	Margarine	Soda, carbonated, lemon–lime type
9	Sugar, beet and cane	Orange juice, frozen, reconstituted
10	Soda, carbonated, fruit flavored	Bread, white
11	Butter	Soft drinks, from powder
12	Tea	Beef, ground, regular
13	Milk, cow, skim	Beef, steak, loin/sirloin

[a] Adapted from Wolf (1981); frequency of consumption based on a 14-day dietary record maintained by 22 subjects aged 14–64.
[b] Adapted from Pennington (1983); weight of consumption calculated as average grams per day for both male and female of the 14–16, 25–30, and 60–65 year age groups. Data obtained from >50,000 participants (in excess of 100,000 daily diets) as part of the 1977–1978 USDA Nationwide Food Consumption Survey, and the Second National Health and Nutrition Examination Survey conducted in 1976–1980.

well as those in the categories of "foods as eaten" and of core foods supply the majority of a nutrient in the diet. The analysis of low levels of nutrients in a food (<5% RDA per serving) may require extensive modification of analytical procedures in order to obtain adequate analytical accuracy and precision. In this case, a decision must be made weighing the value of the analytical data against the cost of obtaining it.

Analysis of nutrients in foods as eaten provides the most meaningful data on the nutrient composition of foods as consumed by human beings. The use of limited distribution centers by many U.S. food suppliers, the widespread consumption of processed foods, and the rigorous quality control systems used by the food industry have reduced the variability of the nutrient content in many foods. Several laboratories are currently collecting data and will soon be able to estimate the variability of the nutrient content in selected foods from nationwide samplings.

SUMMARY

The determination of nutrients in foods is a complex process requiring decisions at several steps. Nutrients selected for analysis should be associated with public health problems; there should be a lack of adequate analytical data; and accurate and precise analytical methods should be available. In the case of some nutrients, accurate and precise analytical procedures, including extraction and preparation schemes, need to be developed before analysis of nutrients in foods can proceed. The analysis of nutrients in those foods for which data are sparse should concentrate on frequently consumed foods that have been prepared by customary procedures; those foods that contribute large amounts (or concentrations) should be the first priority. Compilation of accurate and precise nutrient data bases will permit health professionals to accurately assess nutrient intake and utilization and to improve human health through nutrition education and/or therapy.

ACKNOWLEDGMENTS

The assistance of the staff of the Nutrient Data Research Branch, Consumer Nutrition Division, HNIS, USDA, in updating the information in Figs. 2.2–2.9 is greatly appreciated. The contribution of Fig. 2.14 by Dr. K. K. Stewart, Department of Food Science, Virginia Polytechnic Institute and State University, Blacksburg, VA, is gratefully acknowledged. The preparation of illustrations for this chapter by Mr. Paul Padovano is also greatly appreciated.

BIBLIOGRAPHY

BROWN, J.F., VANDERSLICE, J.T., MAIRE, C.E., BROWNLEE, S.G., and STEWART, K.K. 1981. Control programming without language: Automation of vitamin B-6 analysis. J. Automatic Chem. *3,* 187–190.

CONSUMER AND FOOD ECONOMICS INSTITUTE. 1976. Composition of Foods: Dairy and Egg Products, Raw, Processed, Prepared. U.S. Dep. of Agric. Handb. *No. 8-1.* U.S. Dep. of Agric., Washington, DC.

CONSUMER AND FOOD ECONOMICS INSTITUTE. 1977. Composition of Foods: Spices and Herbs, Raw, Processed, Prepared. U.S. Dep. of Agric. Handb. *No. 8-2.* U.S. Dep. of Agric., Washington, DC.

CONSUMER AND FOOD ECONOMICS INSTITUTE. 1978. Composition of Foods: Baby Food, Raw, Processed, Prepared. U.S. Dep. of Agric. Handb. *No. 8-3.* U.S. Dep. of Agric., Washington, DC.

CONSUMER AND FOOD ECONOMICS INSTITUTE. 1979A. Composition of Foods: Fats and Oils, Raw, Processed, Prepared. U.S. Dep. of Agric. Handb. *No. 8-4.* U.S. Dep. of Agric., Washington, DC.

CONSUMER AND FOOD ECONOMICS INSTITUTE. 1979B. Composition of Foods: Poultry Products, Raw, Processed, Prepared. U.S. Dep. of Agric. Hand. *No. 8-5.* U.S. Dep. of Agric., Washington, DC.

CONSUMER AND FOOD ECONOMICS INSTITUTE. 1980. Composition of Foods: Soups, Sauces and Gravies, Raw, Processed, Prepared. U.S. Dep. of Agric. Handb. *No. 8-6.* U.S. Dep. of Agric., Washington, DC.

CONSUMER NUTRITION DIVISION. 1980. Composition of Foods: Sausages and Luncheon Meats, Raw, Processed, Prepared. U.S. Dep. of Agric. Handb. *No. 8-7.* U.S. Dep. of Agric., Washington, DC.

CONSUMER NUTRITION DIVISION. 1982A. Composition of Foods: Breakfast Cereals, Raw, Processed, Prepared. U.S. Dep. of Agric. Handb. *No. 8-8.* U.S. Dep. of Agric., Washington, DC.

CONSUMER NUTRITION DIVISION. 1982B. Composition of Foods: Fruits and Fruit Juices, Raw, Processed, Prepared. U.S. Dep. of Agric. Handb. *No. 8-9.* U.S. Dep. of Agric., Washington, DC.

CONSUMER NUTRITION DIVISION. 1983. Composition of Foods: Pork Products, Raw, Processed, Prepared. U.S. Dep. of Agric. Handb. *No. 8-10.* U.S. Dep. of Agric., Washington, DC.

CONSUMER NUTRITION DIVISION. 1984. Composition of Foods: Vegetables and Vegetable Products, Raw, Processed, Prepared. U.S. Dep. of Agric. Handb. *No. 8-11.* U.S. Dep. of Agric., Washington, DC.

EXLER, J. 1982. Iron content of food. Home Econ. Res. Rep. *No. 45.* U.S. Dep. of Agric., Washington, DC.

HARNLY, J.M. 1983. Simultaneous multielement atomic absorption spectrometry. 186th Nat. Meet., Am. Chem. Soc., Washington, DC (Abstract No. 165).

KREHL, W.A. 1956. A concept of optimal nutrition. Am. J. Clin. Nutr. *4,* 634–641.

LAWRENCE, J.F. 1981. Organic Trace Analysis by Liquid Chromatography. Academic Press, New York.

NATIONAL RESEARCH COUNCIL, FOOD AND NUTRITION BOARD. 1980. Recommended Dietary Allowances, 9th ed. Natl. Acad. of Sci., Washington, DC.

PENNINGTON, J.A.T. 1983. Revision of the total diet study food list and diets. J. Am. Diet. Assoc. *82,* 166–173.

STEWART, K.K. 1980. Nutrient analysis of foods: State of the art for routine analysis. *In* Nutrient Analysis Symposium. K. K. Stewart (Editor). Assoc. of Off. Anal. Chem., Arlington, VA.

STEWART, K.K. 1981A. Nutrient analyses of food: a review and a strategy for the future. *In* Beltsville Symposia in Agricultural Research [4] Human Nutrition Research. G.R. Beecher (Editor). Allanheld, Osmun Publishers, Totowa, NJ.

STEWART, K.K. 1981B. Flow-injection analysis: a review of its early history. Talanta *28,* 789–797.

VANDERSLICE, J.T., and MAIRE, C.E. 1980. Liquid chromatographic separation and quantification of B-6 vitamers at plasma concentration levels. J. Chromatogr. *196,* 176–179.

VANDERSLICE, J.T., STEWART, K.K.., and YARMAS, M.M. 1979. Liquid chromatographic separation and quantification of B-6 vitamers and their metabolite, pyridoxic acid. J. Chromatogr. *176,* 280–285.

VANDERSLICE, J.T., MAIRE, C.E., DOHERTY, R.F., and BEECHER, G.R. 1980. Sulfosalicylic acid as an extraction agent for vitamin B-6 in food. J. Agric. Food Chem. *28,* 1145–1149.

VANDERSLICE, J.T., BROWN, J.F., BEECHER, G.R., MAIRE, C.E., and BROWNLEE, S.G. 1981A. Automation of a complex high-performance liquid chromatography system. Procedures and hardware for a vitamin B-6 model system. J. Chromatogr. *216,* 338–345.

VANDERSLICE, J.T., MAIRE, C.E., and BEECHER, G.R. 1981B. Extraction and quantification of B-6 vitamers from animal tissues and human plasma: a preliminary study. *In* Methods in Vitamin B-6 Nutrition. J.E. Leklem and R.D. Reynolds (Editors). Plenum Press, New York.

VANDERSLICE, J.T., STEWART, K.K., ROSENFELD, A.G., and HIGGS, D.J. 1981C. Laminar dispersion in flow-injection analysis. Talanta *28,* 11–18.

VANDERSLICE, J.T., BROWNLEE, S.G., CORTISSOZ, M.E., and MAIRE, C.E. 1984. Vitamin B-6 analysis: sample preparation, extraction procedure, and chromatographic separations. *In* Modern Gas and Liquid Chromatography of the Vitamins. A.P. DeLeenheer, W.E. Lambert, and M.G. DeRuyter (Editors). Marcel Dekker, New York.

WATT, B.K., and MERRILL, A.L. 1963. Composition of Foods: Raw, Processed, Prepared. U.S. Dep. of Agric. Handb. *No. 8.,* U.S. Dep. of Agric., Washington, DC.

WOLF, W.R. 1981. Assessment of organic nutrient intake from self-selected diets. *In* Beltsville Symposia in Agricultural Research [4] Human Nutrition Research. G.R. Beecher (Editor). Allanheld, Osmun Publishers, Totowa, NJ.

3

Computers in the Food Analysis Laboratory

Raymond E. Dessy[1]

Today's food analysis laboratory has no choice but to embrace the electronic laboratory. Faced with awesome legal and fiscal imperatives the laboratory manager must make some decisions which involve a synergistic interaction between hardward, operating systems, and software. This ensemble must provide the services required by his beleaguered staff. The implementation must take into consideration the psychological, political, diplomatic, and fiscal patterns of the environment.

In some installations a single large mainframe computer is conceived as the solution to the problem of providing real-time data acquisition, data manipulation, and storage/retrieval. This approach totally disregards the difficulties and impossibilities associated with providing a variety of relevant-time, interactive responses with a central processing unit (CPU) running a classical time-shared operating system.

In other environments, frustrated scientists are active participants in the revolution initiated by microcircuit fabrication. It is both a technological revolution and a true revolution, as people flee the restrictions and inconvenience of a large computing facility. Each user envisages his own personal computer, small, cuddly, and controllable. Fed by the momentum of "microcomputer madness," many scientists acquire hardware which is underkill for their application, and rapidly

[1] Chemistry Department, Virginia Polytechnic Institute and State University, Blacksburg, VA 24061

transform themselves from first class scientists to second class programmers and systems analysts.

This brief chapter is an attempt to convey a rational plan for laboratory automation. The words and philosophies introduced are more completely presented in a series of five longer chapters which have appeared in the A/C Interface columns in *Analytical Chemistry* (Dessy 1982, 1983A–D).

HARDWARE

The instruments of today are microprocessor based, with embedded computers that control the device, collect data, and present final reports to local printers. These intelligent instruments should have RS-232 serial asynchronous ASCII ports which allow them to communicate these reports to larger machines. The alternative is a laboratory in which the scientist becomes a high-priced secretary, copying data from one form to another.

A cluster of these intelligent instruments is attached to a larger microcomputer which has enough disk and memory space to successfully buffer the transmitted data in case of failures further upstream in the network that is being built. Such redundancy is important in laboratories where continued throughput is necessary. This facility also provides control and data acquisition service to more traditionally designed instruments, as well as special equipment that must be constructed in-house.

Architecturally, this processor should have the following features:

1. At least a 24-bit program counter, and two-dimensional (paged and segmented) memory management capabilities.
2. A good two-dimensional interrupt system.
3. Double precision integer math, and single and double precision floating arithmetic in microcoded hardware; or a math coprocessor.
4. An industry standard bus.
5. A removable Winchester disk drive.

Above these nodes exists a larger laboratory computer. Architecturally it should have the following features:

1. A 32-bit CPU with excellent memory management. Noncontiguous memory mapping, high-speed cache page and segment descriptors, and cache look-forward/look-behind are essential.
2. Array processing capabilities.

3. DMA node controller for network communication.
4. Multiplexed serial ports.
5. Register transfer, and DMA analog-to-digital converter subsystems.
6. Printers adequate to the report generation requirements of the facility.
7. High resolution interactive graphics facilities, in black and white and color.

NETWORK ARCHITECTURE

A number of these facilities are threaded together in a local area network (LAN) (Fig. 3.1). This communication medium permits the information resident in each LAN node to be shared with other pro-

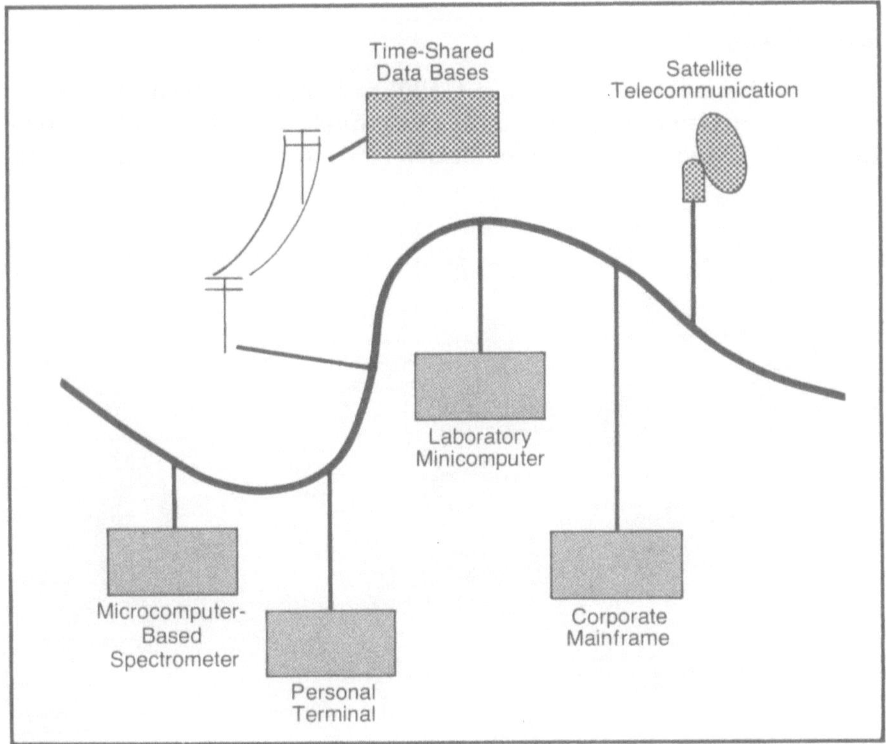

FIG. 3.1. Local area network.
Reprinted with permission from Anal. Chem. (1982) 54, 1167A. Copyright 1982 American Chemical Society.

cessors. In a large facility, with many testing sites and administrative centers, such exchange is essential. Several standards are emerging from the de facto and de jure forces in the electronic marketplace.

CSMA/CD Baseband

The DIX consortium (Digital Equipment Corporation, Intel, and Xerox) have promulgated a baseland local area network that employs CSMA/CD technology (carrier sense, multiple access/collision avoidance). Using Manchester phase encoding, data are placed on the single coaxial line connecting all nodes at rates up to 10 MHz (Fig. 3.2). About 100 users can share the conduit without apparent conflict. At least ten other major instrument vendors have announced their support of this baseband approach. Although variants exist, the general approach is often called Ethernet, derived from the original name given by Xerox to a specific set of protocols.

Broadband

In another approach, midsplit CATV technology is used to create a frequency division multiplexed set of channels on a single coaxial line (Fig. 3.3). Using frequency modulation (FM) or vestigial sideband am-

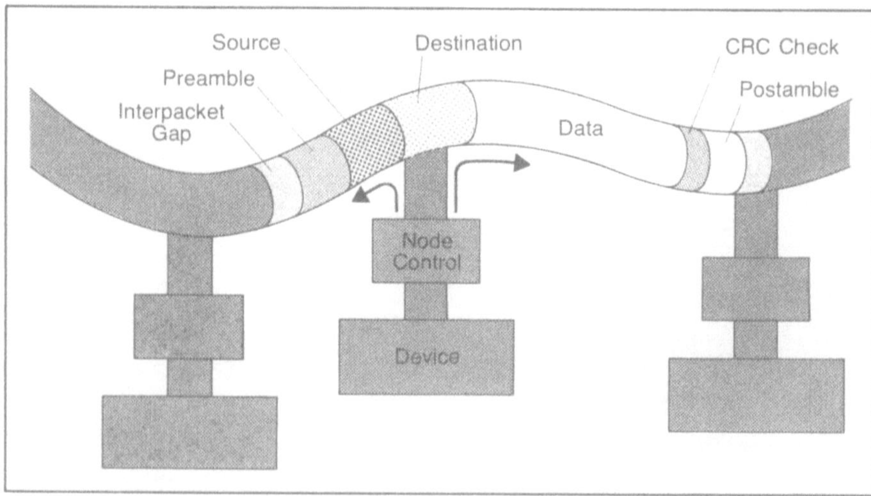

FIG. 3.2. Baseband bus contention network packet construction.
Reprinted with permission from Anal. Chem. (1982) 54, 1167A. Copyright 1982 American Chemical Society.

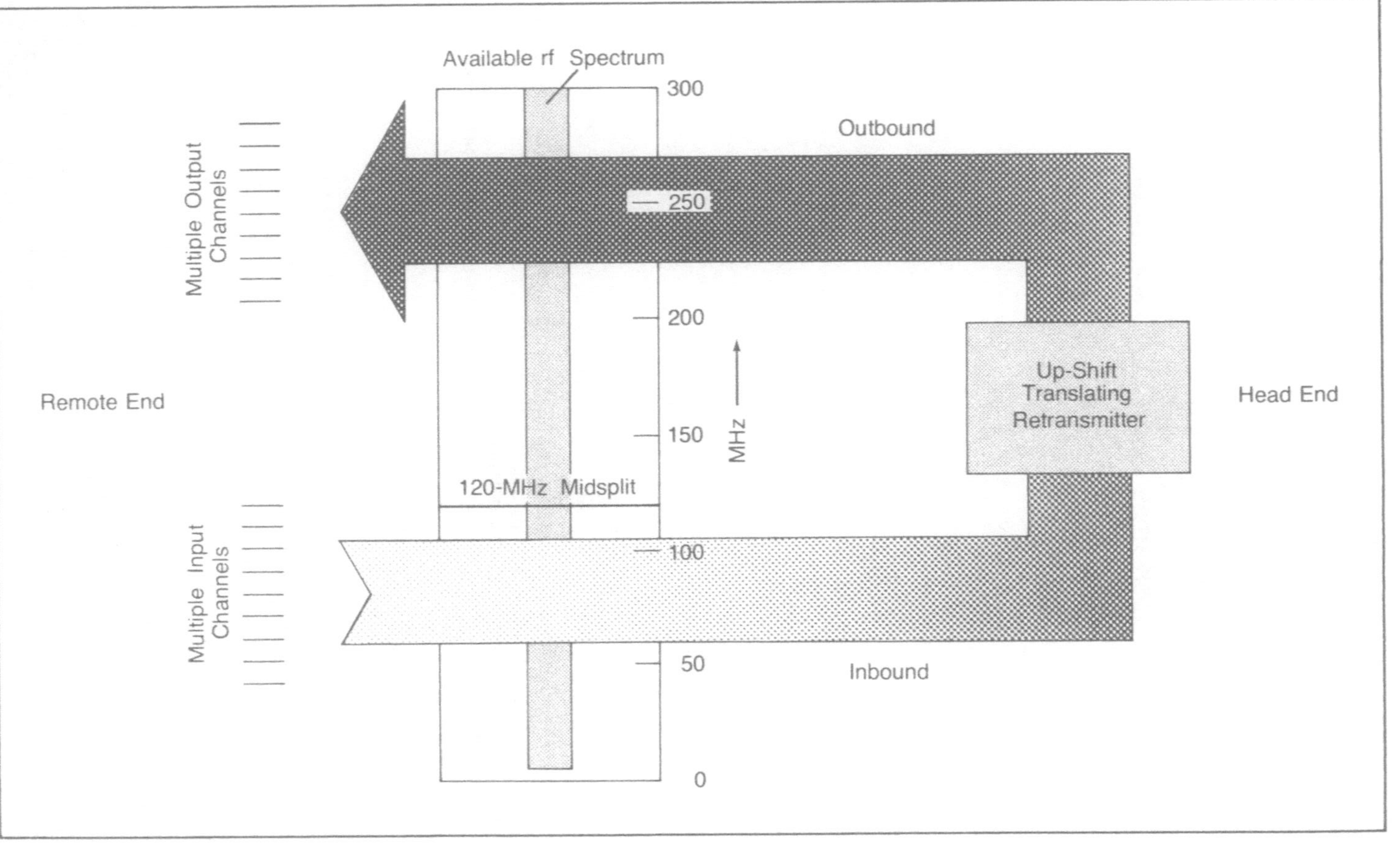

FIG. 3.3. Typical broadband channel allotment.
Reprinted with permission from Anal. Chem. (1982) 54, 1167A. Copyright 1982 American Chemical Society.

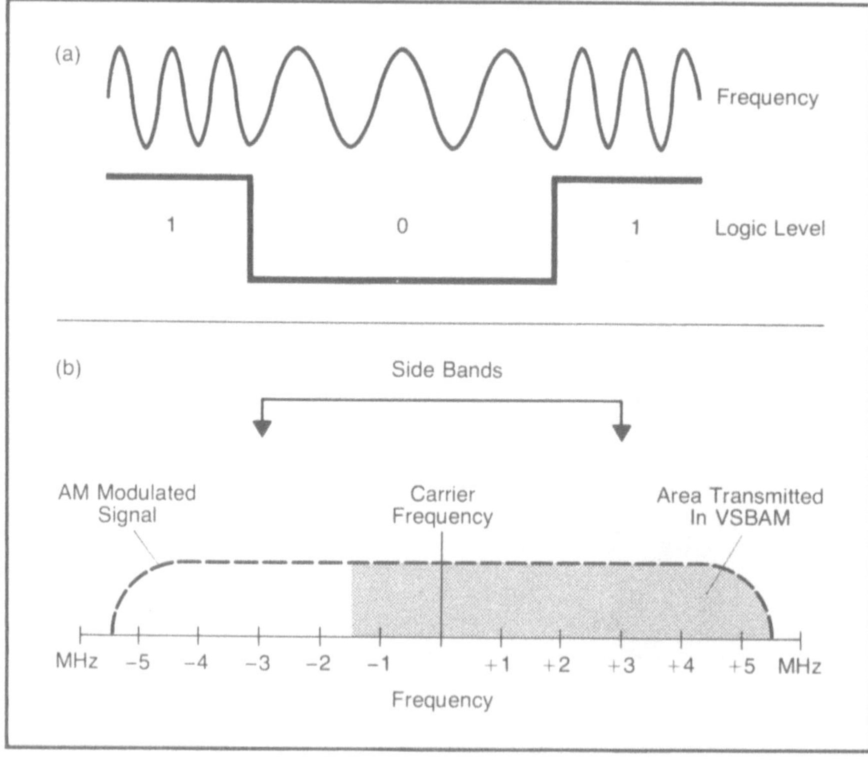

FIG. 3.4. (a) Broadband frequency modulation transmission protocol (frequency shift keying); (b) broadband amplitude modulation transmission protocol (VSBAM—vestigial sideband amplitude modulation).
Reprinted with permission from Anal. Chem. (1982) 54, 1167A. Copyright 1982 American Chemical Society.

plitude modulation (VSBAM) techniques (Fig. 3.4), about 100 channels can concurrently share the facility using CSMA/CD techniques. The cable can thus handle about 100,000 users.

Token Passing Ring

Yet another approach employs a ring architecture, with the ability to have a node make a reservation for another task in the ring (Fig. 3.5). This *token* can assure what the worst case response of the system will be and is required if relevant time response is envisaged.

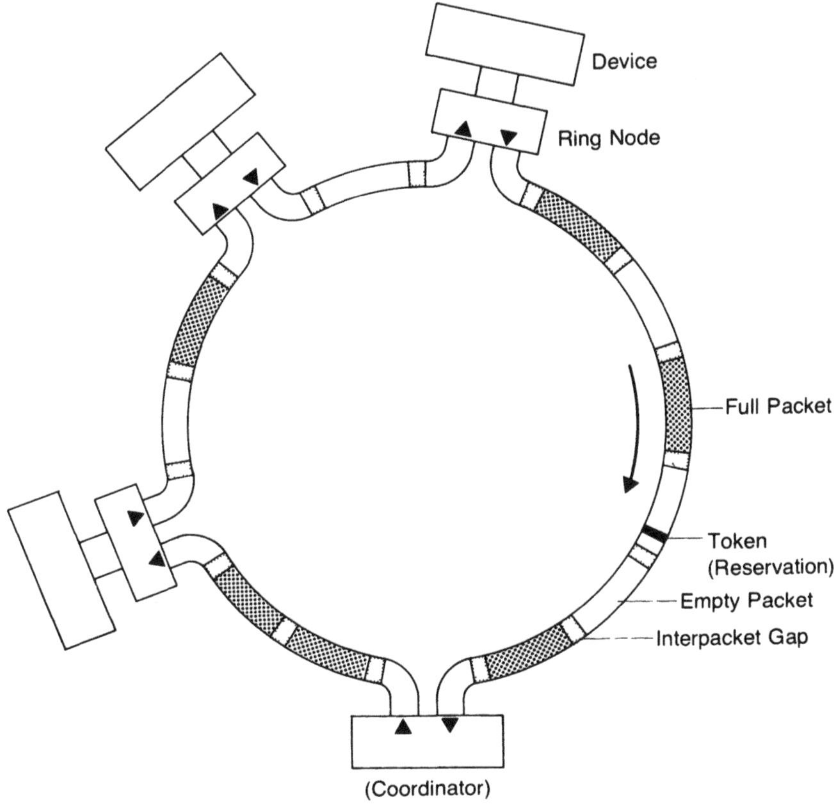

FIG. 3.5. Ring network, packet flow.
Reprinted with permission from Anal. Chem. (1982) 54, 1167A. Copyright 1982 American Chemical Society.

PBX Digital Switching

These technologies are colliding headlong with the PBX digital switching facilities developing rapidly in the aftermath of the Bell dismemberment.

All four approaches are now being installed in various pharmaceutical and petrochemical laboratories.

OPERATING SYSTEMS

The lower instrument cluster nodes in the network should run under a simple multitasking operating system. The upper nodes, attached to the LAN, should run under a multiuser, multitasking system (Fig. 3.6).

The instrument cluster nodes should be typified by

1. A round-robin multitasking scheduler.
2. Direct interrupt servicing via an automatic vectored and queued machine architecture.

The LAN nodes should be typified by

1. A framing-period oriented scheduler.
2. Queued service of system requests, interrupt requests, and operator requests.
3. A hierarchical directory structure.
4. Virtual memory, virtual array handling.
5. Good access control.

Both operating systems should have the following features:

1. The floating point formats on the two types of processors should be identical.
2. File compatibility or file exchange programs should exist.
3. A source code for all software must be available.

Both of the systems should have a disk facility installed that has a common format.

LANGUAGES

The instrument cluster nodes should be programmed in an indirectly threaded code language. Some of the attributes it must have are as follows:

1. Has full extensibility.
2. Is untyped.
3. Is interactive and immediate.
4. Is pointer oriented.
5. Is capable of generating code that is reentrant and recursive.
6. Supports coroutines, and/or rendezvous.
7. Permits intralineal use of assembler and higher level commands.
8. Possesses structured assembly macros.
9. Allows top-down design and bottom-up programming.

The LAN nodes should be programmed in a more conventional language which has the following characteristics:

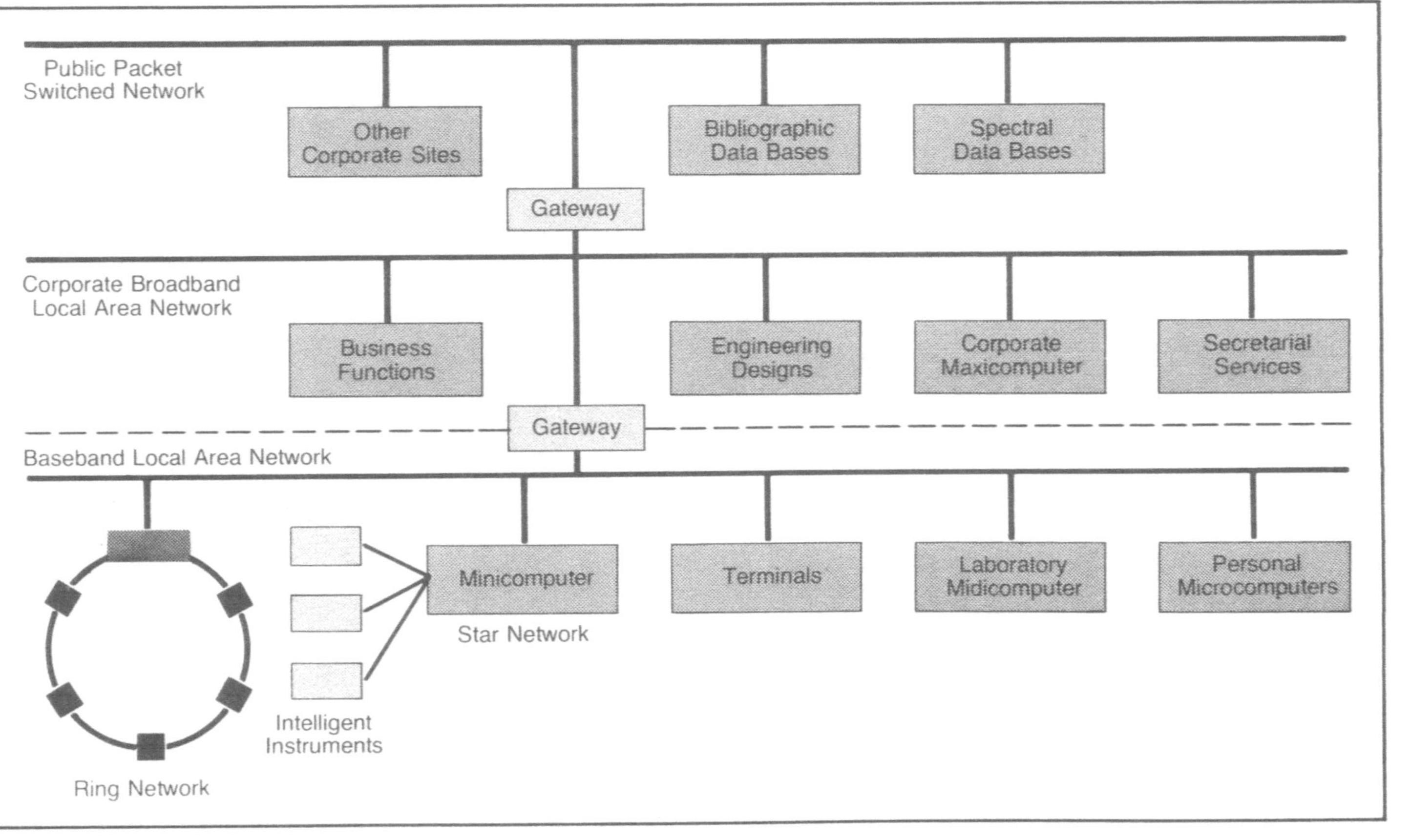

FIG. 3.6. Hierarchial network interconnection.
Reprinted with permission from Anal. Chem. (1982) 54, 1167A. Copyright 1982 American Chemical Society.

1. Is extensible.
2. Has moderate to heavy typing.
3. Is optimizing compiler oriented.
4. Enforces structured programming practices.
5. Allows top-down design, top-down programming.
6. Permits user-defined data structures and variable filling.
7. Supports pipe-lining.

Both language systems should support the following:

1. Arithmetic, logic, and Boolean operations at a high level.
2. Named procedures.
3. A full "if . . . else . . . then, begin . . . end, begin . . . while . . . repeat, case . . . end . . . case" set of conditionals, as well as DO/FOR structures that will handle negative increments, and extended range.
4. String manipulation operators.
5. Overload capabilities.
6. Exception handling service under programmer control.

LABORATORY INFORMATION MANAGEMENT SYSTEMS

The purpose of the LAN and its associated hardware, is to support a suitable laboratory information management system (LIMS).

The typical electronic laboratory starts by adding intelligent instruments. It rapidly increases its throughput of samples, and augments the quality of data produced. But it is rapidly inundated with both additional samples and the increased data that it is capable of generating. This sorcerer's apprentice needs to be controlled by a LIMS system. Concurrent with this is a conduit, the LAN, on which the information can flow. And shortly after these are installed, a good statistics package and scientific word processing support facility become an obvious need. Since the LIMS is the crucial part of this development, in the sense that it must be custom designed for the laboratory, some of the characteristics that any laboratory would require are listed as follows:

1. Sample entry, log-in, and bar-code identification.
2. Worklist assignments to the various instruments involved.
3. Receipts for return to the submitter.
4. Human cordial data entry, with self-editing capabilities.

5. Audit trail.
6. Release of analytical results and sample results by "signature."
7. Separate chemical data bases and information management data bases.
8. Ability to create operations information reports with respect to backlog, instrument load, personnel load, and cost analyses appropriate to the laboratory.
9. Good backup, backout, and backoff facilities.
10. Archiving of primary analytical data, as well as concentrated reports.
11. Quality control and statistics package, with interactive graphics capability.

THE COMPANY ENVIRONMENT

All of the above exist, embedded in the environment of the company's corporate computer facility. Many analysts see a common scenario developing, with a baseband LAN being used to connect the laboratory nodes together. A gateway would connect this LAN to a broadband or PBX LAN incorporating the larger computer facilities used for business applications (Fig. 3.6). Rapid developments are being made in the area of fully relational data bases capable of functioning in a distributed processing environment. This means that it would be possible for the various corporate structures to access "views" of data existing on other systems, where authorization existed. These views are not limited to the structures envisaged at the time of data base installation, as is true of most other data base management systems (DBMS) approaches. Relational data bases thus provide for changes in the questions that are asked by personnel. And with the ability to transmit these relationships to various nodes, it becomes possible to implement the decision support services (DSS) that are becoming necessary in today's laboratory and business environment.

These LAN installations would connect, in turn, to public packet switched networks (Fig. 3.6). Rapidly X.25 is becoming a standard protocol in the United States for such long-haul communication. This provides intracorporation connection, as well as access to the various electronic data bases that are becoming the preferred route into the technical literature. First generation data bases based on bibliographic search techniques are well established. Second generation data bases, where actual chemical and physical data are stored, are becoming more common. An example is the NIH/NBS/EPA consortia generation of the CIS system, which permits access to the wide variety of spectral data bases. Even third generation data bases, providing artifactual intelligence in the search for chemical identity, substructure components, synthetic approaches, or metabolite production, are becoming available.

There is no doubt that the electronic office is here. The laboratory must recognize that it too must change its work pattern.

ROBOTICS—COMPUTERIZATION VS. AUTOMATION

Many laboratories confuse the step of computerizing the laboratory with its true automation. The former only automates the data capture, flow, and condensation/report processes. True automation comes with importation of techniques such as flow injection analysis, which is the subject of another chapter in this treatise (ses Stewart, Chapter 15). Many pharmaceutical and analytical testing laboratories are also installing computer-controlled robots to move and manipulate samples, and even instruments. This is the subject of a separate A/C Interface paper (Dessy 1983D).

IMPLEMENTATION AND INSTALLATION

The technology involved with automation of the food laboratory is really well understood. What makes the job of the laboratory manager difficult is that he must simultaneously cope with a new vocabulary involving both hardware and software, fend off vendors afflicted with shark-frenzy, and enter into the arena of high technology politics. There are some simple rules that are derived from a number of case histories, some successful, others not:

1. It is important to co-opt the fealty of the users at an early date by involving them in the decision-making process.
2. Tasks in the laboratory that have high visibility should be attacked first. Although many of these do not have "glamour," they will create an atmosphere of credibility if the problems involved are quickly solved, with minimum disruption of the work habits in existence. Thus credibility is important as more complicated installations are attempted.

3. Any attempt to implement a full electronic capture of data by a LIMS system should be avoided until a manual entry stage has been fully tested to assure the integrity of the system.

4. One should avoid installation of electronic mail and word processing on the smaller lab computers. They may establish evidence of progress early in the installation, but they are incompatible with good real-time response.

5. The software tools should be developed with an eye towards the human cordiality required of a successful installation. User input as to the normal and abnormal sample flow in the laboratory, and what the actual needs for improved information retrieval are, can be elicited only by patient education and survey. The ergonomics of data entry, report preparation, and distribution must be considered.

6. No arguments based on reducing the number of personnel required in the short run should be used to justify the system. That is usually untrue and certainly creates a tense situation among prospective users. However, some installations report payback periods of 18 months, even for systems costing $500,000 to install.

7. Site visits to similar facilities and extensive presentations and commitments by the vendors selected for the final bid request are mandatory.

8. The personnel responsible for installation of the final system must be highly visible and responsive during the initial phases. Problems must be corrected quickly to avoid loss in credibility.

9. Management must be convinced of the need for in-house education. The individuals performing this function should be recognized for their effort.

We have a choice. We can control our computer system by understanding it. And we can have the system do things for us. Or we can let others do the automation design *for* us, and have the system do things *to* us!

BIBLIOGRAPHY

DESSY, R.E. 1982. Local area networks: Part 1. Anal. Chem. *54,* 1167A.

DESSY, R.E. 1983A. Laboratory information management systems: Part 1. Anal. Chem. *55,* 70A–80A.

DESSY, R.E. 1983B. Languages for the laboratory: Part 1. Anal. Chem. *55,* 650–662A.

DESSY, R.E. 1983C. Operating systems for the laboratory. Anal. Chem. *55,* 883A–892A.

DESSY, R.E. 1983D. Robotics and the laboratory: Part 1. Anal. Chem. *55,* 1100A–1114A.

4

Sample Preparation and Its Role in Nutritional Analysis

Harry G. Lento[1]

INTRODUCTION

Within recent years the growing concern regarding the ability of the world to nourish its ever-increasing population has kindled a keen awareness regarding the nutritional value of foods. As a consequence, a concerted effort is being made by the industrial, governmental, and academic communities to gain more knowledge about the nutritional quality of our food supply and to promulgate this information in a precise and meaningful manner to the consumer through nutritional labeling.

In satisfying this need, the food scientist has as his primary responsibility the task of developing information regarding the nutritional composition of specific foods. The ubiquitous nature of many of the nutrients; their presence in trace amounts; their susceptibility to further reaction and change and the wide variety of food types requiring compositing, blending, and analysis are just a few of the problems confronting the analyst in this work. Here we focus attention on one such critical area: i.e., the role that sample preparation plays in the scheme of developing information regarding the nutritional composition of foods, in general, and specifically in the development of information used in nutritional labeling.

[1]Director of Corporate Analytical Laboratories, Campbell Institute for Research & Technology, Campbell Soup Co., Camden, NJ 08101

REDUCING VARIATION BY UNIT COMPOSITING

In beginning, it is important to consider the problem confronting the food scientist concerned with establishing a label value. Obviously, before any food is labeled, sufficient analytical data must be obtained to establish that the nutritional content is truly representative of that particular food. In an ideal situation, a nutrient would vary only slightly within the product; and under these conditions, reliable label values can be established from very small samplings. Unfortunately, in the real world of the food analyst, this is often not the case and it is well recognized that there can be wide fluctuations in the nutritive content. Seasonal variation, plant locations, climatic conditions, processing conditions, the method of sample preparation, or the actual analytical determination itself may induce wide variation in the value obtained. Under such circumstances, it is not easy to arrive at a truly representative value, and oftentimes this poses some difficulty. Herein lies a problem. On one hand, it would be unfair to the manufacturer or processor of the food who is responsible for labeling to penalize the product by grossly understating its nutritional content to allow for these variabilities. On the other hand, federal regulations regarding nutritional labeling require that each nutrient in the food be within 80% of the label values (Federal Register 1973).

Fortunately, unit composition can often circumvent many of these problems (Compliance Procedures for Nutritional Labeling 1973). Perhaps this point can be better illustrated by Fig. 4.1 which shows two normal distribution curves representing the vitamin A content of a product containing β-carotene as the primary source of this vitamin. The lower curve A shows vitamin A values established from the analysis of individual units. Curve B shows the same data obtained when 12-unit composites are plotted. The data obtained from single-can analysis show a mean RDA value of 60%, with a standard deviation of approximately 20%. Recommended labeling guidelines suggest that the label value be set two standard deviations from the mean to assure 95% compliance. Figure 4.1A shows for single unit analysis that a label value of 20% RDA, although considerably removed from the mean, satisfies these criteria.

Again, referring to Fig. 4.1, curve B shows the same data obtained when 12-unit composites are analyzed and data plotted. As can be seen, the curve is similar to single-can analysis. The effect of 12-unit compositing is to reduce the variability by the square root of 12; that is, by a factor of 3.64 units. A comparison of the two curves demonstrates

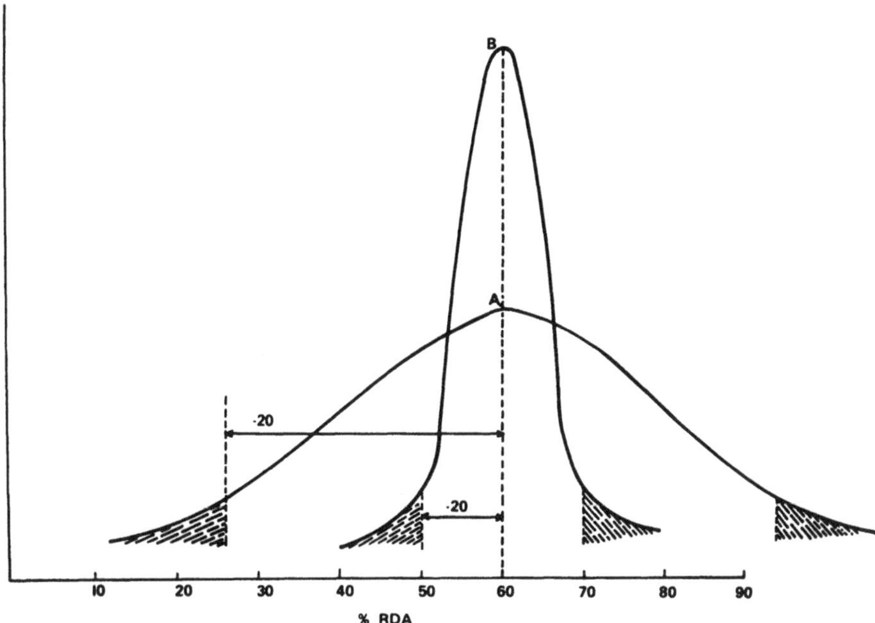

FIG. 4.1. Normal distribution curve for vitamin A. (A) Single can; (B) 12 can.

quite clearly the effect that compositing has in reducing such variability. This figure also shows that the actual label value of 50% of an RDA can be set considerably closer to the mean value of 60% with the same degree of confidence that 95% of product will meet this value. It can be seen from this that unit compositing solves the problem of insuring label compliance, while at the same time presenting a more realistic representation of the true nutritional quality.

PROBLEMS IN UNIT COMPOSITING AND SAMPLE PREPARATION

From a practical standpoint, however, it is easy to imagine the difficulty associated with preparing for nutritional analysis uniform and completely homogeneous 12-unit composites of such diverse samples as pies, lunchmeats, breads, frozen dinners, chickens, institutional size products, etc. To further complicate the problem, oftentimes the methods utilized in the determination of many of the nutrients in-

TABLE 4.1. RELATIONSHIP OF TOTAL COMPOSITE TO PERCENTAGE SAMPLE ACTUALLY ANALYZED

Nutrient	Total sample composite weight (g)[a]	Final analytical determination	
		% Original composite	Concentration of nutrient (μg)
Thiamin	2900	4×10^{-5}	5
Riboflavin	2900	2×10^{-6}	0.2
Niacin	2900	7×10^{6}	1
Vitamin A	2900	9×10^{-6}	0.8
Calcium	2900	9×10^{-6}	5
Iron	2900	8×10^{-6}	25
Sodium	2900	7×10^{-8}	10

[a] Sample weight of 12 units at 8.5 oz.

volved are so sensitive that only a minute quantity of the original composite is subjected to the actual analysis itself. Perhaps this point is best demonstrated from Table 4.1 which shows some of the representative levels of the typical nutrients and the amounts of sample composite required in the actual analytical determination.

When one considers that the individual composites in some instances must be prepared from several thousand grams, it is easy to see the importance that sample preparation and sample uniformity have in implementing any program to analyze for trace quantities of many of these nutrients. It is absolutely essential, therefore, that the analyst employ sampling techniques that will minimize the error involved in selecting and preparing the sample for analysis. More simply stated, the general problem in sampling and sample preparation is to insure that the sample actually analyzed contains the same amounts or concentrations of nutrients that are found in the original. Obviously, the more heterogeneous the material, the more complex the manipulations involved in obtaining a representative portion.

In the case of processed foods, Fig. 4.2 traces schematically the theoretical steps involved in the transport of a sample from a production line to the laboratory. As can be seen, each stage in the preparative scheme involves the sampling of the population with a corresponding reduction in the size: from the production line through unit compositing; subsampling; and finally, to actual physical analysis of the sample itself. This figure also indicates the manner in which the systematic introduction of product $(\sigma_P)^2$, sampling $(\sigma_S)^2$ and analytical $(\sigma_A)^2$ variability combined to affect the total variability $(\sigma_T)^2$ of the system. Because of the way in which incremental variances combine, only reduction in those sources of variance that make major contributions to the total variance can effect a major change in the system. Within re-

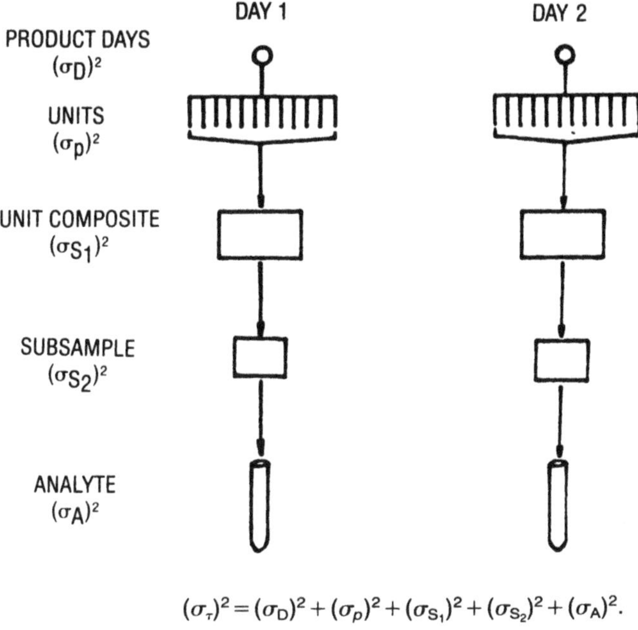

$$(\sigma_T)^2 = (\sigma_D)^2 + (\sigma_p)^2 + (\sigma_{S_1})^2 + (\sigma_{S_2})^2 + (\sigma_A)^2.$$

FIG. 4.2. Theoretical sampling scheme.

cent years, intensive research efforts have been directed toward increasing the level of sophistication of many of the analytical techniques used for nutritional analyses so as to reduce the analytical variability. Such has not been the case with sample preparation. In fact, it has been suggested that the combined error in sample preparation may be two to three times greater than those involved in the actual determination (van der Wal and Snyder 1980).

Remembering that our primary concern is to establish product variability, it is absolutely essential that the total contributions of sample and analytical variances be held to a minimum. Perhaps this can be better illustrated from the data presented in Fig. 4.3. This figure shows the manner in which the total variance of the system increases under conditions in which product variability (10%) and analytical variability (5%) are held constant and the variability due to sample preparation increased. As can be seen, the total variance in the system, for all practical purposes, is directly and almost linearly related to the variability introduced by the error in the preparative step. Further, when this error becomes appreciable, it tends to completely mask the variance of the nutrient in the food itself.

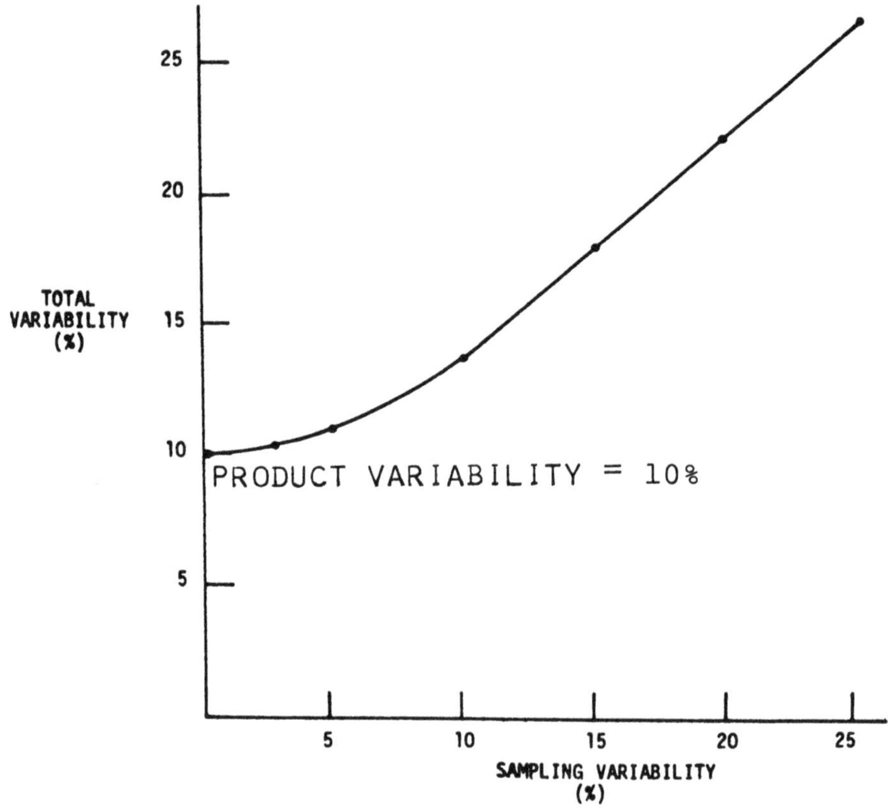

FIG. 4.3. Relationship of sampling variability to variability of system.

Obviously, the experimental approaches that can be utilized to evaluate sampling error can be as varied as the food matrix itself. Since sampling schemes and sample preparation procedures for different food types are available (AOAC 1980; Aulik 1974; Methods of Vitamin Assay 1966), it is not the intention here to provide specific recommendations along these lines. However, Table 4.2 provides an example of the type of experimental design that can be used to evaluate the efficacy of the sample preparative step. This table shows the results obtained from replicate samplings of "homogeneous" blends of 12-unit composites of the particular food matrices in which, prior to the analytical determination, the entire subsample was reblended. By selecting an analytical method with a high degree of precision ($\sigma_A = \pm 0.10$), statistical evaluation of the data can clearly establish whether or not a particular food matrix may pose any blending problem.

TABLE 4.2. EVALUATION OF THE EFFECTIVENESS OF SAMPLE COMPOSITION

Subsample	Sample A % Solids[a]		Sample B % Solids[a]	
1	54.5	54.5	41.7	41.8
2	54.6	54.5	41.6	41.7
3	54.2	54.4	42.1	42.2
4	54.4	54.4	42.1	42.0
5	54.6	54.8	40.6	40.3
6	54.4	54.4	41.7	41.6
	$(\sigma_T)^2 = (0.15)^2$		$(\alpha_T)^2 = (0.59)^2$	
	$(\sigma_A)^2 = (0.10)^2$		$(\sigma_A)^2 = (0.10)^2$	
	By diff. $(\sigma_S)^2 = (0.10)^2$		By diff. $(\sigma_S)^2 = (0.55)^2$	

[a] Duplicate determination.

VARIABILITY AND LABEL VALUE

Any discussion concerning the role of sample preparation in nutritional analysis would not be complete without some consideration being given to the interrelationship between sampling error and serving size and to the effect that such errors may have on the final label value. Federal regulations regarding nutritional labeling clearly stipulate that all nutrients must be expressed on the basis of the amount in one complete serving. Table 4.3 illustrates the manner in which the variability (± 0.01) associated with such factors as nonuniformity of sample preparation, sampling error, etc., can be amplified as a consequence of expressing the nutritive content on a per serving basis.

TABLE 4.3. EFFECT OF SERVING SIZE ON LABEL VALUE (% RDA FOR THIAMIN)

Content/100 g (variability ± 0.01)	Content/serving (0.05 mg ± 0.01 mg)	% RDA/serving
	Cond. soup (142 g)	
	0.057 → 0.028 mg / 0.045 mg	3.8 → 4.5% RDA / 3.0% RDA
	Main dish (270 g)	
0.05 → 0.06 mg / 0.04 mg	0.14 → 0.16 mg / 0.11 mg	9.0 → 10.8% RDA / 7.2% RDA
	Frozen dinner (450 g)	
	0.22 → 0.27 mg / 0.10 mg	15.0 → 18.0% RDA / 12.0% RDA

PREPARATION OF SAMPLE

It is well recognized that many of the nutritive constituents of food products are susceptible to oxidation, thermal, and photochemical decomposition. Table 4.4 lists the stability properties of some of the various substances requiring nutritional analysis. For these reasons, the area in which blending or sample preparation is to be conducted is usually chosen to be one with no outside windows or at least with subdued lighting. Ideal conditions are those in which a series of amber or red fluorescent lights are mounted overhead and used during the blending operation. For the most part, noncanned products should be received and held in the frozen state prior to analysis. The samples should be thawed overnight at 4°C and then stored in the dark and brought to room temperature prior to blending. Experience over the past several years covering an extensive program of nutritional labeling has shown that three main tools for blending will usually suffice for handling practically all types of food material requiring compositing and homogenization. These tools include a huge industrial size blender, a heavy-duty meat grinder, and some sort of high-speed chopper, such as a Hobart or Fitzmill unit. Another interesting blending technique which this author has had occasion to use is that of cryogenic blending. As the name implies, samples are hard frozen by the addition of liquid nitrogen and blended, usually in a Waring Blendor. As the samples return to room temperature, the nitrogen gas volatilizes leaving a fairly homogeneous and representative product. This technique is particularly useful where it is practically impossible to prepare a homogeneous sample by other ordinary techniques.

Table 4.5 shows some examples of the broad spectrum of foods normally encountered in the day-to-day operation of an analytical laboratory whose primary function is the analysis of these nutrients. Also

TABLE 4.4. POTENTIAL NUTRIENT LOSS ON BLENDING

	Thermal	Oxidative	Photodecomposition
Protein	×		
Fat		×	
Carbohydrate	×		
Ascorbic acid	×	×	
Carotene		×	×
Vitamin A		×	×
Niacin			
Riboflavin			×
Thiamin	×	×	

TABLE 4.5. REPRESENTATIVE FOOD TYPES AND EQUIPMENT USED IN SAMPLE PREPARATION FOR NUTRITIONAL ANALYSIS

Product	Sample size (× 12) (g)	Meat grinder	Fitzmill	Blender
Frozen "inst." foods	38,000	×	×	×
Pies	3,800			×
Pie fillings	4,100			×
Frozen cakes	5,400	×	×	
Canned soups	3,600			×
Canned "inst." soups	18,000		×	×
Grains	2,100		×	×
Spice mixes	2,100			×
Main dish	3,700			×

shown in this table are the various types of blending units that can be employed for compositing and homogenization of such sample types.

SUMMARY

Over the last two decades considerable progress has been made in improving the speed, accuracy, and precision of food analysis. The chain is only as strong as its weakest link, and still critical to any analytical chain is the preparation of the sample received prior to the analytical determination. Unfortunately, this is an area frequently overlooked by the food analyst. Within recent years the impact of nutritional labeling regulations and the need for compositing large sample sizes to insure representative samplings only serve to reemphasize the importance of sample preparation in the analytical scheme (see Table 4.3).

BIBLIOGRAPHY

AOAC. 1981. Official Methods of Analysis, 13th Edition. Assoc. Off. Anal. Chem., Washington, DC.

ASSOCIATION OF VITAMIN CHEMISTS, INC. (Editor). 1966. Methods of Vitamin Assay, 3rd ed. Wiley Interscience, New York.

AULIK, D.J. 1974. Sample preparation for nutrient analysis. J. Assoc. Off. Anal. Chem. *57*, 1190–1192.

Compliance Procedures for Nutritional Labeling. 1973. U.S. Govt. Printing Office 0 513–823 (20).

Federal Register. 1973. Food labeling. Federal Register 38 (13), Part III.

VAN DER WAL, R., and SNYDER, L.R. 1980. Precision of high performance liquid chromatographic assay with per sample treatment. Error analysis for the "Technicon" Fast LC system. Clinical Chem. *27*, 1233.

5

NBS Standard Reference Materials for Food Analysis

Robert Alvarez[1]

INTRODUCTION

In recent years, the chemical composition of foods has received increased national attention. Foods are being scrutinized by scientists in the food industry, government agencies, professional societies, and academic institutions. One of the more important aims in nutrition is to determine the effects of food nutrients on health and disease.

The National Research Council (NRC) Committee of the National Academy of Sciences report that offers guidelines for a diet likely to reduce the risk of cancer has led to controversy (see Fox 1982). The NRC findings have been greatly publicized as exemplified by a recent newspaper article (Bishop 1982) and magazine article (Ross 1983). Unaccustomed to hearing reassuring reports about cancer, the public has hastened to learn about the role of vitamins and selenium in preventing cancer.

Selenium is one of six essential trace elements for which the Food and Nutrition Board of the NRC provides estimated safe and adequate daily dietary intakes as narrow ranges (Recommended Dietary Allowances, Revised 1980). For example, the daily intake for selenium is 0.02–0.06 mg for infants (6 months to 1 year) and increases to 0.05–0.2 mg for adults. The daily intake for chromium is the same as that for selenium for all age groups while intakes for the other four elements, copper, manganese, fluorine, and molybdenum, are some-

[1] National Bureau of Standards, Office of Standard Reference Materials, Gaithersburg, MD 20899

what higher. For these six trace elements, the NRC warns: "Since the toxic levels for many trace elements may be only several times usual intakes, the upper levels for the trace elements given in this table should not be habitually exceeded." Yet, trace elements including these six are difficult to determine accurately in foods as shown in the section on Reliability of Food Analyses.

Mertz (1975, 1981) has described the status of essential trace element research and micronutrient deficiencies in humans and animals. He emphasizes the difficulties of arriving at a meaningful trace element analysis and a valid interpretation.

Constituents in foods are also believed to be factors in causing or promoting such degenerative diseases as heart disease, stroke, diabetes, osteoporosis, and hypertension (Sanders 1979). For example, the dietary intake of sodium has become of concern particularly for infants and hypertensives. Reacting to a popular outcry against too much salt in the diet of infants, food processors reduced the salt content of their formulas resulting in a formula deficient in chloride. This deficiency had adverse effects on the health of infants leading to the Infant Formula Act of 1980. At present, there is voluntary labeling of the sodium content of processed foods, but legislation that would mandate sodium labeling has been proposed.

The Food and Drug Administration (FDA) monitors foods for the presence of chemical contaminants unintentionally added to the food supply and also for certain mineral nutrients. The list of constituents determined and individual food items analyzed continues to grow (Worthy 1983). Examples of FDA analytical methodology have been presented (Boyer *et al.* 1979; Kuennen *et al.* 1982).

Interest continues to grow in establishing dietary requirements for nutrients; determining the role of nutrients in health and disease; accumulating baseline concentration data for nutrients in foods; investigating the effects on nutrients of producing and processing foods by different methods; and in monitoring foods over a long term for the presence of environmental pollutants, such as lead, mercury, and industrial chemicals. These analyses must be of demonstrated reliability to avoid conflict between government and industry, and between buyer and seller. Providing additional information on constituents in food products will result in increased testing needs and heavier laboratory workloads for both food producers and regulatory agencies. These increased demands will lead, in turn, to greater use of automated analytical systems. These rapid methods are usually calibrated with laboratory synthetic standards. In these methods, the instrumental response generated by the analytes are compared with those generated by "known" concentrations of the same constituents in the synthetic standards. Consequently, a primary factor affecting the accuracy of the results is the reliability of the standards. If the analytes and matrix components are of questionable purity, the analytical results, especially of micronutrients and potentially hazardous trace elements, will also be questionable. Therefore, primary quality assurance controls should be used to validate the experimental data.

The purpose of this chapter is to consider the current reliability of food analyses, and the use of Standard Reference Materials of certified composition and properties for improving their accuracy.

RELIABILITY OF FOOD ANALYSES

Stewart (1979) has evaluated the state of development of methodology for determining nutrients in foods according to the probability of obtaining a correct value, the speed of analysis, and cost. His classification is based on five descriptive terms: (1) sufficient, (2) substantial, (3) conflicting, (4) fragmentary, and (5) little to none. He has listed selected carbohydrates, lipids, macrominerals, trace elements, protein, vitamins, and available calories under these five terms.

If a constituent is listed under "sufficient," the probability of obtaining a correct value is excellent; if under "substantial," it is good to excellent; if under "conflicting," it is fair; if under "fragmentary," it is poor; and if the constituent is listed under "little to none," the probability of obtaining a correct value is "very low."

The speed of analysis ranges from fast for methods considered "sufficient" to very slow for those listed as "little to none." And, analytical costs are generally low when "sufficient" methodology is available but greatly increase as methodology deteriorates.

For the inorganic micronutrients, only methodology for determining copper and zinc is considered "sufficient." For selenium, it is considered "substantial;" for arsenic, chromium, fluorine, iodine, and manganese, "conflicting"; for molybdenum, "fragmentary"; and for cobalt, heme-iron, nonheme-iron, silicon, tin, and vanadium, the methodology is considered "little to none."

The range of analytical results provided by laboratories analyzing subsamples of the same homogeneous material also indicates the reliability of methodology. In one such interlaboratory study reported by Alvarez *et al.* (1979), the results for chromium determined in brewers' yeast by 14 laboratories ranged from 0.351 to 5.40 μg/g (ppm). The participants used two analytical techniques: atomic absorption spectrometry (AAS) and neutron activation. An analytical method of choice because of its simplicity and low cost, AAS showed more variation in the results than did those provided by neutron activation. This brewers' yeast was also analyzed at NBS using two techniques: stable isotope dilution-thermal source mass spectrometry and radiochemical neutron activation. Based on the close agreement of the results, the yeast was issued by NBS as SRM 1569 with a certified chromium concentration of 2.12 ± 0.05 μg/g.

In a continuation of this investigation, a second interlaboratory study was conducted in which the collaborators analyzed a different brewers' yeast using either Standard Reference Material (SRM) 1569 or other biological SRMs as controls. The reported results showed significantly closer agreement.

In an interlaboratory study reported by Heckman (1979), samples of homogenized foods analyzed for iodine showed a wide discrepancy in the analytical results confirming Stewart's assessment that the probability of obtaining a correct result for iodine was no better than fair. The study included samples of canned corn, dried skim milk, applesauce, whole wheat cereal, and canned tuna. As an example of the results, the iodine concentration reported for the canned corn ranged from 0.0089 to 1.20 μg/g. Or, on the basis of the percentage of recommended daily allowance for a 30-g serving, the range is 0.18 to 24%.

Dybczynski *et al.* (1980) have reported interlaboratory analyses of a milk powder for inorganic constituents by independent methods. In this study, the range of reported laboratory means for some important trace elements was as follows: As, 5–544 ng/g (ppb); Cd, 1–1660 ng/g; Co, 0.004–51 μg/g; Cr, 0.02–52 μg/g; Cu, 0.08–72 μg/g; Fe, 0.70–20 μg/g; Hg, 1–666 ng/g; I, 0.05 (one determination)–3.6 μg/g; Mn, 0.12–55 μg/g; Mo, 0.15–3.6 μg/g; Pb, 0.04–246 μg/g; Se, 25–313 ng/g; V, 0.002–20 μg/g; and Zn, 26–4194 μg/g.

The reliable determination of trace elements is acknowledged to be difficult. But, the determination of major and minor elements in the same milk powder also posed serious problems for some of the laboratories. This was apparent from the range of reported means, in percentage by weight (%), for these important macronutrients: Ca, 0.012–2.6; Cl, 0.11–3.4; K, 1.27–284; and Na, 0.32–0.72.

Unless the concentration of a constituent in a homogeneous material has been established by a definitive or a reference method of known accuracy, an interlaboratory study is best conducted using two or more independent analytical methods to minimize the possibility of method bias. Under these conditions, concordant results from laboratories using these independent methods establish a reliable consensus concentration for a constituent and also identify laboratories reporting errant results. However, by the time a laboratory receives a report of the study and becomes aware of an analytical problem, many important erroneous determinations of nutrients in food products may have been made and published. The use of food certified reference materials as primary controls for monitoring the precision and accuracy of methods can reveal an analytical difficulty at the time of occurrence.

VALIDATION OF METHODS AND ANALYTICAL DATA

Knowledge of the accuracy of analytical data is the foundation of an analytical laboratory's quality assurance system. But unlike precision or repeatability of analytical data, accuracy is difficult to evaluate. Unreliable analytical data may result from poor methodology, improper instrument calibration, impure reagents, faulty manipulations by the analyst, or a combination of these factors.

One method of estimating the accuracy of data and at the same time of validating the method that produced them is to compare these data with those generated on the same homogeneous material by a method of known accuracy (Cali 1974; Uriano and Gravatt 1977). But as Matthews (1976) states, "The need for reliable standardized methods of analysis is acute for many foods and nutrients."

A second approach for testing the reliability of analytical data is to analyze subsamples of the same homogeneous material by two or more independent methods, which differ in theory and practice. If the results obtained by the methods agree, then the probability of method bias has been greatly reduced. However, if the results do not agree, an all-too-frequent, frustrating experience, then the data from either or both methods may be erroneous.

In their paper on quality assurance in the analysis of foods for trace constituents, Horwitz *et al.* (1980) of the Food and Drug Administration observe that the best type of material to verify the correctness of a series of analyses, if it is available, is a Standard Reference Material certified by the National Bureau of Standards. Furthermore, the authors conclude from their interlaboratory quality control studies that all proven causes for deviations greater than two standard deviations can be attributed to standard solutions, gas chromotographic (GC problems, or calculation errors, with *standard solution problems being the culprit about half of the time.*

For use in anayzing foods and beverages, NBS issues matrix SRMs as primary controls for quality assurance applications; high-purity chemicals for preparing standard solutions, ampouled solutions with known concentrations of selected organic constituents for determining GC response factors; and instrument-performance materials for monitoring and checking the performance of laboratory instruments, such as spectrophotometers, polarimeters, thermometers, and pH meters.

NBS STANDARD REFERENCE MATERIALS FOR FOOD ANALYSIS

Biological Matrix SRMs Certified for Chemical Composition

The certificates of analysis for the biological matrix SRMs, used as accuracy controls, list certified concentrations with their estimated uncertainties and noncertified concentrations.

Certified concentrations are based on the results of either a definitive method or two or more independent analytical methods. A definitive method is a method of highest accuracy for the constituent. As an example of the use of a definitive method for certification, lead was determined in most of the food and plant tissue SRMs by isotope dilution mass spectrometry (IDMS). The lowest certified lead concentration in these SRMs is 0.020 ± 0.010 μg/g, the lead content of the wheat flour SRM. Barnes *et al.* (1982) have reported the principles of IDMS methodology and its inherent advantages for providing accurate concentrations of elements, such as lead in the food and plant tissue SRMs. Gramlich *et al.* (1977) have applied IDMS to the determination of zinc, cadmium, and lead in biological and environmental materials and demonstrated that the uncertainty in accuracy is independent of the concentration, provided that the analytical blank is small compared with the element concentration. Moore *et al.* (1975) provided similar data for molybdenum, rubidium, and strontium and cited additional data for lead, nickel, and cadmium.

Table 5.1 lists elements for which a certified concentration is given in the certificate of at least one of the five NBS food SRMs. The table does not show the estimated uncertainties of the concentrations which are given in the Certificate of Analysis. For example, chlorine is certified in Bovine Liver, SRM 1577a, at a concentration of $0.28 \pm 0.01\%$ where 0.01 is the estimated uncertainty. Values listed in parentheses, such as (1.0%) for chlorine in Oyster Tissue, SRM 1566, are not certified but are listed for information only (see Appendix to this chap-

TABLE 5.1. CERTIFIED CONCENTRATIONS OF ELEMENTS IN THE NBS FOOD SRMS[a]

Element	Oyster tissue, SRM 1566	Bovine liver, SRM 1577a	Wheat flour, SRM 1567	Rice flour, SRM 1568	Brewers' yeast, SRM 1569
Arsenic	13.4	0.047	(0.006)[b]	0.41	—
Cadmium	3.5	0.44	0.032	0.029	—
Calcium	0.15%	120	0.019%	0.014%	—
Chlorine	(1.0%)	0.28%	—	—	—
Chromium	0.69	—	—	—	2.12
Cobalt	(0.4)	0.21	—	0.02	—
Copper	63.0	158	2.0	2.2	—
Iron	195	194	18.3	8.7	—
Lead	0.48	0.135	0.020	0.045	—
Magnesium	0.128%	600	—	—	—
Manganese	17.5	9.9	8.5	20.1	—
Mercury	0.057	0.004	0.001	0.0060	—
Molybdenum	(≤0.2)	3.5	(0.4)	(1.6)	—
Nickel	1.03	—	(0.18)	(0.16)	—
Phosphorus	(0.81%)	1.11%	—	—	—
Potassium	0.969%	0.996%	0.136%	0.112%	—
Rubidium	4.45	12.5	(1)	(7)	—
Selenium	2.1	0.71	1.1	0.4	—
Silver	0.89	0.04	—	—	—
Sodium	0.51%	0.243%	8.0	6.0	—
Strontium	10.36	0.138	—	—	—
Sulfur	(0.76%)	0.78%	—	—	—
Uranium	0.116	0.00071	—	—	—
Vanadium	2.3	—	—	—	—
Zinc	852	123	10.6	19.4	—

[a] Concentrations in μg/g or where noted percentage by weight. Certificates of Analysis provide estimated uncertainties for the certified concentrations.
[b] Values in parentheses are not certified.

ter). The same considerations apply to Table 5.2 for NBS plant tissue SRMs.

The plant tissue SRMs were developed primarily for validating plant nutrients and potentially phytotoxic elements in foliar analyses. Although not foods, these SRMs are nevertheless useful for validating important elemental determinations in foods, especially plant tissue foods. For example, Citrus Leaves (SRM 1572), which is certified for iodine, is valuable for establishing more reliable concentrations of this analytically difficult trace element. The certified concentration, 1.84 ± 0.33 μg/g, is based on the results of an IDMS method for iodine developed recently at NBS by Gramlich and Murphy (1982) and on the results of neutron activation analysis (NAA) reported by Allegrini (1982). Although iodine is a mononuclidic element and ordinarily not amenable to IDMS determination, it was determined by spiking samples with a long-lived radioactive isotope ^{129}I, to which the ^{127}I in the sample could be compared after chemical processing.

TABLE 5.2. CERTIFIED CONCENTRATIONS OF ELEMENTS IN THE NBS PLANT TISSUE SRMS[a]

Element	Orchard leaves, SRM 1571	Citrus leaves, SRM 1572	Tomato leaves, SRM 1573	Pine needles, SRM 1575
Aluminum	—	92	(0.12%)[b]	545
Antimony	2.9	(0.04)	—	(0.2)
Arsenic	10	3.1	0.27	0.21
Barium	(44)	21	—	—
Beryllium	0.027	—	—	—
Boron	33	—	(30)	—
Cadmium	0.11	0.03	(3)	(<0.5)
Calcium	2.09%	3.15%	3.00%	0.41%
Chromium	2.6	0.8	4.5	2.6
Copper	12	16.5	11	3.0
Iodine	(0.17)	1.84	—	—
Iron	300	90	690	200
Lead	45	13.3	6.3	10.8
Magnesium	0.62%	0.58%	(0.7%)	—
Manganese	91	23	238	675
Mercury	0.155	0.08	(0.1)	0.15
Molybdenum	0.3	0.17	—	—
Nickel	1.3	0.6	—	(3.5)
Nitrogen	2.76%	(2.86%)	(5.0%)	(1.2%)
Phosphorus	0.21%	0.13%	0.34%	0.12%
Potassium	1.47%	1.82%	4.46%	0.37%
Rubidium	12	4.84	16.5	11.7
Selenium	0.08	(0.025)	—	—
Sodium	82	160	—	—
Strontium	37	100	44.9	4.8
Sulfur	(1900)	0.407%	—	—
Thorium	0.064	0.15	0.17	0.037
Uranium	0.029	(≤0.15)	0.061	0.020
Vanadium	—	0.24	—	—
Zinc	25	29	62	—

[a] Concentrations in μg/g or where noted percentage by weight. Certificates of Analysis provide estimated uncertainties for the certified concentrations.
[b] Values in parentheses are not certified.

The mean concentration of iodine in SRM 1572 determined by NAA was within the estimated limits of uncertainty of the certified concentration. The NAA procedure described by Allegrini *et al.* (1981) is a modification of the one described by Rook (1977).

Tables 5.3 and 5.4 list elements that have not been certified in the food and plant tissue SRMs, respectively. These noncertified values provide additional information on the composition of the SRM. They may be certified at a later date if a need is indicated.

A certificate is revised when additional characterization has been performed. As an example, the certificate for Oyster Tissue (SRM 1566) shown in the Appendix to this chapter had been revised to include a certified value for vanadium. This is the first time that a certified value

TABLE 5.3. NONCERTIFIED CONCENTRATIONS OF ELEMENTS, IN μg/g, IN THE NBS FOOD SRMS

Element	Oyster tissue, SRM 1566	Bovine liver, SRM 1577a	Wheat flour, SRM 1567	Rice flour, SRM 1568	Brewers' yeast, SRM 1569
Aluminum	—	(2)	—	—	—
Antimony	—	(0.003)	—	—	—
Bromine	(55)	(9)	(9)	(1)	—
Iodine	(2.8)	—	—	—	—
Tellurium	—	—	(≤0.002)	(≤0.002)	—
Thallium	(≤0.005)	(0.003)	—	—	—
Thorium	(0.1)	—	—	—	—

TABLE 5.4. NONCERTIFIED CONCENTRATIONS OF ELEMENTS, IN μg/g, IN THE NBS PLANT TISSUE SRMS

Element	Orchard leaves, SRM 1571	Citrus leaves, SRM 1572	Tomato leaves, SRM 1573	Pine needles, SRM 1575
Bismuth	(0.1)	—	—	—
Bromine	(10)	(8.2)	(26)	(9)
Cerium	—	(0.28)	(1.6)	(0.4)
Cesium	(0.04)	(0.098)	—	—
Chlorine	(690)	(414)	—	—
Cobalt	(0.2)	(0.02)	(0.6)	(0.1)
Europium	—	(0.01)	(0.04)	(0.006)
Fluorine	(4)	—	—	—
Gallium	(0.08)	—	—	—
Lanthanum	—	(0.19)	(0.9)	(0.2)
Lithium	(0.6)	—	—	—
Samarium	—	(0.052)	—	—
Scandium	—	(0.01)	(0.13)	(0.03)
Tellurium	(0.01)	(0.02)	—	—
Thallium	—	(≤0.01)	(0.05)	(0.05)
Tin	—	(0.24)	—	—

for this essential element has been listed in a biological SRM. The concentration is based on an IDMS method developed by Fassett and Kingston (1983).

High-Purity Organic and Inorganic SRMs

Compounds of known high purity are necessary for preparing standard solutions. These solutions, which have accurate concentrations of analytes, can then be diluted to prepare laboratory synthetic standards with a range of accurate concentrations for the analytes. The standards are used with instrumental techniques to establish analytical curves relating the instrumental response to the known concentration of the analyte. Therefore, if the purity of the compound is questionable, the analytical results will be likewise.

Examples of available high purity SRMs are: cholesterol (SRM 911) with a certified purity of $99.8 \pm 0.1\%$; D-glucose (SRM 917) with a certified purity of 99.9%; and calcium carbonate (SRM 915) with a certified purity of 99.9+%.

Miscellaneous Chemical Composition SRMs

Stabilized Wine, SRM 1590. Stabilized Wine is certified for its ethanol content based on the concordant results of two independent methods—gas chromatography and titration of the ethanol after distillation from the wine. In addition to the certified concentration of ethanol, the Certificate of Analysis provides information on the concentrations of copper, iron, potassium, sodium, volatile acidity, and total acidity. These noncertified values are based on a collaborative study conducted by the American Society of Enologists.

Constituents in Water. Water is a major constituent of many beverages and canned foods and therefore its purity must be assured to provide wholesome products. Several SRMs are available for validating water analyses. They include: Trace Elements in Water (SRM 1643a) and Halocarbons in Methanol for Water Analysis (SRM 1639). The SRM 1643a is certified for the concentrations of 17 elements at ng/g (ppb) levels in acidified water. The elements include arsenic, cadmium, chromium, lead, and selenium.

The SRM 1639 is certified for the concentrations of chloroform, chlorodibromomethane, bromodichloromethane, bromoform, carbon tetrachloride, trichloroethylene, and tetrachloroethylene at ng/mL levels in methanol.

The SRM 1639 was developed primarily for adding known accurate amounts of the certified halocarbons to water and other samples. It is also intended for use in calibrating chomatographic instrumentation and determining instrument-response factors.

Human Serum, SRM 909. Human Serum is a lyophilized material which is certified for selected electrolytes and organic constituents at normal levels in serum. It is useful for metabolic studies of nutrients.

At present, the certificate lists certified concentrations of calcium, chloride, cholesterol, glucose, lithium, magnesium, potassium, sodium, urea, and uric acid. These concentrations are based on the high accuracy definitive methods. The certificate also includes noncertified values for seven enzymes determined collaboratively using "best available methodology" and a value for total protein determined at NBS using a candidate reference method. In progress are analytical determinations of trace elements which are of special interest to nutritionists.

SRMs For Polarimetry

Two SRMs have been developed for calibrating polarimeters. They are Sucrose, SRM 17c, and Dextrose, SRM 41b.

The certificate for SRM 17c provides certified values for the optical rotation, index of refraction, density, and their associated standard errors of a "normal solution" at 20.00 ± 0.01°C. A "normal" sugar solution is defined as 26.0160 g of "pure" sucrose weighed in vacuum and dissolved in pure water and diluted to 100.000 cm^3 at 20.00°C. This corresponds to 23.7017 g sucrose per 100.00 g of aqueous solution.

The certificate for SRM 41b, which is certified as a chemical of known purity, provides values for α-D-glucopyranose, β-D-glucopyranose, moisture, ash, and values for the specific rotation of aqueous and methyl sulfoxide solutions.

SRMs for Thermometry, Spectrophotometry, Fluorimetry, and pH Measurements

Chemical reactions involving enzymes are very sensitive to small temperature changes. Therefore, accurate measurement of reaction temperature is an important step in the analytical procedure. For validating temperature measurements, a gallium melting point thermometric device (SRM 1968) is available with a certified melting point of 29.7723°C. A clinical laboratory thermometer with calibrated points at 0°, 25°, 30°, and 37°C is also available. Nearing completion is a rubidium melting point thermometric device for calibrating thermometers accurately near 37°C.

Spectrophotometers are used in analytical methods for constituents in foods. Proper calibration is necessary to obtain satisfactory performance and reliable analytical results. A Glass Filter for Transmittance Measurement, SRM 2030, was developed for use in determining color values of fat-soluble resins by spectrometric techniques. The filter is individually calibrated and certified for transmittance at a wavelength of 465.0 nm and for a spectral bandpass of 2.7 nm.

In addition, other SRMs are available to verify the accuracy of the transmittance and absorbance scales of spectrophotometers. They consist of another glass filter (SRM 930d) having additional calibration points, a metal-on-quartz filter (SRM 2031), liquid filters (SRM 931c), and crystalline potassium dichromate for use in solution form as an ultraviolet absorbance calibrator (SRM 935). A crystalline potassium iodide of known purity (SRM 2032) is available for assessing heterochromatic radiant energy (stray light) in ultraviolet spectrophotometers below 260 nm. The specific absorbance of an aqueous solution of KI is certified at eight wavelengths, in 5-nm intervals from 240 to 275 nm. Stray light in UV spectrophotometers is assessed by comparing the certified values of specific absorbance to the absorbance values of aqueous solutions of SRM 2032 measured under conditions of known concentrations, pathlength, and temperature.

For calibrating fluorescence spectrometers, a quinine sulfate dihydrate material (SRM 936) was analyzed for impurities, and a solution of SRM 936 in perchloric acid was certified for the relative molecular emission spectrum, from 375 to 675 nm for an excitation wavelength of 347.5 nm.

The SRMs for preparing solutions of known hydrogen ion concentrations are available. The certificates give directions for preparing the solutions and tables of pH values at various temperatures.

SRMS UNDER DEVELOPMENT

A Nonfat Dry Milk, Proposed SRM 1549. This is being analyzed for constituents of nutritional and toxicological signifiance. The lot of material was tested for homogeneity and found to be satisfactory.

Chlorinated Pesticides in Isooctane, Proposed SRM 1583. This has been prepared and is being analyzed for six pesticides at μg/mL levels. They are: δ-BHC, γ-BHC (Lindane), p,p′-DDE, p,p′-DDT, HHDN (Aldrin), and heptachlor epoxide. The SRM will enable accurate amounts of these pesticides to be added to samples of foods and other agricultural products being analyzed for determination of the pesticides by the method of additions.

Trace Elements in Urine, Proposed SRM 2670. This is being analyzed for trace elements of metabolic and toxicological significance. This material is expected to be issued in 1983.

Phenols in a Solvent SRM. This is being prepared for adding known amounts of these compounds to beverages and foods for determination by the method of additions.

APPLICATIONS OF NBS STANDARD REFERENCE MATERIALS

Horwitz (1976) describes a case where high concentrations of lead, cadmium, copper, and zinc were reported in food produced in an area of smelting and metallurgical activities. But after using FDA validated methods of analysis and SRMs, the levels of lead and cadmium obtained were reduced by as much as a factor of ten.

The SRMs are used in other countries as well as the United States. In an interlaboratory trace element survey involving 48 laboratories from Australia and New Zealand, Koh *et al.* (1981) found that laboratories using a certified SRM produced more accurate results than those which did not. The authors cited the use of NBS SRM 1571 Orchard Leaves and NBS SRM 1577 Bovine Liver in this survey.

FUTURE SRMS FOR FOOD ANALYSIS

Certified concentrations for many elements of nutritional and toxicological importance in animal and plant tissue SRMs are available. Lacking are certified values for the important micronutrients: tin, fluorine, and silicon. Development of reliable methodology would enable these essential elements to be certified.

In his paper on analytical methods for foods in the 1980s, Tanner (1982) refers to the need for certified organic constituents in food matrices as indicated in an NBS Workshop on Reference Materials for Organic Nutrient Measurements.

The proceedings of this workshop, available from NBS as Special Publication 635 (1982), considered reference materials that were needed and could be prepared with reasonable hope of success. They included the following: (a) corn syrup certified for glucose, maltose, maltotriose, and water content; (b) partially hydrogenated vegetable oil certified for cholesterol; (c) nonfat dry milk certified for fortified amounts of niacin, thiamin, and riboflavin; (d) either nonfat dry milk or a sugar-containing dry drink mix certified for vitamin C; and (e) a nonfat dry milk certified for galactose and lactose.

A reference material that was needed, but presented purity or stability problems, was vitamin A palmitate and β-carotene in partially hydrogenated vegetable oil.

As summarized by Uriano (1982) in the proceedings of the previously cited workshop, certification of organic constituents in a food SRM will depend on (1) assuring the homogeneity and long-term stability of both constituents and matrix, and (2) the development of methods having the requisite precision and accuracy.

In addition to the foregoing technical considerations, NBS funding for the development of methods to certify nutrients depends on the priority assigned to this activity compared to other proposed SRM projects. The author would appreciate learning of the readers' needs for SRMs in the food SRM category.

BIBLIOGRAPHY

ALLEGRINI, M. 1982. Personal Communication. Univ. of Pavia, Pavia, Italy.

ALLEGRINI, M., BOYER, K.W., and TANNER, J.T. 1981. Neutron activation analysis of total diet food composites for iodine. J. Assoc. Off. Anal. Chem. *64,* 1111–1115.

ALVAREZ, R., WOLF, W.R., and MERTZ, W. 1979. Biological reference materials certified for chromium content. *In* Chromium in Nutrition and Metabolism. D. Shappcott and J. Hubert (Editors). Elsevier North Holland, New York.

BARNES, I.L., MURPHY, T.J., and MICHIELS, E.A.I. 1982. Certification of lead concentration in Standard Reference Materials by isotope dilution mass spectrometry. J. Assoc. Off. Anal. Chem. *65,* 953–956.

BISHOP, J.E. 1982. Diet studies. Scientists conduct research on nutrients that may block cancer. The Wall Street Journal, November 15.

BOYER, K.W., TANNER, J.T., and GAJAN, R.J. 1978. Multielement analytical techniques at FDA. Am. Lab. *10*(2), 51–65.

CALI, J.P. 1974. A systematic approach to accuracy in clinical chemistry. Med. Instrum. *8,* 17–21.

DYBCZYNSKI, R., VEGLIA, A., and SUSCHNY, O. 1980. Report on the Intercomparison Run A-11 for the Determination of Inorganic Constituents of Milk Powder, Int. Atomic Energy Agency, Vienna, Austria.

FASSETT, J.D., and KINGSTON, H.M. 1983. NBS Report of Analysis. (Method is being written for publication.)

FOX, J.L. 1982. Diet/cancer recommendations debated. Chem. Eng. News, *60* (26), 26–28.

GRAMLICH, J.W., and MURPHY, T.J. 1982. In Annu. Rep. 1982, Center for Analyt. Chem. Rep. NBSIR 82–2620. Natl. Bur. Stand. Gaithersburg, MD.

GRAMLICH, J.W., MACHLAN, L.A., MURPHY, T.J., and MOORE, L.J. 1977. The determination of zinc, cadmium, and lead in biological and environmental materials by isotope dilution mass spectrometry. *In* Trace Substances in Environmental Health, XI. A Symposium. D.D. Hemphill (Editor). Univ. of Missouri, Columbia.

HECKMAN, M.M. 1979. Analysis of foods for iodine: interlaboratory study. J. Assoc. Off. Anal. Chem. *62,* 1045–1049.

HORWITZ, W. 1976. Problems of sampling and analytical methods. J. Assoc. Off. Anal. Chem. *59,* 1197–1203.

HORWITZ, W., KAMPS, L.R., and BOYER, K.W. 1980. Quality assurance in the analysis of foods for trace constituents. J. Assoc. Off. Anal. Chem. *63,* 1344–1354.

KOH, T.S., BENSON, T.H., and JUDSON, G.J. 1981. Quality control in trace element analysis: an interlaboratory study in Australia and New Zealand. Chem. Aust. *48,* 139–140.

KUENNEN, R.W., WOLNIK, K.A., and FRICKE, F.L. 1982. Pressure dissolution and real sample matrix calibration for multielement analysis of raw agricultural crops by inductively coupled plasma atomic emission spectrometry. Anal. Chem. *54,* 2146–2150.

MATTHEWS, R.H. 1976. Nutrient content of wheat and wheat products. *In* Proc. of the 9th Natl. Conf. on Wheat Utilization Res. Rep. ARS-NC-40 available from Northern Regional Res. Center, ARS, U.S. Dep. of Agric., Peoria, IL.

MERTZ, W. 1975. Trace element nutrition in health and disease: contributions and problems of analysis. Clin. Chem. *21,* 468–475.

MERTZ, W. 1981. The essential trace elements. Science *213,* 1332–1338.

MOORE, L.J., GRAMLICH, J.W., and MACHLAN, L.A. 1975. Application of isotope dilution to the high accuracy trace analysis of environmental and health standards. *In* Trace Substances in Environmental Health, IX. A Symposium, D.D. Hemphill (Editor). Univ. of Missouri, Columbia.

RECOMMENDED DIETARY ALLOWANCES. 1980. Food and Nutrition Board, Natl. Acad. Sci., Washington, D.C.

ROOK, H.L. 1977. The determination of iodine in biological and environmental Standard Reference Materials. J. Radioanal. Chem. *39,* 351–358.

ROSS, W.S. 1982. At last, an anti-cancer diet. The Readers Digest, February, 78–82.

SANDERS, H.J. 1979. Nutrition and health. Chem. Eng. News. *57* (13) 27–46.

STEWART, K.K. 1979. Nutrient analysis of foods: State of the art for routine analysis. *In* Nutrient Analysis Symp. Proc. Assoc. Off. Anal. Chem., Arlington, VA.

TANNER, J.T. 1982. Analytical methods for foods in the next decade. J. Assoc. Off. Anal. Chem. *65,* 531–534.

URIANO, G.A. 1982. The process and requirements for the development of a Standard Reference Material. *In* Reference Materials for Organic Nutrient Measurements. NBS Spec. Pub. *635.* S.A. Margolis (Editor). Natl. Bur. Stand., Gaithersburg, MD.

URIANO, G.A., and GRAVATT, C.C. 1977. The role of reference materials and reference methods in chemical analysis. CRC Critical Rev. Anal. Chem. *6,* 361–411.

WORTHY, W. 1983. Federal food analysis program lowers detection limits. Chem. Eng. News *61* (10) 23–24.

APPENDIX

The National Bureau of Standards is responsible under Federal statute for issuing Standard Reference Materials (SRMs) (see pp. 97–99) to assist investigators in improving the accuracy of their laboratory tests. For the food science laboratory, these well-characterized, certified materials are available to serve as accuracy-control materials; to prepare primary standard solutions; and to evaluate and monitor the performance of instruments and devices, such as polarimeters and spectrophotometers. Of the approximately 900 different SRMs listed in the current catalog, the biological matrix materials are especially suitable for long-term quality assurance of food analyses. Examples of these are Oyster Tissue (SRM 1566); Bovine Liver (SRM 1577a); Wheat Flour (SRM 1567); Rice Flour (SRM 1568); and a Non-Fat Milk Powder (Proposed SRM 1549), expected to be issued in late 1983. The Certificates of Analysis for these SRMs include certified concentrations of nutritionally and toxicologically important elements. Other SRMs for food and beverage analysis include a stabilized wine and compounds of certified high purity, such as cholesterol. Additional SRMs have been developed for metabolic studies, such as Human Serum (SRM 909).

National Bureau of Standards

Certificate of Analysis

Standard Reference Material 1566

Oyster Tissue

This Standard Reference Material is intended primarily for use in calibrating instrumentation and validating methodology for the chemical analysis of marine animal tissue.

Certified Values of Constituent Elements: The certified values for the constituent elements are shown in Table 1. Certified values are based on results obtained by reference methods of known accuracy; or alternatively, from results obtained by two or more independent and reliable analytical methods. Non-certified values are given for information only in Table 2. All values are based on a minimum sample size of 250 mg of the dried material.

NOTICE AND WARNINGS TO USERS

Expiration of Certification: This certification is invalid after 5 years from the date of shipping. Should it become invalid before then, purchasers will be notified by NBS.

Storage: The material should be kept tightly closed in its original bottle and stored in a desiccator at temperatures between 10-30 °C. It should not be exposed to intense sources of radiation, including ultraviolet lamps or sunlight.

Use. A minimum sample weight of 250 mg of the *dried* material (see Instructions for Drying) is necessary for any certified value in Table 1 to be valid within the stated uncertainty. The bottle should be shaken well before each use, and closed tightly immediately after use.

The statistical analysis of the data was performed by K. R. Eberhardt and H. H. Ku of the Statistical Engineering Division.

The overall direction and coordination of the analytical chemistry measurements leading to this certificate were performed in the NBS Center for Analytical Chemistry by P. D. LaFleur.

The technical and support aspects involved in the preparation, certification, and issuance of this Standard Reference Material were coordinated through the Office of Standard Reference Materials by R. Alvarez.

Washington, D.C. 20234
February 22, 1983
(Revision of Certificate
Dated 12-12-79)

George A. Uriano, Chief
Office of Standard Reference Materials

(over)

Instructions for Drying: Before weighing, samples of SRM 1566 should be dried to constant weight by one of the following procedures:

1. Reduced-pressure drying at room temperature for 48 hours over $Mg(ClO_4)_2$ in a vacuum desiccator at approximately 1.3×10^4 Pa (100 mm Hg).
2. Vacuum drying at room temperature for 24 hours at a pressure of approximately 30 Pa (0.2 mm Hg) using a cold trap.
3. Freeze drying for 20 hours at a pressure of approximately 3 Pa (0.02 mm Hg).

Source and Preparation of Material: The oysters for this reference material were obtained by the FDA Bureau of Shellfish Sanitation from a commercial source. They had been shucked, frozen, and packaged in sealed plastic bags. The oyster material was ground, freeze-dried, and powdered at the U.S. Army Natick Research and Development Command, Natick, Mass., under the direction of L. Hinnegardt and G. C. Walker. At NBS, preliminary analyses of the material homogeneity indicated that an improvement in homogeneity would be required to establish more reliable certified values for a minimum sample size of 250 mg. Accordingly, the material was cryogenically ground by J. R. Moody and J. Matwey. It was then blended and bottled at NBS, after which it was again freeze-dried at the Natick, Mass., laboratory.

Homogeneity Assessment: Randomly selected bottles of SRM 1566 were sampled and tested for homogeneity by neutron activation and atomic absorption spectrometry. No inhomogeneity was observed for the following elements determined by neutron activation: Na, Cl, V, and Mn. The values for Mg, K, Cu, Zn, and Cd determined by atomic absorption spectrometry were within the imprecision of the method; however, Ca does exhibit some inhomogeneity--approximately 4% relative standard deviation.

Analysts:

Center for Analytical Chemistry, National Bureau of Standards:

1. J. V. Bailey
2. C. Blundell
3. T. J. Brady
4. M. Diaz
5. L. P. Dunstan
6. M. S. Epstein
7. J. D. Fassett
8. M. Gallorini
9. E. L. Garner
10. T. E. Gills
11. J. W. Gramlich
12. R. R. Greenberg
13. S. Hanamura
14. S. Harrison
15. E. F. Heald
16. G. M. Hyde
17. W. R. Kelly
18. H. M. Kingston
19. W. F. Koch
20. R. M. Lindstrom
21. G. J. Lutz
22. L. A. Machlan
23. W. A. MacCrehan
24. E. J. Maienthal
25. J. Maples
26. O. Menis
27. J. D. Messman
28. J. R. Moody
29. L. J. Moore
30. T. J. Murphy
31. P. J. Paulsen
32. T. C. Rains
33. H. L. Rook

Cooperating Analysts:

34. University of Tokyo, Tokyo, Japan; Y. Dokiya (NBS Guest Worker).
35. Division of Chemistry, National Research Council of Canada, Ottawa, Canada; S. Berman, A. Desaulniers, J. McLaren, A. Mykytiuk, D. Russell, and S. Willie.
36. Ibaraki Electrical Communication Laboratory, Nippon Telegraph and Telephone Public Corporation, Tokai, Ibaraki, Japan; K. Kudo and K. Kobayashi.
37. Food Research Division, Health Protection Branch, Tunney's Pasture, Ottawa, Ontario, Canada; R. W. Dabeka, A. D. McKenzie, and H. B. S. Conacher.

Table 1. Certified Values of Constituent Elements

Element[1]	Content[2], Wt. Percent	Element[1]	Content[2], Wt. Percent
Calcium[b,d]	0.15 ± 0.02	Potassium[d]	0.969 ± 0.005
Magnesium[a,d]	0.128 ± 0.009	Sodium[b,f]	0.51 ± 0.03
Element[1]	**Content[2], μg/g**	**Element[1]**	**Content[2], μg/g**
Arsenic[a,f,g,h]	13.4 ± 1.9	Nickel[a,e,h]	1.03 ± 0.19
Cadmium[a,d,e,f,h]	3.5 ± 0.4	Rubidium[d,f]	4.45 ± 0.09
Chromium[d,e,f]	0.69 ± 0.27	Selenium[a,e,f]	2.1 ± 0.5
Copper[a,c,e,f]	63.0 ± 3.5	Silver[a,f]	0.89 ± 0.09
Iron[b,c,e,f]	195 ± 34	Strontium[b,d]	10.36 ± 0.56
Lead[a,d,e,h]	0.48 ± 0.04	Uranium[d]	0.116 ± 0.006
Manganese[a,c,f]	17.5 ± 1.2	Vanadium[d]	2.3 ± 0.1
Mercury[a,f]	0.057 ± 0.015	Zinc[a,c,d,e,f,h]	852 ± 14

1. Analytical Methods:
   [a] Atomic absorption spectroscopy
   [b] Atomic emission spectroscopy, flame
   [c] Atomic emission spectroscopy, inductively coupled plasma
   [d] Isotope dilution mass spectrometry, thermal ionization
   [e] Isotope dilution mass spectrometry, spark source
   [f] Neutron activation
   [g] Photon activation
   [h] Polarography

2. Based on dry weight. (For drying instructions, see the section of this certificate on Instructions for Drying.) The estimated uncertainty is given as 95 percent tolerance limits for coverage of at least 95 percent of the measured values of all bottles of SRM 1566. For a given element, the following statement can be made at a confidence limit of 95 percent. "If the concentrations were measured for all bottles, at least 95 percent of these measured values should fall within the indicated limits." The concept of tolerance limits is discussed in Chapter 2, Experimental Statistics, NBS Handbook 91, 1966, and page 14, The Role of Standard Reference Materials in Measurement Systems, NBS Monograph 148, 1975.

Table 2. Non-certified Values of Constituent Elements

Element	Content[1] (Wt. Percent)
Chlorine	(1.0)
Sulfur	(0.76)
Phosphorous	(0.81)
	(μg/g)
Bromine	(55)
Cobalt	(0.4)
Fluorine	(5.2)
Iodine	(2.8)
Molybdenum	(≤0.2)
Thallium	(≤0.005)
Thorium	(0.1)

[1] Based on dry weight. (For drying instructions, see the section of this certificate on Instructions for Drying.)

6

Quality Assurance of Analysis of Inorganic Nutrients in Foods

James M. Harnly,[1] *Wayne R. Wolf,*[1]
and Nancy J. Miller-Ihli[1]

INTRODUCTION

The most important aspects of any analytical method are the steps taken to ensure the quality of the results. Quality assurance is the system of activities whose purpose is to provide assurance that the overall quality control limits are being met (Taylor 1981). The quality control limits (established by the analytical laboratory, or an outside organization) define the expected accuracy and precision of the results. Quality assurance must cover all aspects of the analytical process, from sampling, through sample treatment, standards preparation, and method validation, to the data handling and evaluation. Without a defined quality assurance program, all analytical results must be suspect.

With recent advances in modern analytical techniques, the need for quality assurance is even more critical. The current trend towards multielement determinations, automation, and computerization has resulted in the analyst being less directly involved in the actual analytical measurement process. All too frequently the analyst accepts computerized results without question.This point is well made by Frank and Kowalski (1982) in a review of chemometrics, which should be re-

[1]Beltsville Human Nutrition Research Center, Nutrient Composition Laboratory, U.S. Department of Agriculture, Beltsville, MD 20705

quired reading for everyone involved with automated and computerized methods. Without the proper quality assurance, the latest advances in analytical methodology only result in erroneous results being turned out at a faster rate.

This chapter considers quality assurance for the analysis of inorganic nutrients in foods by atomic spectrometry. The first section provides a general overview of quality assurance and the second section an example of how these concepts have been applied specifically to a large-scale analytical project.

GENERAL ASPECTS OF QUALITY ASSURANCE FOR ATOMIC SPECTROMETRY

Sampling

Sampling is the selection, from a population, of a finite number of units to be analyzed. The object is to choose the units so that the results obtained for the analyses are an accurate characterization of the entire population. A bias in sampling can produce a bias in the characterization of the population.

The problems in establishing a valid sampling of foods are considerable because of the many variables. For processed foods the analyst must consider the wide variety of brand names available, geographical location, seasonal variation, and the effect of shelf life. There are also longer-term variations associated with the development of new processing methods and new sources of raw materials. The number of variables associated with fresh foods and commercially prepared foods are equally complex. Fresh foods will vary with locale, season, weather, fertilization, and so on. Commercially prepared foods will have the variables associated with fresh and processed foods plus the variations associated with the method of preparation. Unless valid statistical methods are used, the results of the study may not be representative beyond the limited population analyzed.

Statistically valid sampling is a complex science. The reader is referred to the many excellent publications on the topic for detailed discussions in areas of interest (Youden and Steiner 1967; NBS #422 1976; Kratochvil and Taylor 1981).

Validation of Analytical Data

There are three general aspects to validating analytical data. These are (1) developing a method of analysis, (2) validating the method, and (3) ensuring that the method remains valid throughout its use.

To develop a reliable routine method of analysis, the whole analytical process must be critically examined and understood. The strengths, weaknesses, limitations, and ruggedness of the analytical aspects of a procedure must be documented by appropriate research and development. Potential matrix effects must be identified for each type of sample. Each type of food is basically a different chemical matrix. For a particular nutrient, a method that is valid for one type of food may not be valid for another. Different instrumental parameters and/or sample preparation methods may be necessary for different matrices and for different elements. These analytical variables must be identified and taken into account in the development of an accurate procedure.

Once the method has been developed and is under control (i.e., reproducible results with high precision can be obtained), the accuracy must be validated. The method must be shown to give results in agreement with the "true" value of the analyte in the samples.

The final step, after a method has been developed, validated, and put in routine use, is to establish appropriate monitoring procedures to ensure that the data obtained remain valid. The key to this quality control (QC) is the availability and use of the appropriate control samples.

Analytical Methodology. The analytical methodology will be considered under four broad headings: (a) sample treatment, (b) standards preparation, (c) signal measurement, and (d) optimization of analytical parameters. A more detailed discussion of analytical methodology has been previously presented by the authors (Harnly and Wolf 1983).

The main goal of the sample treatment, for atomic spectrometry, is to convert the sample to a liquid form (suitable for analysis) without adulterating the metal content. To obtain accurate results, all aspects of the sample treatment (collection, storage, homogenization, subsampling, and digestion) must be designed to avoid either contamination or loss of any of the elemental components. For intermediate and major constituents (greater than 1 $\mu g\ g^{-1}$), this aspect is important, but can be accomplished without extreme measures. For trace elements (less than 1 $\mu g\ g^{-1}$), however, every phase of the sample treatment is critical. An attitude of "useful paranoia" towards contamination is a prerequisite for accurate trace metal determinations.

Laboratory techniques and equipment for maintaining the trace metal integrity of the sample have been well reviewed by a number of authors. General laboratory techniques have been covered by Thiers (1957), Zeif and Mitchell (1976), Veillon and Vallee (1978), Moody

(1982), and Harnly and Wolf (1983). Comparison of specific digestion or solubilization methods can be found in several texts (Thiers 1957; Christian and Feldman 1970; O'Haver 1976B). To summarize these sources, all aspects of the analytical process must be questioned. Contamination can arise from laboratory equipment, glass and plasticware, reagents, and airborne particulate matter. Metal losses can occur during the sample digestion and through contact with sample containers and equipment. Ideally, as little sample treatment as possible is desired.

Atomic spectrometric methods determine the concentration of samples by comparing their analytical signals to those of a series of calibration standards. As a result, the analytical determinations are only as accurate as the standards. A number of schemes are employed for calibration. Matrix matching is the most accurate but is time consuming and requires detailed knowledge of the sample matrix. Dilution of the samples and standards into a common matrix can eliminate the most obvious interferences and is suitable for large numbers of samples. The method of standard additions uses the sample matrix as the standard but assumes equilibrium between the added standard and the endogenous elements. Pure standards can sometimes be used depending on the nature of the sample and the method of analysis.

The complexity of the calibration standards is dependent on the number of elements to be analyzed and the method of analysis. Multielement standards require compatibility of the elements in solution (no coprecipitation or opposing pH requirements), additional precautions (generally higher acid concentrations) to ensure the long-term stability of the solution, and high purity of each of the components (especially when one component is several orders of magnitude less concentrated than another). By comparison, single element standards are much simpler. The method of analysis will determine the nature and severity of interferences. Low temperature (<3000°C) atomization sources make matrix matching or a common diluent necessary to avoid chemical interferences, whereas high temperature (>5000°C) sources require high purity of the standards to avoid spectral interferences.

Accurate and precise measurements of the analytical signal are crucial to every determination. There are two major sources of measurement error that are of concern to the analyst; those errors which arise from the detection system and those which arise from the chemical and physical nature of the sample (O'Haver 1976B). Detection system errors arise from uncertainties inherent in the atomization and detection processes and are usually random in nature. Errors due to the chemical and physical nature of the sample are systematic in nature and are usually called interferences. There are two types of interferences: additive interferences and multiplicative interferences. Spectral interferences, broad band and line overlap interferences, are examples of the additive type while chemical and physical interferences associated with the sample matrix, matrix effects, are examples of the multiplicative type. The types and severity of interferences differ according to the method of analysis and the nature of the sample. Interferences must be corrected in order to obtain accurate determinations, whereas random errors are evaluated using statistics.

Detection, or random, errors are inherent in the instrument operation and are present in the measurement of every analytical signal, regardless of whether it is a blank, standard, or sample. The uncertainty of a measurement noise can be evaluated by making repeat measurements of the signal and computing the standard deviation. The signal-to-noise ratio, or inversely, the percentage coefficient of variation (the noise-to-signal ratio expressed as a percentage) can be used to establish the reliability of a signal. The random noise in the baseline determines the lowest analytical signal which can be detected, the detection limit. Detection limits are useful for evaluating a method's suitability for a particular analysis, for comparing instruments and methods, and for evaluating instrumental performance between experiments. The detection limit is defined as three times the standard deviation of the baseline (IUPAC 1976), a coefficient of variation (CV, or relative standard deviation) of 33%. The quantitation limit is generally defined as ten times the standard deviation of the baseline (ACS 1980), a CV of 10%.

Optimization of the analytical parameters is straightforward for the single-element mode of analysis. All aspects of the sample treatment, the standards preparation, and the signal measurement process (instrumental parameters) can be optimized for the element of interest. Thus the single-element mode of analysis is well suited for samples which require high precision, whose concentrations fall close to the detection limit, or which are subject to interferences.

Multielement determinations on the other hand require compromises in almost all analytical parameters: the sample treatment, standards preparation, the signal measurement process, and the handling and critical evaluation of the data. The sample treatment must be carried out in a manner that is either the most suitable for all elements or the best suited for the most sensitive element, i.e., the most volatile element determines the maximum temperature for a dry ashing method. The sample size, the dilution factor, and the instrumental parameters cannot be optimized for all elements simultaneously. The analyst can select parameters that compromise all elements equally, favor the element with the worst signal-to-noise ratio, or are weighted in the manner desired. Finally, handling and critical evaluation of the data are compromised with respect to time. As the data are produced at faster rates, it becomes impossible for the analyst to individually compute the concentration for each element for each sample and critically examine each result for accuracy. Instead, these tasks are turned over to a computer.

Multielement methods are best suited for the analysis of large numbers of samples provided the compromises in the analytical parameters can be tolerated. The extent of compromise is dependent on the method of analysis and the elements to be analyzed. The success of inductively coupled plasma-atomic emission spectrometry can be attributed to the minimum compromise required for the analysis of almost all of the metals. The analyst must assess the laboratory's requirements and evaluate the ability of individual or multielement methods to meet those needs.

Method Validation. Having established an analytical method, the analyst must validate the accuracy of the method. This is best done by verifying the results using an independent method or by obtaining accurate results for a certified reference material of the same composition. Other, less conclusive methods are the comparison of results to values reported in the literature and the comparison of results with other laboratories for the analysis of an exchange sample.

The establishment of two independent methods for determining the same elements is the ideal method for validating analytical results. Independent methods, each based on a different physical principle, seldom suffer from the same systematic biases, or interferences. If different sample preparation methods are employed, then analytical agreement of the two methods allows the analyst to place a great deal of confidence in the result. This approach is not usually considered since it is either beyond the capability of most laboratories or is not usually economically feasible. This is especially true for the analysis of trace elements. The development of a single state-of-the-art method can be quite a drain on time, money, and resources. Development of a second method is not usually considered unless the purpose of the laboratory is to produce highly reliable values for an appropriate reference material.

The most common means of validating results is to obtain accurate determinations for a certified reference material of the same chemical composition. Certified reference materials are reference materials accompanied by a certificate issued by a recognized official standards agency. These certificates give the reference value of the component plus confidence limits. The National Bureau of Standards (NBS) is the United States' source of certified reference materials (see Alvarez, Chapter 5). The NBS Standard Reference Materials (SRMs) are carefully prepared for homogeneity and stability and are characterized by at least two independent analytical methods (NBS # 492 1977). Currently NBS lists nine biological SRMs which may be appropriate for foods.

The International Atomic Energy Agency also has several biological matrices certified for a number of inorganic elements including trace elements (Parr 1980). Several other laboratories have issued reference materials and a directory of these materials is available (ISO 1982). There is promise for several new materials in the near future (Ihnat *et al.* 1982; Muntau 1980).

The major problem associated with reference materials is finding one similar to the samples to be analyzed. The limited number of currently available food reference materials means that, in many cases, it is necessary to use a reference material that only superficially resembles the sample. Cluster analysis procedures have been used to show that for metals there is relatively little composition overlap between the available biological reference materials (from NBS and the IAEA) and 160 of the most commonly consumed foods in the United States (Wolf 1982). In addition, the range of concentrations of the elements in the reference materials is not fully compatible with the range of concentrations found in foods. There is currently a great need for a wider range (both of food types and of elemental concentrations) of food reference materials.

In the absence of a second, independent method or a suitable reference material, results can be compared to previously reported values in the literature. Agreement with the published values lends strong support to the accuracy of the method, but disagreement can be very dissatisfying. In most cases, there is no way to resolve differences between the data without the development of an independent method. It is possible that the analytical differences arise from the samples being different. This difference would be best resolved by the analysis of a common sample.

Analysis of a common exchange sample by a group of laboratories is another means of evaluating analytical results. Agreement with other laboratories lends strong support to the validity of the method, but differences may be difficult to resolve. If the group exchanging samples is not sufficiently large, or, if each laboratory employs the same analytical method, the consensus result may not be the most accurate value. The agreement between laboratories is more significant if more than one independent method is employed.

Quality Control Standards. After a method has been developed, validated, and implemented, it is necessary to ensure that the results continue to be valid. It is necessary to periodically analyze a sample of known value. Primary or certified reference materials can be expensive and are frequently not available in large quantities. These materials are generally used only for the initial validation of a method. For routine usage, it is necessary for laboratories to develop secondary reference materials or quality control (QC) samples. The QC samples usually consist of a particularly large amount of an individual sample or several samples pooled together. The QC samples are carefully characterized using certified materials as standards (if possible) and the best possible quality assurance methods to determine the "true" value of their components and the normal ranges of variation.

The QC samples are used in a number of ways. They may be analyzed in conjunction with the samples in a random or systematic pattern. Atomic spectrometric methods require recalibration at frequent intervals. It is desirable to analyze at least one QC sample for each calibration. Quality control samples may be labeled or analyzed "blind" (unidentified to the analyst). The results of the QC samples are compared with the predetermined "true" values and ranges of variation. Poor accuracy or precision of a QC sample throws doubt on the validity of the results of all the samples analyzed during that calibration, and usually leads to rejection of the results for those samples. Rejection criteria must be established by the analyst ahead of the time of actual analysis.

Data Handling and Evaluation

After an analysis, the raw data must be used to convert the sample readings into concentrations. It is necessary to determine a calibration function that best fits the calibration standards; the sample concentrations must be computed; and the results must be evaluated with respect to their reliability and with respect to the experimental design. The recent advances in computer technology have had a dramatic impact on all aspects of data handling and evaluation.

Calibration Functions. The historical approach to calibration was to restrict the analysis to the linear range, plot the calibration standards on graph paper, throw out those points that did not look "right," draw the best straight line to fit the remaining points, and then convert the sample signals to concentrations using the "linear calibration." With the advent of tabletop calculators, straight lines were fit to the calibration data using the least squares method. With today's computer technology, much more sophisticated mathematical operations are possible. Complex equations can now be used for calibration functions (Hornbeck 1975; Rowe and Routh 1977; Stineman 1980) and statistical methods of analysis can be applied to the raw data and the computed concentrations.

The first step in any calibration procedure is to inspect the calibration data. The standards must show good agreement between repeat determinations. A systematic change in the standard values usually indicates drift of one of the analytical parameters. The standards must also have an acceptable signal-to-noise ratio. Standards whose signals have a poor signal-to-noise ratio can deviate considerably from the true value unless determined a sufficient number of times to obtain an accurate mean value. Finally, the standard signals must progress logically in the order and proportion of their concentrations. Although a standard may appear in error by visual inspection, elimination on the basis of a statistical test may prove difficult. Many times the best approach is to redetermine the standard or make up a new standard.

The most accurate means of calibration, in either the absorption or emission mode, is to use a pair of standards whose concentrations bracket that of the sample as tightly as possible. This approach eliminates any concern about linearity and reduces the time interval between the determination of the standards and the sample, minimizing fluctuation, or drift, errors. Of course this approach is not practical for large numbers of samples, when the samples include a wide range of concentrations or when operating in the multielement mode.

Data Evaluation. Computation of sample concentrations is never the final step in the analytical process. Even though the analyst may have no further involvement in the project, the results will always be used (or misused) by someone. All too often, the end users of the data lack an analytical background and tend to interpret the results as absolute, forgetting the inherent uncertainty in each value. It is, therefore, the duty of the analyst to assign a value of uncertainty to the data in order to ensure their proper usage.

Repeated determinations of the same sample can be used to compute a standard deviation that reflects the measurement error. The significance of the standard deviation can be expanded by including more variables. If the repeat determinations include different preparations of the sample then the standard deviation will reflect measurement and sample preparation uncertainty. By performing the repeat measurements on different days, the standard deviation will reflect the uncertainties listed above as well as those uncertainties arising from daily variations in instrumental performance and calibration. The best method of characterizing analytical errors is to include multiple determinations made on different days by different analysts of different sample preparations. For this reason, the standard deviations of the QC sample are usually the most accurate measurement of analytical uncertainty.

The use of the signal-to-noise ratio to define "reliable" data has found wide acceptance since specific ratios, the detection limit, and the quantitation limit are now defined by the IUPAC (1976) and ACS (1980). These limits divide all analytical data into three categories: undetectable (coefficient of variation, CV>33%), detected or semi-quantitative (33%>CV>10%), and quantitative (CV<10%). Such a classification of the data can create problems when repeat determinations of a sample fall into different categories. If the true value of a sample falls close to either the detection limit or the quantitation limit, then the normal distribution expected for repeated determinations will yield results falling on both sides of the limit. If this limit is applied as a filter to determine whether data are to be reported, then a fraction of the determinations will be discarded. Averaging the remaining fraction will produce a biased result. This bias will increase as the discarded fraction increases.

An intuitive limit is zero. Most analysts will throw out zeros or negative values. Horwitz *et al.* (1980) have shown, however, that when the CV is 100%, 17% of the data will fall below zero for a normal distribution. It is intuitive that negative concentrations have no meaning. Yet, failure to include negative and zero values would produce an average which is biased high.

The signal-to-noise ratio for individual measurements is extremely poor below the quantitation limit. However, for a sufficiently large number of determinations a reliable estimate of the mean value can be obtained. The standard deviation of the average value improves with the square root of the number of determinations. It is seldom worthwhile to make the necessary number of determinations to obtain an acceptable standard deviation of the mean, but, in all cases, all positive, negative, and zero values must be used to obtain an unbiased average of the sample concentration. The average and the standard deviation of the average can then be used to calculate a coefficient of variation. The same filters (CV equal to 33 or 10%) can then be applied without biasing the results although they are no longer strictly "detection" or "quantitation" limits. These limits do allow a categorization of the reliability of the data. However, if the data are to be subjected to further statistical analysis, filtering must be avoided.

Depending on the purpose of the study and the sampling scheme employed, the data may be tested using a wide variety of statistical methods. Some of the simpler tests are (1) the t test to determine whether data or groups of data are significantly different, (2) the F test to determine whether the variances of the data are different, (3) one-way and two-way analysis of variance to test for one- and two-dimensional patterns in the data, and (4) multiple regression analysis to determine whether the data are dependent on preselected variables. These tests are described in most statistics texts (Wine 1964; Zar 1974). In addition, statistics packages are available for almost every micro and minicomputer and for every large computer.

With the current emphasis on multielement determinations, automation, and computerization, it is now possible to generate data faster than ever before. The weakest link in the analytical process is increasingly becoming the interpretation of the data. Development of the field of study of "chemometrics" has offered great potential for strengthening this data evaluation link. This discipline uses mathematical and statistical methods to design an optimal measurement procedure and to provide maximum chemical information by analyzing chemical data (Frank and Kowalski 1982). A number of chemometric methods are available for evaluation of multicomponent data, such as those generated in multielement atomic spectroscopic studies. The techniques can be very useful in determining and sorting out multifactorial relationships within groups of samples. These chemometric concepts are only recently being applied to the development of automated computerized analytical atomic spectroscopic systems to generate trace element data in foods.

Computerization. Data which previously kept the analyst busy for weeks can now be processed in a matter of minutes using computerized mathematical approaches which until now were too complex and/or time consuming to be used. However, computerization does create additional quality assurance problems for the analyst. Since a computer is restricted to a predesigned program, deviations from the expected data patterns can lead to erroneous results. This is not because of inaccuracies in the number handling of the computer but because of the rigid nature of the computer logic. Once established, a program will treat each datum in an identical fashion. Whereas the analyst might note that one of the standards in a calibration was reading slightly high or low, the computer, unless specifically programmed to check the standards (which would be difficult), will accept the data as absolute. This problem is compounded in multielement spectrometers where most of the collected data have never been seen by the analyst. Unless the computer is programmed to detect a wide variety of errors and to evaluate the validity of the data, significant analytical errors can result.

To completely automate the data reduction process, each step, which was previously performed by the analyst, must be translated into a computer algorithm, or a series of algorithms. This is an extremely complex undertaking. As programs become more versatile, they become more complex and cumbersome. The logic of complex programs is difficult to follow and more susceptible to errors. On the other hand, conceptually simple programs are often too narrow in their application. As a result, most programs are a compromise; logically as simple as possible but able to handle common problems.

It is imperative that the analyst be familiar with how the data are handled by a computer program. This knowledge allows the analyst to inspect the data and results, detect errors, and anticipate conditions for which the program algorithms are not appropriate. Obtaining listings of commercially available programs is not always possible. Most manufacturers do not make original programs, or "source" listings, available to the users. Thus, the analyst is able to obtain only a general idea of how the data are being handled. Certainly, the analyst must try to evaluate the program's performance under as many different circumstances as possible. In most cases, analyses of reference materials and quality control samples are the best methods for evaluating the accuracy of the analytical methods and the computer programs.

CASE STUDY—DETERMINATION OF INORGANIC NUTRIENTS IN U.S. FRUIT AND VEGETABLE JUICES

A study was undertaken by the Nutrient Composition Laboratory at the U.S. Department of Agriculture to characterize the nutrient content of fruit and vegetable juices sold in the United States. The inorganic phase of the study consisted of the determination of 11 elements of nutritional interest. The main goal was to identify the range of elemental concentrations found for all juices and for each generic juice. This section presents, in detail, the quality assurance steps employed to ensure valid data for this study.

Sampling

To ensure the analyzed samples were representative of the national population, a carefully prepared sampling scheme was implemented (Moss *et al.* 1982). The most commonly consumed brands of fruit and vegetable juices (apple, cranberry, grape, grapefruit, lemonade, orange, pineapple, prune, and tomato, and mixed vegetable juices), according to dietary surveys and market shares, were purchased in two different major supermarkets in each of five major metropolitan areas: Atlanta, Boston, Chicago, Houston, and Los Angeles. A total of 190 juice samples was collected.

Validation of Analytical Data

Analytical Methodology

Sample treatment. Upon arrival at the laboratory, the samples were processed as follows:

1. Ready to drink juices were opened and poured into polyethylene Mason jars (Fisher Scientific, Pittsburgh, PA).
2. Concentrated juices were opened and poured into polyethylene Mason jars with the specified amount of high purity, deionized water (Milli-Q Water Purification System, Millipore Corp., Bedford, MA).
3. Samples were blended for 10 sec on medium high speed, to avoid foaming, using an Oster Blendor (Oster, Milwaukee, WI) with Teflon-coated blades.
4. Five to seven aliquots of the juice were transferred to 15-mL polyethylene tubes (Falcon, Oxnard, CA).
5. Tubes were capped and stored in a −29°C freezer until used.

All plasticware and equipment had been checked for contamination during the method development. Plasticware is routinely cleaned using a chelating detergent, present as the sodium salt, and then thoroughly rinsed with deionized water.

Prior to analysis, the samples were digested using a scaled-down version of the acid hydrolysis method reported by McHard *et al.* (1976A) for the preparation of orange juice samples. It was shown (McHard *et al.* 1976B) that this digestion procedure is suitable for the determination of eight inorganic nutrients: Ca, Cu, Fe, K, Mg, Mn, Na, and Zn. This procedure was found to yield results comparable to dry and wet ashing techniques and was much more suitable for preparing a large number of samples. The modified digestion procedure is as follows:

1. The samples were thawed by sitting at room temperature overnight.
2. One gram of juice was weighed into a 15-mL polyethylene tube.
3. One milliliter of concentrated HNO_3 (Ultrex brand, Baker Chemical Co., Phillipsburg, NJ) was added to the tube.
4. The samples were allowed to sit overnight.
5. The next day approximately 8 mL of Milli-Q water were added.
6. The samples were then cooled to room temperature, diluted to 14 mL, and filtered into clean 15-mL polyethylene tubes.
7. The sample digest was aspirated directly for analysis.

The concentrated nitric acid is routinely checked for contamination each time a new bottle is opened. Contamination of high purity reagents is infrequent but has been found.

The amount of sample employed in the sample digestion procedure was chosen to provide the desired level of detection for each element. An a priori decision was made that the method should be able to detect 5% of the recommended daily allowance (RDA) or 5% of the lower limit of the recommended safe and adequate daily dietary intake (SADDI) occurring in a 6-oz (170-mL) serving. In Table 6.1, the RDAs, the SADDIs, and the concentration corresponding to 5% of these values in 170 mL are listed. If a 1-g sample is digested and diluted to 14 mL, as described in the experimental section, the detection limits listed in the third column (routine atomic absorption detection limits multiplied by 14, the sample preparation factor) can be achieved.

In general, it is desirable that the sample concentrations be at least five times the detection limits to ensure good signal-to-noise ratios. Quantitation at the detection limit is poor by definition (a signal-to-noise ratio of 3 to 1, a relative standard deviation of 33%). It can be seen that, with the exception of Cr, all concentrations exceed the detection limit by a factor of 5. The reliable determination of Cr would require a 25-fold increase in the sample size. Consequently, a 1-g sample size was used despite the loss of the Cr values.

Standards preparation. From preliminary studies, the ranges of expected concentrations of the major elements and some of the minor

TABLE 6.1. FLAME ANALYSIS OF JUICE SAMPLES

Element	Adult RDA (mg/day)	5% RDA/170 mL (μg/mL)	Detection limit ($\times$ 14) (μg/mL)[a]
Mn	2.5–5.0[b]	0.7–1.4	0.08
Zn	15	4.4	0.8
Fe	18	5.3	0.8
Cu	2–3[b]	0.6–0.9	0.1
Mg	350	103	0.004
Ca	1200	352	0.04
Na	1100–3300[b]	325–975	0.04
Cr	0.05–0.2[b]	0.02–0.06	0.1
Ni	—	—	6.0
Co	—	—	1.0

[a] Detection limits based on the use of 1 g of sample into 14 mL in the sample preparation.
[b] Safe and adequate daily dietary intake.

elements were determined. A series of eight calibration standards, containing all 11 elements, was prepared to cover these ranges and an additional concentration factor of two to five at each end (Table 6.2). For those elements which could not be detected, calibration standard concentrations were selected which covered the range from the detection limit to at least 15 times the detection limit. A mixed element stock solution, which was also the highest standard, was prepared in 7% HNO_3 from commercially available single-element standards (Spex Industries, Inc., Metuchen, NJ). The stock solution contained the following: 200 μg mL^{-1} Ca, 4 μg mL^{-1} Co, 2 μg mL^{-1} Cr, 2 μg mL^{-1} Cu, 4 μg mL^{-1} Fe, 500 μg mL^{-1} K, 200 μg mL^{-1} Mg, 4 μg mL^{-1} Mn, 50 μg mL^{-1} Na, 10 μg mL^{-1} Ni, and 2 μg mL^{-1} Zn.

TABLE 6.2. MULTIELEMENT STANDARD CONCENTRATIONS FOR JUICE ANALYSES

Element	Standards concentration range (μg/mL)
Mn	0.016–4.0
Zn	0.008–2.0
Fe	0.016–4.0
Cu	0.008–2.0
Cr	0.008–2.0
Ni	0.04–10.
Co	0.016–4.0
Ca	0.8–200.
Mg	0.8–200.
Na	0.2–50.
K	2.0–500.

The mixed stock was diluted by factors of 2.5, 5, 10, 25, 50, 100, and 250 in 7% HNO_3 to provide the range of concentrations shown in Table 6.2.

Signal measurement. The analytical measurements were made with a simultaneous multielement atomic absorption continuum source (SIMAAC) spectrometer using an air–acetylene flame as an atomization source. The SIMAAC was developed at USDA and has been previously described in detail (Harnly *et al.* 1979). The SIMAAC is composed of a xenon arc continuum source, an echelle polychromator modified for wavelength modulation, and a dedicated minicomputer. Wavelength modulation with a continuum source provides state-of-the-art background correction (Snelleman 1968; Harnly and O'Haver 1977) and detection limits comparable to conventional AAS and is compatible with either flame or electrothermal atomization. In addition, this technique allows absorbances of varying sensitivity to be computed, extending the calibration curves to 4–6 orders of magnitude of concentration for each element (Harnly and O'Haver 1981). The echelle polychromator permits up to 20 elements to be determined simultaneously. The automated sampling and computerized nature of the instrument permit a high sample throughput rate (Harnly *et al.* 1982) as well as long-term data storage and report generation.

The computer programs for SIMAAC have been described in detail previously (Harnly *et al.* 1982). There are three essential programs: (1) the data acquisition and file storage program; (2) the data modification program (permits descriptive parameters, such as run type and standard concentration, to be modified; computed absorbance readings cannot be changed); and (3) the data averaging, calibration, sample concentration calculation, and report generation program. Only the first of these programs is involved in the signal measurement. This program is not interactive, i.e., the program takes intensity data, computes absorbances, and displays the absorbance for one or all elements but does not test the data in any way or perform any inspection tasks.

All analytical measurements were made using the same protocol. First, a blank atomization was repeated five to ten times to allow the baseline standard deviation to be computed by the computer. Next, a complete set of calibration standards was atomized. The samples were then atomized with a blank and a standard repeated at regular intervals between samples. Finally, a second set of calibration standards was atomized.

The juice samples were analyzed in "batches." Each "batch" consisted of triplicate preparations of 14 samples and a quality control sample. This batch of 45 solutions was analyzed using the protocol in the preceding paragraph. Following the analysis of each batch, a report was generated by the computer. The analysis of a "batch" of samples took 1–1½ hours from the start of atomization to the report generation. As many as three batches were analyzed in a single day.

The only systematic interferences of concern for atomization with an air–acetylene flame using SIMAAC are matrix effects. The relatively low (as compared to high-energy plasmas) temperature of the air–acetylene flame is well documented as to its susceptibility to chemical interferences. Consequently, it is necessary to verify the accuracy of the measurement of each element. Although such interferences should properly be discussed here, it is more convenient to consider the problem in conjunction with the optimization of analytical parameters in the next section.

Spectral interferences are rare since wavelength modulation with a continuum source eliminates broad band interferences and line overlap interferences are virtually nonexistent. It has been suggested that wavelength modulation widens the wavelength region viewed and increases susceptibility to spectral lines falling within the viewed region. However, the modulation ranges (peak to peak) fall between 0.06 and 0.16 nm. In every case, this interval is less than the spectral bandpass recommended for line source AAS when employing a deuterium source background corrector. To eliminate any doubt, a program was developed which prints out the spectra of each element (Miller-Ihli *et al.* 1982). Using this program, it was verified that no line overlap interferences occurred within the modulation interval, for any of the elements in any of the nine major types of juices.

Optimization of analytical parameters. The critical parameters for the accurate multielement determination in an air–acetylene flame are those which control the flame: the air and acetylene flow controls, which regulate the air–acetylene ratio, and the burner height control, which regulates the viewing height (the height at which the continuum source beam passes through the flame). In the single-element mode, drastically different air–acetylene ratios and viewing heights are used for each element. In the multielement mode, however, suitable compromise settings are necessary which provide accurate determinations for each element.

The flame parameters were optimized for the simultaneous determinations of Ca, Co, Cr, Cu, Fe, K, Mg, Mn, Na, Ni, and Zn. Previous studies for a 5% HCl matrix (Harnly *et al.* 1982) showed that a viewing height low in a lean air–acetylene flame gave good recoveries (100 ± 5%) for a variety of NBS standard reference materials and USGS standard rocks for all the elements examined except Ca and Mg. A similar optimization was repeated in this study for the 7% HNO_3 matrix. Spike recoveries were determined for a quality control sample of pineapple juice since there are no appropriate standard reference materials. Tables 6.3 and 6.4 show the recovery matrices for Mn and Ca. Similar to Co, Cu, Fe, K, Mg, Na, and Zn, the Mn showed little variation in recovery as a function of the flame parameters. The Ca, Cr, and Ni showed distinct recovery patterns with Ca (Table 6.4) the most severe. For Cr and Ni, good recoveries were observed at low viewing heights (3 mm) but the recoveries decreased higher in the flame. The recoveries for Ca were just the opposite, improving at higher viewing heights, in a lean air–acetylene flame. Consequently, a viewing height low (3 mm) in a lean flame (air–acetylene ratio of 4.2) was optimum for all elements except Ca, which was accurately determined at a high viewing height (12 mm) in the same flame.

The usefulness of these two sets of flame parameters for all the juices was checked by determining spike recoveries for each element in each of the nine general types of fruit and vegetable juices. Each of the nine juices was prepared as described in the sample treatment section, spiked, and analyzed, for all 11 elements, low (3 mm) and high (12 mm) in a lean (4.2 air–acetylene ratio) flame. The average recoveries

TABLE 6.3. PERCENTAGE SPIKE RECOVERY VS. FLAME CHARACTER[a]

	Height in flame (mm)			
Air–acetylene (v/v)	3	6	9	12
4.9	99	101	97	99
4.2	101	104	101	99
3.7	101	97	99	99
3.4	101	99	101	97
3.1	104	101	104	106

[a] For Mn (279.5 nm).

TABLE 6.4. PERCENTAGE SPIKE RECOVERY VS. FLAME CHARACTER[a]

	Height in flame (mm)			
Air–acetylene (v/v)	3	6	9	12
4.9	64	86	94	95
4.2	52	79	92	100
3.7	47	70	82	95
3.4	43	50	70	78
3.1	36	55	68	72

[a] For Ca (422.6 nm).

TABLE 6.5. PERCENTAGE SPIKE RECOVERIES OF NINE TYPES OF FRUIT AND VEGETABLE JUICES

	Height in flame (mm)	
Element	3	12
Ca	79	100
Co	99	83
Cr	106	84
Cu	104	102
Fe	85	90
K	104	105
Mg	101	105
Mn	101	99
Na	109	103
Ni	85	53
Zn	58	102

and standard deviations are shown in Table 6.5. Good recoveries (100 ± 15%) were found at a 3-mm height, for all elements except Ca and, surprisingly, Zn. Good recoveries were obtained for Ca and Zn at 12 mm. Based on these results, the 11 elements in fruit and vegetable juices were determined routinely in two experiments. The Co, Cr, Cu, Fe, K, Mg, Mn, Na, and Ni were determined at a viewing height of 3 mm in a lean (4.2 air–acetylene ratio) flame and the Ca and Zn were determined at a 12-mm height in the same flame.

Method Validation. There were no suitable certified reference materials available for the fruit and vegetable juice samples with which to validate the method. In addition, no second method was available to the laboratory for most of the elements. Fortunately, McHard *et al.* (1980) have shown that orange juices from different geographic locations (Florida, California, Brazil, and Mexico) have surprisingly distinctive and consistent metal concentrations. Consequently, since commercial orange juices in this country are almost exclusively from Florida, McHard's AAS data for Florida orange juice (McHard *et al.* 1976B) made an excellent reference. Three cans of a single brand of orange juice were selected, prepared in triplicate, and analyzed simultaneously for all 11 elements using the flame parameters selected in the previous section. The results are shown in Table 6.6. In general, the results are in excellent agreement with those of McHard *et al.* Close inspection reveals that Mn and Na are factors of two and three times too high, respectively. The Co, Cr, and Ni were not detect-

TABLE 6.6. ANALYSIS OF ORANGE JUICE SAMPLES

	McHard *et al.* (8 Brands)[a]		SIMAAC Results (1 Brand)[b]	
Element	Mean conc. (μg/g)	Range (mg/g)	Mean conc. + S.D. (μg/g)	Range (μg/g)
Mn	0.12	0.11–0.13	0.25 ± 0.04	0.20–0.30
Zn	0.37	0.33–0.38	0.34 ± 0.10	0.20–0.50
Fe	1.22	0.73–2.45	1.19 ± 0.27	0.82–1.65
Cu	0.29	0.25–0.33	0.32 ± 0.06	0.24–0.44
Mg	106.	98–115	115 ± 6	103–123
Ca	92.3	85–105	95.8 ± 11.8	77–104
Na	4.3	2.10–8.95	12.1 ± 0.4	11.5–12.4
K	1863.	1775–1925	1861 ± 103	1726–2000

[a] McHard *et al.* (1976B).
[b] $n = 9$, three samples of one brand prepared in triplicate.

able. More recent data (McHard et al. 1980) list an average Mn concentration of 0.18 μg/g in Florida orange juice determined by a variety of methods (ICPAES, DCPAES, FAFS, and FAAS). The high Na results from the authors' laboratory arise from contamination during the sample preparation. Na is an extremely difficult contaminant to eliminate. No effort was made to minimize Na contamination in the sample preparation procedure and to obtain accurate Na values since the Na levels (<5 μg/g) represent only 0.05% of the lower SADDI limit (Table 6.1) in a 6-oz serving. It was not deemed worthwhile to clean up the procedure.

Quality Control Samples. The quality control sample was prepared as follows: A case of 24-oz cans of pineapple juice was purchased at a local supermarket. Five cans were selected at random, blended, and combined. Five- to seven-mL aliquots were transferred to 250 of the 15-mL polyethylene tubes. The tubes were capped and stored in a −29°C freezer until used.

The pineapple juice quality control sample was analyzed with each batch of samples over the 12 months of the study. The results for the pool sample (Table 6.7) reflect the day-to-day fluctuation of the routine instrument operation and sample treatment providing a means of evaluating the results for each batch of samples and assessing the long-term precision of the method.

For elements well above the quantitation limit (Ca, Mg, K, and Mn), the long-term CVs ranged from 3.0 to 6.9%. Elements whose mean concentrations fell between the quantitative limit and the detection limit (Cu, Fe, and Zn) had CVs from 28 to 40%. The expected range

TABLE 6.7. ANALYSIS OF THE PINEAPPLE QUALITY CONTROL SAMPLE[a]

Element	Mean ± σ (μg/g)	Relative standard deviation (%)
Ca	98.5 ± 4.6	4.7
Cu[b]	0.45 ± 0.15	28.
Fe[b]	1.64 ± 0.65	40.
K	723. ± 42.	5.8
Mg	80.8 ± 2.4	3.0
Mn	4.35 ± 0.30	6.9
Na	2.72 ± 1.58	63.
Zn[b]	1.42 ± 0.42	30.

[a] $n = 18$.
[b] Mean results fall between the detection limit (3σ) and the quantitation limit (15σ)

of CVs is from 33%, at the detection limit, to 6.7%, at the quantitative limit. The Na results fell above the quantitation limit, but variations in the contamination level (discussed in the previous section) resulted in an unusually large CV. The signals for Co, Cr, and Ni fell below the detection limit for every determination.

Prior to the start of the juice analysis program, the pineapple sample was analyzed 90 times (three batches of 10 samples determined in triplicate on three different days). These values were used as a baseline value to which the quality control sample was compared (Table 6.8). If the results were unsatisfactory, all the results, or just the results for the element for which poor recovery was obtained, were thrown out and the samples were rerun.

The use of the quality control sample for evaluating the accuracy of each batch analysis is illustrated in Table 6.8 for Mn. In each case, the triplicate analyses are averaged to give a single value. It can be seen that results were satisfactory for all runs except 12–14, run on January 14, 1981. For runs 12 and 13, high recoveries were obtained only for Mn. Samples high in Mn (approximately 200 times higher than the Mn concentrations in the juices) had been determined the day before and the nebulizer and burner assembly had not been properly cleaned. As a result, the values for Mn in the juices were biased high by 89 and 37%, respectively. The Mn in the juices results were thrown out and the samples were rerun for Mn only. Batch 14, run the same day, showed low recoveries for Mn and every other element. A check of the original data revealed an apparent sensitivity shift during the analysis. All the results were thrown out for batch 14 and the samples were rerun for each element.

TABLE 6.8. Mn DETERMINATIONS IN PINEAPPLE JUICE QUALITY CONTROL SAMPLE

Batch	Date	n	Concentrations (μg/g)
—	08-12-80	10	4.35 ± 0.38
—	08-21-80	10	4.48 ± 0.25
—	08-25-80	10	4.88 ± 0.24
		30	4.57 ± 0.30
1	09-10-80	1	4.46
2	09-18-80	1	4.47
3	09-23-80	1	4.40
4	10-14-80	1	3.84
5	10-31-80	1	4.53
6	11-07-80	1	4.48
8[a]	11-19-80	1	4.14
10[a]	12-10-80	1	3.77
11	12-10-80	1	4.35
12	01-14-81	1	8.17[b]
13	01-14-81	1	5.95[b]
14	01-14-81	1	2.70[c]
15	01-29-81	1	4.31
16	01-29-81	1	4.58
17	02-05-81	1	4.03
18	02-05-81	1	4.35
19	08-11-81	1	4.66
20	08-11-81	1	4.86
		15[d]	4.35 ± 0.30[d]

[a] No quality control standard was run with batches 7 and 9.
[b] Samples high in Mn (approximately 200 times the Mn concentration in juices) were run the previous day.
[c] Low recoveries for all elements. Sensitivity shift during the experiment.
[d] Batches 12, 13, and 14 omitted from average.

Data Handling and Evaluation

Calibration and Result Computation. For the large numbers of samples analyzed in this study, the only reasonable method of quantitation was the use of the traditional calibration curve. The calibration standards (discussed earlier), because they are made up approximating the elemental ratios found in orange juice, are more closely matrix matched to the samples than single-element standards would be. The standards were prepared in a common diluent, the 7% HNO_3 solution which results from the digestion procedure. Still, matrix effects were anticipated and extensive studies were undertaken to determine the atomization parameters which produced accurate results (see the section on Optimization of Analytical Parameters).

The calibration, result computation, and report generation are performed by the third of the three programs listed in the Signal Measurement section. This program first averages the absorbances of all the similarly identified blanks, standards, and samples. For the blank type designated by the analyst, the standard deviation is also computed. The standard deviation of the blank (σ_B) is used to determine the absorbance detection limit ($3\sigma_B$) and quantitation limit ($10\sigma_B$). Standards, whose absorbances fall below the quantitation limit, are not used since they are usually not determined enough times to obtain an accurate mean value. Results are computed for all samples (even if their absorbance is less than the blank) but the results are flagged as "less than the detection limit" or "less than the quantitation limit" in the report.

The SIMAAC computer report lists the concentration (in micrograms per milliliter) for each sample determined in that batch. A disk copy of the report is currently hand carried to the main laboratory computer (a hard wire link is currently being purchased) where it is incorporated into the main juice file which contains the background information (weight of the sample, final digestion volume, dilution factors, and sample identification). The elemental concentrations (in micrograms per gram) are then computed and stored for further evaluation.

The SIMAAC computer report lists the detection limit and characteristic concentration for each element. These values permit the instrumental operation to be compared for the analysis of each batch (Table 6.9). It can be seen that for Cu, as for the other elements, the figures of merit were reproducible between batches.

A number of curve fitting routines have been investigated (Miller-Ihli *et al.*, 1984) but a linear fit between each set of standards has been found to be as accurate as much more sophisticated functions, provided at least three standards are used per order of magnitude. Thus each sample absorbance is considered separately. The standards which bracket the sample are then used to compute the sample concentration. At the low end of the calibration curve, if no standard exists at a lower absorbance than the sample, a result is computed by extrapolating from the lowest standard through zero, providing the lowest standard is in the linear range. This result is also flagged in the report. At the upper end, if no standard exists at a higher absorbance than the sample, no result is computed and, again, a flag is set in the report.

As a result of the flags in the report, the analyst knows the signal-to-noise ratio of the sample result and can objectively evaluate its reliability. Unfortunately, systematic errors cannot be characterized statistically and can only be detected by inspection of the individual data by the analyst. Mislabeling of a standard, sample, or blank, en-

TABLE 6.9. DETECTION LIMITS AND CHARACTERISTIC CONCENTRATIONS OF Cu

Batch	Date	Characteristic concentration (μg/mL)	Detection limit (μg/mL)
1	09-10-80	0.12	0.002
2	09-18-80	0.12	0.002
3	09-23-80	0.14	0.002
4	10-14-80	0.043	0.001
5	10-31-80	0.094	0.002
6	11-07-80	0.12	0.004
7	11-19-80	0.079	0.001
8	11-19-80	0.17	0.006
9	11-19-80	0.056	0.001
10	12-10-80	0.16	0.007
11	12-10-80	0.15	0.008
12	01-14-81	0.17	0.01
13	01-14-81	0.19	0.01
14	01-14-81	0.24	0.007
15	01-29-81	0.10	0.005
16	01-29-81	0.14	0.003
17	02-05-81	0.16	0.001
18	02-05-81	0.12	0.003
19	08-11-81	0.12	0.006
20	08-11-81	0.090	0.004

tering an incorrect dilution factor for a standard, or an erroneous reading, which is not detected, can result in inaccurate computed results for one or more samples. The results for quality control samples will sometimes reflect these errors, depending on the error. The only sure way of avoiding these errors is inspection of the data by the analyst. The computer programs permit the data for each element to be listed at any stage in the processing.

Data Evaluation. The best statistics on the reliability of the analytical method are provided by the quality control sample results (Table 6.7). The relative standard deviation for each element includes uncertainties arising from the sample treatment, variations in instrument performance and calibration, between experiments, and the signal measurement. These data were collected over a 12-month period. In general, the precision of the data is ±5% for elements whose concentrations fall above the quantitation limit. At lower concentrations (below the quantitation limit) the precisions are worse, as expected.

It was initially decided that only quantitative data (a CV less than 6.7%) would be reported so that the reliability of each value would be known. This classification, or filtering, of the data was done based on the flags in the computer report. However, it was soon discovered that filtering of the data led to systematic errors. This is illustrated in Ta-

TABLE 6.10. ANALYSIS OF Fe IN FRUIT JUICE SAMPLES

Juice	Analytical conc. (μg/g)[a]	
	Quantitative data only	All values
1	2.8 ($n = 18$)	2.4 ($n = 28$)
2	4.7 (20)	3.5 (27)
3	2.8 (7)	1.6 (32)
4	1.7 (1)	0.6 (18)
5	2.7 (4)	1.6 (23)
6	1.6 (4)	0.9 (24)
7	<2.0 (−)	0.3 (15)
8	1.7 (7)	1.1 (18)
9	2.0 (4)	1.2 (15)
10	1.8 (2)	0.9 (27)
11	3.3 (11)	2.1 (21)
12	<1.8 (−)	0.5 (15)
13	<2.1 (−)	1.5 (6)

[a] Samples were diluted by factor of 14 in the digestion procedure, i.e., μg/mL = 1/14 × μg/g.

ble 6.10 for the determination of Fe in thirteen different juices. Filtering of the data led to systematic errors ranging from 17% (juice #1) to 183% (juice #4). In addition, the filtering process invalidates further statistical analysis of the data.

Based on these results, the data storage programs of the laboratory were modified. Now all computed concentrations are stored, whether quantitative, semiquantitative, or below the detection limit. These files are for laboratory use only. At the completion of a study, these files are used for statistical analysis. Data leaving the laboratory are now filtered using the CV computed from the mean and standard deviation obtained for all the determinations. Consequently reliable data can be reported without introducing a bias.

The analytical data for the fruit and vegetable juices were transferred by phone line, as a file, from the laboratory computer to a larger computer (Washington Computer Center, Washington, DC) where the data were analyzed using SAS (SAS Institute, Inc., Cary, NC), a statistical analysis system with extensive array handling capabilities. The SAS was used to perform all statistical evaluation of the analytical data.

Computerization. The computer power at our disposal makes possible the analytical approach our laboratory has chosen. The dedicated minicomputer of SIMAAC is the heart of the instrument, performing high-speed data acquisition, fitting calibration curves, computing sample concentrations, generating a final report, and creating a data file for the laboratory computer. The laboratory computer provides large volume file merging and storage capabilities. And finally, the large commercial computer facility provides large-scale file handling and statistical analyses. These three computers provide the power and speed to process large volumes of data in a short period of time.

The programs for SIMAAC and the laboratory computer were written in the laboratory. The SAS programs were also written in the laboratory using the instruction set and statistical sub-routines available in the SAS system. In each case, a complete listing of every program is available to the analyst. These programs were thoroughly tested after being written and it is possible at any time to recheck the accuracy of the algorithms.

BIBLIOGRAPHY

ACS. 1980. Guidelines for data acquisition and data quality evaluation. *In* Environmental Chemistry. Anal. Chem. *52,* 2242.

CHRISTIAN, G.D., and FELDMAN, F.J. 1970. Atomic Absorption Spectroscopy. Wiley (Interscience), New York.

FRANK, I.E., and KOWALSKI, B.R. 1982. Chemometrics. Anal. Chem. *54*(5) 232R–234R.

HARNLY, J.M., and O'HAVER, T.C. 1977. Background correction for the analysis of high-solids samples by graphite furnace atomic absorption. Anal. Chem. *49,* 2187–2193.

HARNLY, J.M., and O'HAVER, T.C. 1981. Extension of analytical curves in atomic absorption spectrometry. Anal. Chem. *53,* 1291–1298.

HARNLY, J.M., and WOLF, W.R. 1984. Quality assurance of analysis of inorganic nutrients in foods. *In* Instrumental Analysis of Foods and Beverages—Modern Techniques. G. Charalambous (Editor). Academic Press, New York.

HARNLY, J.M., O'HAVER, T.C., WOLF, W.R., and GOLDEN, B.M. 1979. Background-corrected simultaneous multielement atomic absorption spectrometer. Anal. Chem. *51,* 2007–2014.

HARNLY, J.M., MILLER-IHLI, N.J., and O'HAVER, T.C. 1982. Computer software for a simultaneous multi-element atomic absorption spectrometer. J. Autom. Chem. *4,* 54–60.

HORNBECK, R. W. 1975. Numerical Methods. Quantum Publishers, Inc., New York.

HORWITZ, W., KAMPS, L.R., and BOYER, K.W. 1980. Quality assurance in the analysis of foods for trace constituents. J. Assoc. Off. Anal. Chem. *63,* 1344–1354.

INHAT, M., CLOUTIER, R.A., and WOLF, W.R. 1982. Preparation of a corn agricultural biological reference material. 9th Annu. Meet., Fed. Anal. Chem. Spectrosc. Soc., Sept. 1982. Philadelphia, PA (abstract).

ISO. 1982. Directory of Certified Reference Materials. International Standardization Organization/Remco, Geneva, Switzerland.

IUPAC Commission of Spectrochemical and Other Optical Procedures. 1976. Nomenclature, symbols, units, and their usage in spectrochemical analysis—II. Data interpretation. Pure Appl. Chem. *45,* 99–123.

KRATOCHVIL, B., and TAYLOR, J.K. 1981. Sampling for chemical analysis. Anal. Chem. *53,* 924A–938A.

McHARD, J.A., WINEFORDNER, J.D., and ATTAWAY, J.A. 1976A. A new hydrolysis procedure for preparation of orange juice for trace element analysis by atomic absorption spectrometry. J. Agric. Food Chem. *24,* 41–45.

McHARD, J.A., WINEFORDNER, J.D., and TING, S.-V. 1976B. Atomic absorption spectrometric determination of eight trace metals in orange juice following hydrolytic preparation. J. Agric. Food Chem. *24,* 950–953.

McHARD, J.A., FOULK, S.J., JORGENSEN, J.L., BAYER, S., and WINEFORDNER, J.D. 1980. Analysis of trace metals in orange juice. *In* Citrus Nutrition and Quality. J. Attaway (Editor), pp. 363–392. Am. Chem. Soc., Washington, DC.

MILLER-IHLI, N.J., O'HAVER, T.C., and HARNLY, J.M. 1982. Direct observation of flame atomic absorption spectral interferences. Anal. Chem. *54,* 799–803.

MILLER-IHLI, N.J., O'HAVER, T.C., and HARNLY, J.M. 1983. Time-resolved electrothermal atomic absorption spectra. Appl. Spectrosc. *37,* 429–432.

MILLER-IHLI, N.J., O'HAVER, T.C., and HARNLY, J.M. 1984. Calibration and-curve fitting for extended range AAS. Spectrochim. Acta (in press).

MOODY, J.R. 1982. NBS clean laboratories for trace element analysis. Anal. Chem. *54,* 1358A–1367A.

MOSS, M.K., LI, B.W., HARNLY, J.M., MILLER-IHLI, N.J., HOLDEN, J.M., and WOLF, W.R. 1982. Nutrient composition of juice purchased at retail. Paper #480, 42nd Annu. Meet. Inst. Food Technol., Las Vegas, NV, June 22–25.

MUNTAU, H. 1980. Measurement quality improvements by application of reference materials. *In* Trace Element Analytical Chemistry in Medicine and Biology. P. Brätter and P. Schramel (Editors), pp. 707–726. de Gruyter and Co., Berlin.

NBS Spec. Publ. 422. 1976. Accuracy in Trace Analysis: Sampling, Sample Handling, Analysis. P.D. La Fleur (Editor). U.S. Government Printing Office, Washington, DC.

NBS Spec. Publ. 492. 1977. Procedures Used at the Natl. Bur. of Stand. to Determine Selected Trace Elements in Biological and Botanical Materials. R. Mavrodineau (Editor). U.S. Government Printing Office, Washington, DC.

O'HAVER, T.C. 1976A. Analytical considerations. *In* Trace Analysis: Spectroscopic Methods for Elements, J.D. Winefordner (Editor), pp. 15–62. John Wiley, New York.

O'HAVER, T.C. 1976B. Chemical aspects of chemical analysis. *In* Trace Analysis: Spectroscopic Methods for Elements, J.D. Winefordner (Editor), pp. 63–78. John Wiley, New York.

PARR, R.M. 1980. The reliability of trace element analysis as revealed by analytical reference materials. *In* Trace Element Analytical Chemistry in Medicine and Biology, P. Brätter and P. Schramel (Editors), pp. 631–655. de Gruyter and Co., Berlin.

ROWE, C.T., and ROUTH, M.W. 1977. Ultimate detection limit barrier in furnace AAS. Res./Dev. Nov., 24–30.

SNELLEMAN, W. 1968. An a.c. scanning method with increased sensitivity in atomic absorption analysis using a continuum primary source. Spectrochim. Acta *23B,* 403–411.

STINEMAN, R.W. 1980. A consistently well-behaved method of interpolation. Creative Computing *6* (7), 54–57.

TAYLOR, J.K. 1981. Quality assurance of chemical measurements. Anal. Chem. *53,* 1588A–1596A.

THIERS, R.E. 1975. Separation, concentration, and contamination. Trace Analysis, pp. 636–660. John Wiley, New York.

VEILLON, C. and VALLEE, B.L. 1978. Atomic spectroscopy in metal analysis of enzymes and other biological materials. *In* Methods in Enzymology *54* (25), 446–484. Academic Press, New York.

WINE, R.L. 1964. Statistics for Scientists and Engineers. Prentice-Hall, Inc., Englewood Cliffs, NJ.

WOLF, W.R. 1982. Evaluation of available reference materials for potential use in analysis of foods. Eastern Analytical Symposium. November. New York (abstract).

YOUDEN, W.J. and STEINER, E.H. 1967. Statistical Manual of the Association of Official Analytical Chemists. AOAC, Washington, DC.

ZAR, J.H. 1974. Biostatistical Analysis. Prentice-Hall, Englewood Cliffs, NJ.

ZEIF, M. and MITCHELL, J.W. 1976. Contamination Control in Trace Element Analysis. Wiley (Interscience), New York.

7

Atomic Absorption and Plasma Atomic Emission Spectrometry[1]

M. Ihnat[2]

INTRODUCTION AND HISTORY OF ANALYTICAL ATOMIC ABSORPTION AND EMISSION SPECTROMETRY

Historical Outline of Analytical Atomic Spectrometry

Observations in the mid-eighteenth and early nineteenth centuries by Melvill (1756), Herschel (1823), and Talbot (1826) regarding colors imparted to flames by salts and other materials, studies of the solar spectrum by Wollaston (1802) and Fraunhofer (1817), and studies of spark- and arc-excited spectra by Wheatstone in 1835 and Foucault in 1848, respectively (Schrenk 1975), were the early beginnings of atomic emission and absorption. It is, however, the investigations on emission in about 1860 by Kirchhoff and Bunsen (1860, 1861) that are generally regarded as the foundations of analytical spectrometry. Modern analytical flame emission spectrometry dates to the work reported by Lundegårdh (1934) on the determination of a number of metallic elements in biological samples using air–acetylene flame excitation, a prism spectrograph, and photographic recording. It is considered that quantitative (arc/spark) atomic emission spectrochemical analysis originated about 1882 with Hartley using a spark excitation source. Arc and spark emission spectrometry was the method of choice

[1] Contribution No. M-1397 from the Chemistry and Biology Research Institute

[2] Chemistry and Biology Research Institute, Agriculture Canada, Ottawa, Ontario K1A 0C6

for simultaneous multielement determinations during the three decades (1930–1960) in many fields of analysis notably in metallurgy and geology. Flame emission spectrometry gained rapidly in popularity following the introduction of commercial instruments in 1937–1945.

A century after Wollaston (1802) reported what are believed to be the Fraunhofer lines, due to atomic absorption, and Foucault (Schrenk 1975) made the first observations of atomic absorption by a sodium flame of radiation from an electric arc, Woodson (1939) utilized atomic absorption analytically to determine mercury. It was, however, the independent researches of Walsh (1955) and Alkemade and Milatz (1955) demonstrating that most elements giving free atoms in flames could be detected by atomic absorption that led to the concrete establishment of the technique as a viable analytical tool. Attributes of flame atomic absorption spectrometry are instrument operation simplicity, excellent element specificity and detectivity, capability of measuring about 70 elements, and moderate cost of basic instrumentation. These have led to wide acceptance and phenomenal growth of the method since the first commercial instrument made its appearance in about 1960. Of the numerous technical developments made over the years, one benchmark achievement is the development of electrothermal (flameless) techniques for sample atomization. Early work by King (Fuller 1977) at the turn of the 20th century on arc-heated and resistively heated carbon furnaces for emission spectrometry was followed in 1959 and succeeding years by the work of several groups of investigators, notably L'vov (Fuller 1977), Massman (1967, 1968), and Woodriff (Woodriff and Ramelow 1968; Woodriff *et al.* 1968) on the design of primarily graphite furnaces for atomic absorption. Much fundamental and applications research has followed demonstrating the technique to be analytically useful, particularly for analysis of materials available in small quantities and/or containing low concentrations of sought-for constituents.

Pioneering studies were carried out by Babat in the early 1940s (Fassel 1979) on atmospheric pressure inductively coupled plasmas (ICP). Work by Reed (Fassel 1979) some 20 years later was catalytic in independently initiating studies on the properties of these plasmas for analytical atomic spectrometry by Greenfield and co-workers (1964) and Fassel and co-workers (Wendt and Fassel 1965). Concurrent developments were made on direct current plasma arcs (DCP) as atomization–excitation sources for spectrochemical analysis stemming from the groundwork laid by Margoshes and Scribner (1959) and Korolev and Vainshtein (Keirs and Vickers 1977). Research and commercial developments during the past decade have established the technique of ICP-atomic emission spectrometry as a viable complementary tool to atomic absorption.

Literature on Applications of Atomic Absorption and Emission to Food Analysis

Since the establishment of spectrochemical techniques of analysis, the scientific literature has been inundated by tens of thousands of reports dealing with studies of fundamental aspects and applications of these techniques to the measurement of elements in a vast array of materials. Only a small fraction of publications, however, is devoted to the analysis of foods. Annual rates of publication of reports on atomic spectrometry applied to food analysis during the 1970 decade are summarized in Table 7.1. Atomic absorption has occupied, and still does, a preeminent position with flame methods (FAAS) being the most popular. Although somewhat a more technically demanding technique, electrothermal atomization has increased in usage during the second half of the decade. The low levels of mercury commonly encountered in foodstuffs are most amenable to determination by cold vapor atomic absorption. With respect to emission spectrometry, both flame emission spectrometry and arc/spark atomic emission have resulted in a small steady stream of papers throughout the period, whereas reports dealing with ICP have begun to appear since 1975.

Detailed compendia of methodologies on the application of atomic spectrometry to the analysis of foods do not abound among volumes

TABLE 7.1. ANNUAL PUBLICATION RATES OF REPORTS ON ATOMIC SPECTROMETRY APPLIED TO FOOD ANALYSIS

	Number of reports by technique[a,b]					
	Atomic absorption			Atomic emission		
Year	FAAS	EAAS	Hg–cold vapor	FES	ICP	Arc and spark
1971	30	4	15	3	—	6
1972	59	4	12	10	—	3
1973	85	9	8	8	—	5
1974	66	19	13	4	—	—
1975	49	20	4	4	2	—
1976	41	11	9	1	—	3
1977	49	19	4	8	9	3
1978	40	20	4	5	4	3
1979	32	26	10	2	6	2

[a] Estimated number of publications including journal articles, reviews, books and conference presentations based on *Annual Reports on Analytical Atomic Spectroscopy* (1971–1980).
[b] See Table 7.2 for explanation of nomenclature.

devoted to food analysis. None has yet been published on plasma emission and only a handful are available dealing with atomic absorption. Official and recommended atomic absorption methods for several of the common elements are available in the authoritative and comprehensive methods manuals of the Association of Official Analytical Chemists (AOAC) (Horwitz 1980), the Society for Analytical Chemistry (SAC) (Hanson 1973), and in Technical Bulletin 27 of the Ministry of Agriculture, Fisheries and Food (UK) (1973). Other sources of validated flame atomic absorption spectrometric methods are Approved Methods Committee, W.C. Schaefer (Chairman) (1976), Link (1977), Subcommittee on Procedures of the Chemistry Task Force of the National Shellfish Sanitation Program (1975), The Institute of Brewing Analysis Committee (1973, 1974, 1977), and Mercury Analysis Working Party of the Bureau International Technique du Chlore (1976).

Monographs on food analysis have been very weak on atomic spectrometric applications. Lees (1975), Pearson (1976), and Hart and Fisher (1971) present only brief treatments with the latter authors giving a procedure only for copper in milk. Joslyn (1970) mentions AAS; MacLeod (1973) and Pomeranz and Meloan (1978) present pedagogical treatments of the flame technique but neither give analytical details. A volume on AAS by Pinta (1975) includes an excellent chapter (9) on the FAAS analysis of vegetable materials for major and trace elements and two other sections (16.1 and 16.2) on foods and wines. A recent monograph on atomic absorption spectrometry (Cantle 1982) contains chapters on a methods manual approach to the analysis of a variety of materials including biologicals. The chapter by Ihnat (1982) presents comprehensive procedures for the analysis of a variety of foods by mainly FAAS but occasionally EAAS techniques for 21 of the major and trace elements, gathered from validated methods manuals and other reliable sources. A methods manual by Allen *et al.* (1974) on the chemical analysis of ecological materials documents AAS procedures for 13 elements which can be extended to food and agricultural commodities.

Numerous other monographs, chapters therein, and reviews on AAS by Welz (1976), Robinson (1966), Elwell and Gidley (1966), Price (1974, 1979), Slavin (1968), Ramirez-Muñoz (1968), Rubeška and Moldan (1969), Mavrodineanu (1970), Christian and Feldman (1970), Reynolds and Aldous (1970), Dvořak *et al.* (1971), Kirkbright and Sargent (1974), Koirtyohann and Pickett (1975), Fuller (1977), Varju (1971), Saarloos (1972), Morre (1974), Sakai (1975), Schuller and Egan (1976), Teper (1977), Crosby (1977), and Seiler (1972) provide discussion, reviews, and guides to the literature regarding applications of AAS in the food and allied fields. Excellent reviews appear annually in Annual Reports on Analytical Atomic Spectroscopy and biennially in odd years in Analytical Chemistry.

Scope of Chapter

Following a brief introduction to the basic principles of atomic absorption and emission, an insight will be presented into some of the fundamentals of instrumentation and techniques. The suitability and scope of applicability of absorption and emission spectrometry to the determination of nutritionally essential major and minor elements and those generally considered as toxic in foodstuffs will be assessed. Finally, a description will be given of general practical analytical approaches to the elemental analysis of foods. The aim is to equip the food analyst with an understanding of the applications and potential of atomic spectrometric methods with the hope that these powerful techniques of analysis will enjoy increased widespread acceptance in food analytical laboratories and in the literature on food analysis.

BASIC PRINCIPLES OF THEORY AND INSTRUMENTATION

Theory of Atomic Absorption and Emission

Atomic absorption and emission methods of spectrochemical analysis deal with the excitation of atoms of the analyte of interest and the subsequent measurements of absorbed or emitted energy. Energy in the form of heat within the atomization cell serves to vaporize and atomize the sample producing atoms in the ground or unexcited energy state as well as some in the many allowed excited energy states possessed by the element.

The basic processes depicted in Fig. 7.1 are illustrated with reference, for simplicity, to only one (the lowest) excited energy level. In atomic absorption, the population of ground state atoms is important as a fraction of these is raised to the lowest excited level by absorption of resonance radiation of a specific wavelength from a source external to the flame or electrical atomization unit. In atomic emission, the atomization flame or plasma source is further relied upon to excite ground state atoms to the higher energy levels from which radiation is emitted as the atoms revert to the ground state. It is the mea-

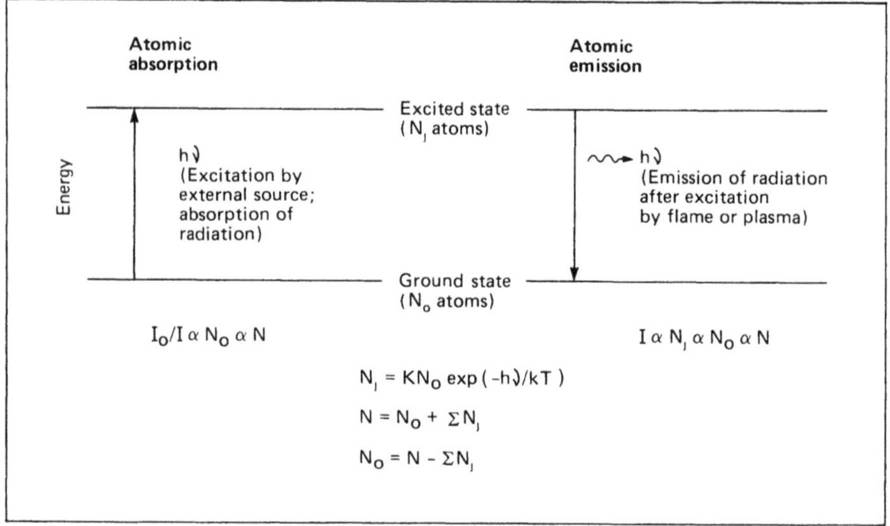

FIG. 7.1. Basic concepts of atomic excitation and emission pertinent to the techniques of atomic absorption and emission spectrometry. N, N_0, and N_j denote the number of atoms in total, in the ground state and in the excited state, respectively, in a vapor at thermal equilibrium; h is Planck's constant, ν is the resonance frequency, k is the Boltzmann constant, T is the absolute temperature, and K is a proportionality constant. In absorption, I_0 and I refer to the intensity of the external radiation source and the transmitted intensity, respectively; in emission, I represents the intensity of emission by the atom.

surement of the absorbed and emitted radiation that is central to atomic absorption and emission spectrometry, respectively. These measurements are proportional to ground state and excited state atomic concentrations, hence to the total number of atoms in the region of observation and in turn to the concentration of the element in the sample of interest. Absorbed and emitted radiation is characteristic of the element and has a wavelength typically in the 200–800-nm optical (ultraviolet and visible) region.

The fundamental physical, chemical, and radiative processes occurring during the transformation of a sample solution in spectrochemical analysis are depicted in Fig. 7.2. Aspiration of the solution into the flame or plasma typically via a nebulizer results in a fine spray or wet aerosol. Heat from the chemical flame or electrical plasma first desolvates then vaporizes or volatilizes the aerosol. Additional absorption of thermal energy serves to convert the vapor into free atoms

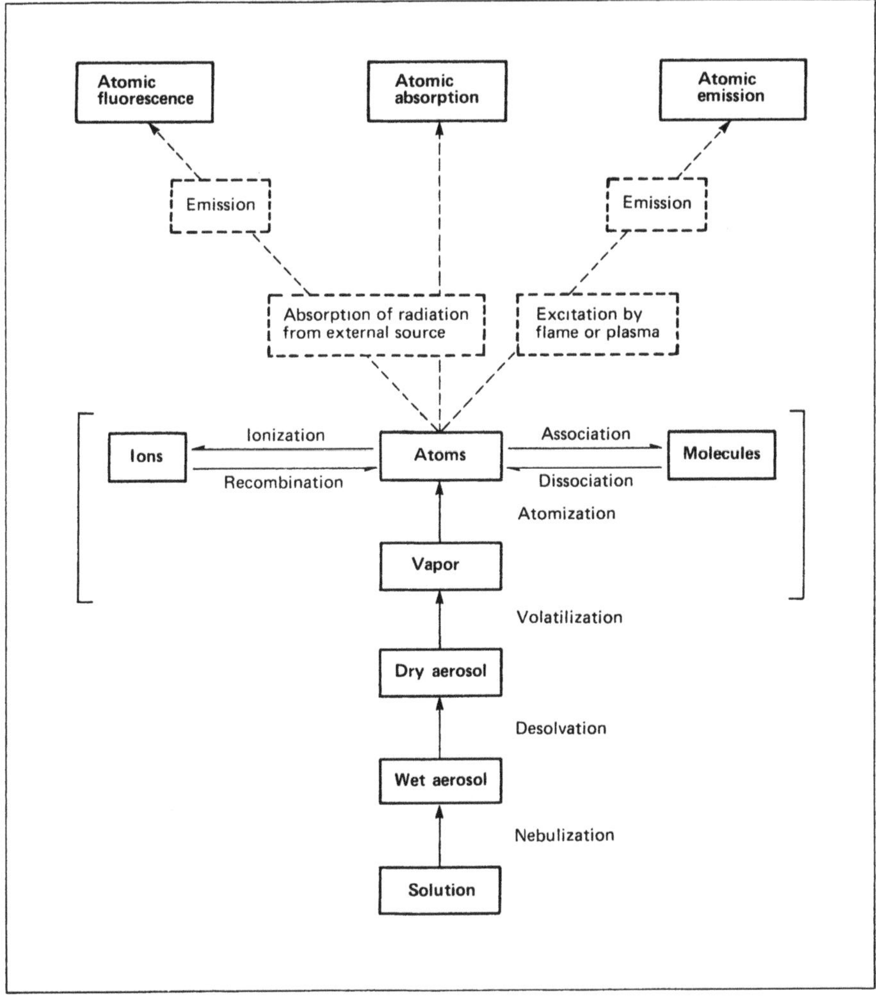

FIG. 7.2. Physical and chemical transformation of a sample solution and related light absorption and emission processes during atomic spectrometry.

which exist in equilibrium with ions and molecules formed by ionization and association, respectively. Absorption of radiation from an external source leads to atomic absorption. Excitation of the ground state atoms by the atomization source and subsequent emission of radiation lie behind the technique of atomic emission. For completeness, the phenomenon of atomic fluorescence is also shown, whereby detection is made of the reemitted radiation previously absorbed from an external source. For in-depth discussion of the principles of atomic absorption and emission the reader is referred to the excellent comprehensive treatments by Mavrodineanu (1970), Dean and Rains (1969, 1971, 1975), Kirkbright and Sargent (1974), and Schrenk (1975).

Definitions and Nomenclature

A multitude of terms and expressions have been used and continue to be applied to describing the techniques of absorption and emission spectrometry and related terms. Table 7.2, including nomenclature of some of the more pertinent terms, and the discussion below should serve to define and characterize the terminology used in this chapter. In spectrochemical analysis, the general term (optical) *spectroscopy* refers to the technique whereby light, emitted from a sample of interest, passes through a dispersing device with one or more exit slits with which measurements of light intensity are made at selected wavelengths or by scanning over the spectral range. *Spectrography* is reserved for techniques where emission spectra are recorded on a photographic plate; *spectrometry* is used if signal detection is performed with a photoelectric device. *Spectral bandpass* (or bandwidth) of the spectrometer is related to the dispersion of the monochromator and

TABLE 7.2. NOMENCLATURE OF ATOMIC SPECTROMETRY

Method and Abbreviation	
Absorption	Emission
AS—Absorption spectrometry[a]	ES—Emission spectrometry[a]
AAS—Atomic absorption spectrometry	AES—Atomic emission spectrometry
FAS—Flame atomic spectrometry	FES—Flame emission spectrometry
FAAS—Flame atomic absorption spectrometry	FAES—Flame atomic emission spectrometry
EAAS—Electrothermal atomization atomic absorption spectrometry	EAES—Electrothermal atomization atomic emission spectrometry
	PAES—Plasma atomic emission spectrometry[a]
	ICPAES—Inductively coupled plasma atomic emission spectrometry
	DCPAES—Direct current plasma atomic emission spectrometry
	DC arc AES/HV spark AES—Direct current arc/high voltage spark atomic emission spectrometry

[a] Generic terms in spectrochemical analysis; the remaining terms more definitively describe the specific techniques. Spectroscopy would be the general term; spectrometry, referring to detection of signal with a photoelectric detector is used as this is the typical current mode of detection.

the mechanical slit width, denotes the width of the spectrum selectable, and is a measure of instrumental resolution of spectral lines.

Flames were early sources for atomization and excitation and in various forms continue to be dominant. In *chemical flames* thermal energy to atomize the sample and excite the atoms results from the combustion of the fuel in the fuel–oxidant mixture. *Electrical or plasma flames* are flamelike electrical discharges classified according to whether discharges occur between two electrodes or a radiofrequency discharge is used to transfer energy to a gas from an electrical power source. An example of the first is the *direct current plasma* (DCP) jet whereas a *radiofrequency inductively coupled plasma* (ICP) is a commonly used radiofrequency discharge plasma flame. Nonflame atomization–excitation sources popular in emission spectrometry are the *direct current arc* (DC arc) and *high voltage spark* (HV spark). In atomic absorption, nonflame sources used most commonly are those based on the *electrothermal atomization* graphite furnace.

With regard to the processes which occur as the liquid sample passes through the nebulizer/flame the distinction between *nebulization* and *atomization* should be noted. Nebulization refers to the mechanical breakup of the liquid to produce a fine spray which eventually is decomposed or atomized into atoms. The term *solvent extraction*-flame or electrothermal atomization atomic absorption spectrometry (SE-FAAS, SEEAAS) pertains to partition of the analyte into an organic phase prior to introduction into a flame or electrothermal device for quantitation. Some elements such as arsenic, selenium and mercury can be converted into volatile hydrides (As, Se) or into elemental atomic vapor (Hg) for introduction into the spectrometer. The pertinent terms for these techniques are *hydride generation atomic spectrometry* and *cold vapor atomic spectrometry* whereby a flame or an electrothermal cell is used with the former and a simple quartz cell for the latter, and measurements of atomic absorption or emission are performed.

Finally, a word about the terms detection limit and sensitivity used to quantitatively describe method performance. *Detection limit* or limit of detection refers to the minimum amount or concentration of an element the method can measure with a given degree of reliability. It has been variously defined, with a common definition being the amount or concentration of an element required to produce a signal equal to twice the magnitude of the background noise. Values reported in subsequent tables include data pooled from the literature using various definitions and specifications; regardless of these inconsistencies, the reader should be able to get reasonable estimates of method detection limits. *Sensitivity* in analytical chemistry is generally taken to refer to the slope of the response-concentration calibration curve. The usual definition historically ascribed to this term in atomic absorption is the amount or concentration of the element required to produce an absorbance of 0.0044 absorbance units. The term *characteristic concentration* is used here instead of the commonly used AAS sensitivity definition in order to prevent confusion with the general analytical definition of sensitivity.

Basic Instrumental Requirements

Basic components constituting instrumentation for atomic absorption and emission spectrometric measurements are shown in Fig. 7.3. For atomic absorption, radiation from a hollow cathode or electrodeless discharge lamp, producing a sharp line spectrum of the element of interest, is focused onto the burner flame containing the atomized sample. Light passing through the flame travels through the entrance slit into the monochromator, set to select and direct the spectrum in the vicinity of the narrow resonance line through the exit slit. This isolated line falls on the detector, typically a photomultiplier tube whose

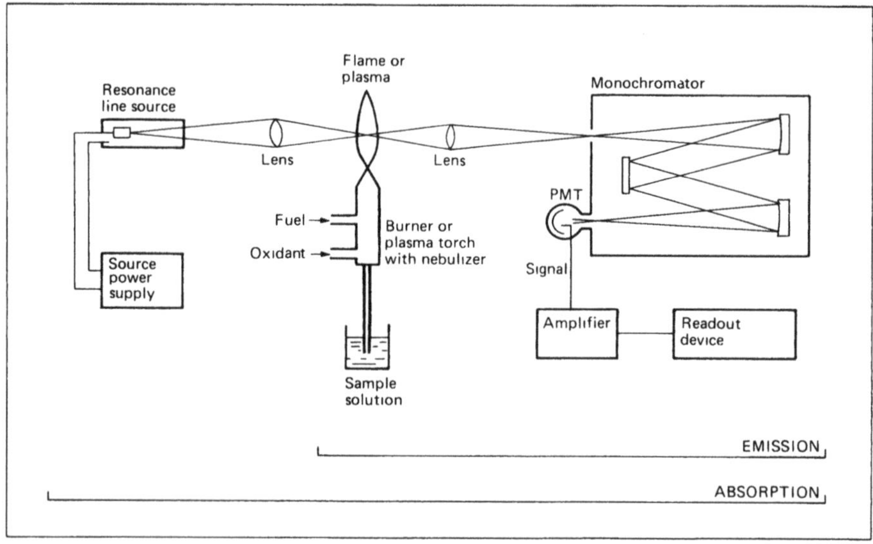

FIG. 7.3. Basic instrumental configurations for atomic absorption and emission spectrometry.

electrical output is amplified and presented on a readout device such as a digital readout meter or recorder. Modulating the resonance line source and tuning the amplifier to the same frequency precludes interference by emission at the same wavelength from flame-excited atoms. Attenuation of the radiation passing through the flame is related to the concentration of the element in the original sample.

In principle, similar instrumental components may be used for emission spectrometry with the important exception that the line source is not required. Emission from flame- or plasma-excited atoms modulated by a vibrating reed placed between the atomizer and monochromator is directed through the monochromator to the detector, amplified, and read. The monochromator can be replaced by a polychromator with a separate photomultiplier tube at each of several exit slits permitting simultaneous multielement determinations.

Although in practice variations and modifications of these components occur, the components described constitute the basic configuration for atomic absorption and emission spectrometry. Notable accessory or peripheral components in modern instruments are hydride and cold vapor generators, electrothermal atomizers, radiofrequency generators to sustain the plasma, automated sample introduction devices, and computers for instrument control and data processing.

SCOPE OF APPLICATIONS TO FOOD ANALYSIS

Elements of Interest and the Elemental Composition of Foods

The nature and concentration of chemical elements found in foods is related to the biological role played by the elements in tissue structure and physiology, and to external factors. Hence both natural processes and deliberate addition and inadvertent contamination during growth, processing, and preparation influence element levels in food materials. Some two dozen of the 90 naturally occurring elements are considered essential for animal and plant life. A number of other elements are found in foods as environmental contaminants at times at concentrations toxic to the food plant or to the consumer of the food product. Essential and toxic elements are presented in Fig. 7.4 in context of the periodic table. Division has been made into three principal groups separately including nutritionally essential major and micro elements and those of toxicological significance. Although a number of elements such as C, Na, K, Mg, and Ca are essential and others such

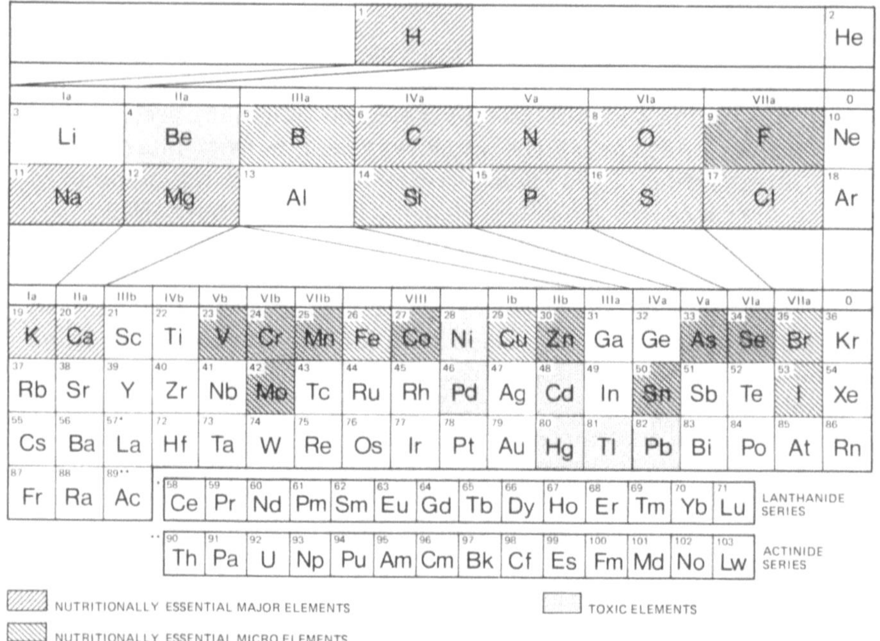

FIG. 7.4. Essential and toxic chemical elements. Elements essential for animal and/or plant nutrition have been subdivided into groups denoted major and micro reflecting concentrations greater and less than 1000 mg/kg, respectively. It should be noted that these are typical divisions and that in some commodities, an element usually considered a macro constituent may be present at trace (micro) levels. Note also the essential/toxic duality exhibited by several elements.

FIG. 7.5. Composition of foodstuffs in respect to chemical elements generally regarded as nutritionally essential and toxic. The bars reflect typical total concentration ranges in the edible portions of foodstuffs within the 12 classes of foods: (1) cereal products, (2) dairy products, (3) eggs and egg products, (4) meat and meat products, (5) fish and marine products, (6) vegetables, (7) fruits and fruit products, (8) fats and oils, (9) nuts and nut products, (10) sugar and sugar products, (11) beverages, and (12) spices and condiments. Arrows on bars indicate that the lower limit is beyond the scale of the figure. A point (●) reflects an average value due to limited information.

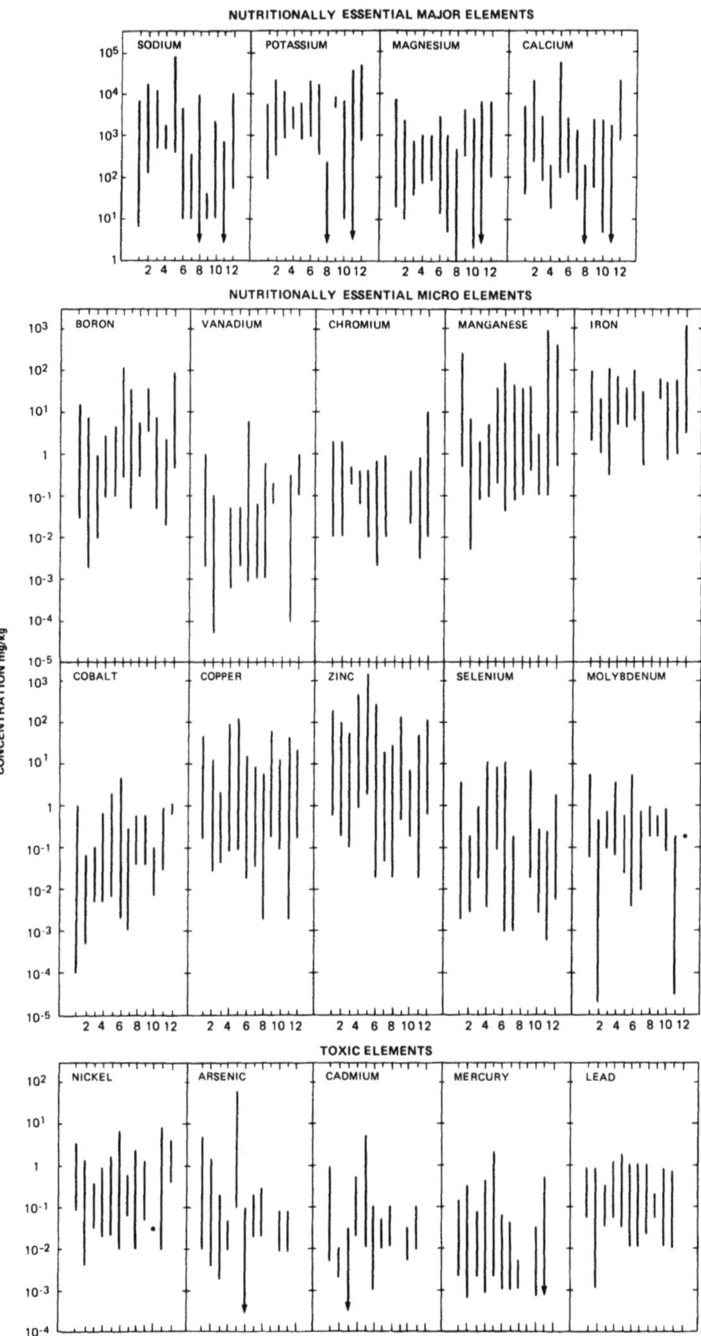
NUTRITIONALLY ESSENTIAL MAJOR ELEMENTS
SODIUM
POTASSIUM
MAGNESIUM
CALCIUM
NUTRITIONALLY ESSENTIAL MICRO ELEMENTS
BORON
VANADIUM
CHROMIUM
MANGANESE
IRON
COBALT
COPPER
ZINC
SELENIUM
MOLYBDENUM
TOXIC ELEMENTS
NICKEL
ARSENIC
CADMIUM
MERCURY
LEAD
CONCENTRATION mg/kg
2 4 6 8 10 12

as Be, Pd, Cd, Hg, Tl, and Pb are regarded as toxic, several elements, F, V, Cr, Mn, Co, Ni, Zn, As, Se, Mo, and Sn exhibit an essential/toxic duality. This population of elements thus encompasses those of interest to the food analyst or researcher dealing with nutrition/toxicology research or conducting analyses for compliance with legislation pertaining to nutrient value and adventitious contamination of foods.

A diagrammatic presentation of typical concentration ranges of some of the essential macro and micro elements and toxic elements found in 12 different classes of foodstuffs is given in Fig. 7.5. These data, reflecting total concentrations in the edible portions of foods reported on a fresh, dry, or processed weight basis, will serve to give the analyst an overview of the elemental composition with respect to some of the more important metallic elements of a variety of food classes. Evident are the extremes of concentration, a factor of about 10^9 between the highest and lowest levels over all elements and commodities, large variations as well within commodity groups, and also indications of levels characteristic of the various food classes.

Applicability of Absorption and Emission Spectrometry to Food Analysis

Analytical atomic absorption and plasma emission spectrometries can detect and measure just about every naturally occurring metallic element and some nonmetals as well. With reference to Fig. 7.4, virtually all elements listed there are amenable to determination by atomic spectrometry with the exception of the inert gases and halogens on the right-hand side of the table and N and O. With these techniques then one has comprehensive tools applicable to most if not all elements of interest to the food analyst.

Detection limits for the elements of interest in the domain of food analysis, achievable by flame and electrothermal atomic absorption and ICP atomic emission are listed in Table 7.3. Three main points are immediately apparent upon scanning the data therein: (1) nonmetals such as B, P, and S exhibit relatively poor limits of detection by FAAS; (2) ICP has a much superior detection limit for these important nonmetallic elements; (3) EAAS generally excels as the most detective technique. Atomic absorption is generally applicable to the estimation of metallic elements although the hydride generation technique has extended its usefulness to the metalloids As and Se. Although nonmetals such as S have been determined indirectly by FAAS at concentrations of interest, this is not the preferred approach; EASS and ICP have decided advantages in this regard. Considering FAAS/EAAS and

TABLE 7.3. DETECTION LIMITS FOR ATOMIC ABSORPTION AND INDUCTIVELY COUPLED PLASMA EMISSION TECHNIQUES[a]

	Detection limit, ng/mL			Ratio of detection limits	
Element	FAAS	EAAS[b]	ICPAES[c]	FAAS/EAAS	FAAS/ICPAES
Al	40	0.5	2	80	20
As	200	2	40	100	5
(Hydride)	0.3				
B	2000	—	3	—	670
Be	2	0.05	0.2	40	10
Ca	2	2	0.2	1	10
Cd	1	0.01	2	100	0.5
Co	5	0.5	4	10	1
Cr	10	1	5	10	2
Cu	5	0.2	3	25	2
Fe	5	0.5	3	10	2
Hg	20	10	20	2	1
(Cold vapor)	0.04				
K	5	0.1	—	50	—
Mg	0.3	0.002	0.05	150	6
Mn	3	0.02	1	150	3
Mo	30	0.5	3	60	10
Na	2	0.02	10	100	0.2
Ni	5	2	10	3	0.5
P	30,000	0.4	100	75,000	300
Pb	10	0.2	10	50	1
Pd	20	2	30	10	0.7
S	5000	50	—	100	—
Se	250	10	50	25	5
(Hydride)	0.3				
Si	100	2	8	50	13
Sn	50	10	20	5	3
Sr	10	0.5	0.1	20	100
Tl	200	2	30	100	7
V	40	10	3	4	13
Zn	2	0.005	3	400	0.6

[a] Typical good detection limits for elements in aqueous solution at typically the most detective lines, adapted from Robinson (1974), Kirkbright and Sargent (1974), Slavin (1968), Varian Techtron (1972B), Dean and Rains (1975), Fuller (1977), Ihnat (1976), Ihnat and Miller (1977), Boumans (1980), Boumans and Bosveld (1979), Scott *et al.* (1974), Dickinson and Fassel (1969), Winge *et al.* (1979), McQuaker *et al.* (1979), and Boumans and de Boer (1972).

[b] Absolute detection limits have been converted to concentration units assuming a 10-μL charge in the electrothermal atomizer.

[c] ICP detection limits reflect both calculated and measured values.

FAAS/ICPAES detection limit ratios and excluding the extreme values for B, P and S, we arrive at means of about 60 and 5, respectively, indicating the superior detection capability of EAAS over FAAS and the comparability of FAAS and ICPAES.

How these spectrometric detection limits for elemental aqueous solutions translate into applicability to the analysis of foodstuffs is summarized in Table 7.4. Presented here, individually for each element

TABLE 7.4. ATOMIC ABSORPTION AND EMISSION DETECTION LIMITS IN RELATION TO ELEMENT CONCENTRATIONS IN FOODS

Element	Ratio of median element concentration in analytical solution to detection limit of technique[a]		
	FAAS	EAAS	ICPAES
Al	0.5	50	15
As	3[b]	0.5	0.03
B	0.01	—	7
Ca	1000	1000	10,000
Cd	0.2	15	0.1
Co	0.2	2	0.3
Cr	0.3	3	0.6
Cu	3	30	4
Fe	40	440	70
Hg	1[c]	0.004	0.002
K	6000	300,000	—
Mg	14,000	2,000,000	80,000
Mn	4	600	15
Mo	0.05	4	0.5
Na	22,000	220,000	500
Ni	1	2	0.4
Pb	0.1	6	0.1
Se	7[b]	0.2	0.04
Sn	1	7	3
V	0.03	0.1	0.3
Zn	25	10,000	20

[a] Concentrations of the elements in the analytical solution presented to the spectrometer are based on estimated median concentrations in foods representing the 12 classes listed in Fig. 7.6, calculated from data tabulated by Ihnat (1982), assuming 1-g sample in a final volume of 100 mL. Concentration detection limits taken from Table 7.3. As a rule of thumb, a ratio of 10 or greater indicates the technique can be expected to be applicable to the analysis of foods with typical levels of the element. It should be noted that use of larger sample weights and resort to other concentration procedures can enhance the ratios making techniques with current ratios in the vicinity of one potentially good performer.

[b] Hybride generation method in conjunction with FAAS.

[c] Cold vapor generation method in conjunction with FAAS.

and spectrometric technique, are ratios of median element concentrations in the analytical food sample solution to method detection limits. Adopting the criterion that for reliable analyses the concentration of the element in solution should be 10 times the detection limit, a ratio of 10 or greater indicates the technique to be suitable for the listed element at the typical level found in a foodstuff. Of the 21 elements, FAAS and ICPAES techniques are each applicable to about $\frac{1}{3}$ of the elements, the four major essential elements Ca, K, Mg, and Na, together with several of the essential micro elements. A somewhat greater proportion of elements (about $\frac{1}{2}$) can be determined by the more detective EAAS technique. These observations are based on generalizations regarding typical analyte levels in analytical samples undergoing normal treatments. It should be noted that variations in the analytical procedure tailored to the sample and its elemental composition, such as using larger sample weights or small final volumes and resorting to concentration steps, will enhance the ratios listed in Table 7.4 making techniques with tabulated ratios in the vicinity of one potentially good performers. In this case, the number of elements determinable rises to about 13, 16, and 10 for FAAS, EAAS, and ICPAES, respectively.

Data in Table 7.4 for FAAS measurements of As, Se, and Hg are presented for the techniques of hydride generation (As, Se) and cold vapor generation (Hg) whether or not a flame is used as an atomizer. This is because the vapor evolution methods are substantially more detective than ordinary flame methods and have become fairly well established. These modifications to sample introduction make the FAAS technique superior to EAAS and ICPAES for determinations of these elements. Toxic elements Cd and Pb present in samples and toxicologically effective at very low concentrations are best amenable to detection by electrothermal atomic absorption. Proceeding down the list of elements in Table 7.4, all elements with the sole exception of vanadium have a ratio of one or greater by at least one of the techniques demonstrating the absorption and emission techniques to be complementary. It is evident, then, that a suitable atomic spectrometric technique commensurate with detectivity necessary for the analyte level of interest is available for the majority of elemental analyses the analyst might be called upon to conduct, making atomic spectrometry a collection of extremely powerful techniques of analyses.

Established Methodologies

As mentioned previously, in spite of the preeminence of absorption and emission atomic spectrometry in analytical chemistry, there exists a dearth of publications on approved methodologies with respect to food analysis applications of atomic spectrometry. Several atomic absorption methods for a limited number of elements have appeared from official organizations and a spark atomic emission spectrographic method has been adopted by the AOAC. Understandably, none has yet appeared for the newer technique of plasma emission. The relatively small number of validated procedures centering on atomic absorption results from the necessary slow progression in the validation and acceptance deliberations of organizations entrusted with this task and the apparently slow acceptance of this technique by food analysts as evidenced by the scant mention of AAS in volumes on food analysis. A summary of current official and other methods found in methodology manuals geared to the bench chemist is presented in Table 7.5.

TABLE 7.5. OFFICIAL AND OTHER ATOMIC ABSORPTION AND EMISSION SPECTROMETRIC METHODOLOGIES FOR FOOD ANALYSIS[a]

Elements	Method source and reference
	Official methods
Al, Ba, B, Ca, Cu, Fe, Mg, Mn, Mo, P, K, Na, Sr, Zn	Association of Official Analytical Chemists (Emission Spectrography) (Horwitz 1980)
As, Ca, Cd, Cu, Fe, Hg, K, Mg, Mn, Na, Ni, Pb, Se, Sn, Zn	Association of Official Analytical Chemists (Horwitz 1980, AOAC)
Cd, Cu, Zn	Analytical Methods Committee of the Society for Analytical Chemistry (Hanson 1973)
Ca, Cu, Mg, Mn, Na, Zn	Ministry of Agriculture, Fisheries and Food (1973)
Ca, Cu, Fe, Mg, Mn, Zn	Approved Methods Committee, American Association of Cereal Chemists, W.C. Schaefer (Chairman) (1976)
Cr, Cu, Fe, Ni	American Oil Chemists' Society (W.E. Link 1977)
Cd, Cu, Pb, Zn	Subcommittee on Procedures of the Chemistry Task Force of the National Shellfish Sanitation Program (1975)
Cu, Fe, Pb, Zn	The Institute of Brewing Analysis Committee (1973, 1974, 1977)
Hg	Mercury Analysis Working Party of the Bureau International Technique du Chlore (1976)
Cu	International Union of Pure and Applied Chemistry (Schuller and Coles 1979)
	Other sources[b]
Al, As, B, Ca, Cd, Co, Cr, Cu, Fe, Hg, K, Mg, Mn, Mo, Na, Ni, Pb, Se, Sn, V, Zn	Ihnat (1982)
Ca, Cd, Cu, Mg, Pb	Lees (1975)
Ca, Cd, Cu, Hg	Pearson (1976)
Cu	Hart and Fisher (1971)
Al, Ca, Cu, Co, Fe, K, Mg, Mn, Mo, Na, Se, Si, Sr, Zn	Pinta (1975)
Al, Ca, Cd, Cr, Cu, Fe, Hg, K, Mg, Mn, Na, Ni, Pb, Sb, Zn	Allen *et al.* (1974)

[a] With the exception of the AOAC emission spectrography method and flame emission as well as atomic absorption methods used for Na and K, all other methods center on atomic absorption spectrometry. Aspiration of solution samples into a flame is the commonest approach, with hydride generation (As and Se) and cold vapor (Hg) techniques of analyte introduction into the spectrometer and electrothermal atomization specified as well.

[b] Books on food analysis or atomic spectrometry giving step by step instructions for use by the bench analyst.

Of all organizations listed in Table 7.5, the AOAC appears to be the most active, getting to official status both atomic absorption spectrometric and atomic emission spectrographic methods for up to 15 elements in a variety of foodstuffs and plants via validation through interlaboratory testing. Absorption spectrometric procedures for 1–6 elements are also available from several other organizations. Other "nonofficial" sources of methods listed in the bottom half of Table 7.5 have been selected from those published as detailed laboratory instructions in volumes devoted to food and related analysis or atomic spectrometry. As the usefulness of a methodology manual is related to the reliability of methods described, the methods therein should have been tested for satisfactory performance. This criterion was taken to be fulfilled in the work of Ihnat (1982) by including, as far as possible, methods promulgated and tested by analytical organizations. To round out the treatment there and to expand the list of elements, methods were also selected from the journal literature. Even at this stage of development in atomic spectrometry, it is evident that only one half of the 28 essential and toxic elements of potential interest to the food analyst (Fig. 7.4) are covered by detailed descriptions of analytical methodology.

FUNDAMENTAL CONSIDERATIONS OF THE ABSORPTION AND EMISSION TECHNIQUES

Principal Lines for Atomic Absorption and Plasma Emission

Transitions of electrons between ground and excited energy levels of the atom as depicted in Fig. 7.1 involve absorption and emission of electromagnetic radiation. This radiation, highly characteristic of the element, corresponds to specific wavelengths in the ultraviolet and visible regions of the spectrum. In reality, many more atomic energy levels exist than simplistically depicted in the figure. Electronic transitions between various combinations of ground and excited state atomic and ionic levels give rise to a multitude of spectral lines for every element producing rich and complex emission spectra as attested by the large number of lines characterized by, for example, Meggers *et al.* (1975) and Boumans (1980) and demonstrated in spectrograms by Gilbert (1959). Important spectral lines for atomic absorption and inductively coupled plasma emission are listed in Table 7.6. In flame atomic absorption and emission, the characteristic wavelength usu-

TABLE 7.6. IMPORTANT SPECTRAL LINES FOR ATOMIC ABSORPTION AND INDUCTIVELY COUPLED PLASMA EMISSION[a]

	Wavelength, nm			Wavelength, nm	
Element	Atomic absorption[b]	Plasma emission[c]	Element	Atomic absorption[b]	Plasma emission[c]
Al	309.27	309.27(I)	Mo	313.26	202.03(II)
	309.28	309.28(I)			379.83(I)
	396.15	396.15(I)			386.41(I)
As	193.70	193.70(I)	Na	589.00	589.00(I)
		197.20(I)		330.24	589.59(I)
		228.81(I)		330.30	330.24(I)
B	249.68	249.77(I)	Ni	232.00	221.65(II)
	249.77	249.68(I)		341.48	232.00(I)
		208.96(I)			231.60(II)
Be	234.86	313.04(II)	P	177.50	213.62(I)
		234.86(I)		178.29	214.91(I)
		313.11(II)		178.77	253.57(I)
Ca	422.67	393.37(II)	Pb	283.31	220.35(II)
	239.86	396.85(II)		217.00	217.00(I)
		317.93(II)			261.42(I)
Cd	228.80	214.44(II)	Pd	247.64	340.46(I)
		228.80(I)		244.79	363.47(I)
		226.50(II)			229.65(II)
Co	240.73	238.89(II)	S	180.7	180.7
		228.62(II)			
		237.86(II)			
Cr	357.87	205.55(II)	Se	196.03	196.03(I)
	359.35	206.15(II)			203.99(I)
		267.72(II)			206.28(I)
Cu	324.75	324.75(I)	Si	251.61	251.61(I)
		224.70(II)			288.16(I)
		327.40(I)			212.41(I)
Fe	248.33	238.20(II)	Sn	235.48	189.98(II)
		239.56(II)		286.33	235.48(I)
		259.94(II)		224.61	242.95(I)
Hg	253.65	194.23(II)	Sr	460.73	407.77(II)
		253.65(I)			421.55(II)
		404.66(I)			216.60(II)
K	766.49	404.41(I)	Tl	276.79	190.86(II)
	404.41	766.49(I)			276.79(I)
		769.90(I)			351.92(I)
Mg	285.21	279.55(II)	V	318.34	309.31(II)
		280.27(II)		318.40	310.23(II)
		285.21(I)		318.54	292.40(II)
Mn	279.48	257.61(II)	Zn	213.86	213.86(I)
		259.37(II)			202.55(II)
		260.57(II)			206.20(II)

[a] Principal lines used in atomic absorption and three lines for each element in inductively coupled plasma emission (except K), giving the best detection limits, according to Boumans (1980); adapted from Robinson (1974), Kirkbright and Sargent (1974), Varian Techtron (1972A), Dean and Rains (1975), Meggers *et al.* (1975), Boumans (1980), Boumans and Bosveld (1979), and Winge *et al.* (1979).

[b] The first line for each element normally gives the best detection limit; the second and third line is an alternative either for similar good detectivity or reduced detectivity for high analyte level samples. Wavelengths within braces are not resolvable with typical spectral bandpasses used in absorption and emission.

[c] Symbols in parentheses refer to state of ionization; (I) and (II) indicate that the spectral lines originate from the atom and singly charged ion, respectively.

ally utilized is the resonance line corresponding to transitions between the ground and lowest excited state. Occasionally the same line is suitable for plasma emission analysis, but due to the different characteristics of this atomization/excitation source, other lines have been observed to give superior detectivities and are thus preferred. Whereas in AAS all principal lines originate from the ground atomic state, in plasma emission, it is often lines resulting from transitions between energy levels within the ion (singly ionized atom) that lead to the best detection limits.

Fundamental Aspects of Sample Decomposition, Atomization, and Excitation

Some of the basic processes occurring during the transformation of a sample solution into an optically active atomic vapor have been diagrammatically presented in Fig. 7.2. In flame or plasma emission, the liquid is usually introduced as a spray of fine micrometer-size droplets from a pneumatic nebulizer, carried by the oxidant and fuel gases producing the flame or plasma supported on a laminar burner or torch, respectively. The high temperature of the flame vaporizes the solvent (usually water) leaving residual salt particles which then volatilize from the solid or molten state into the gaseous state. Decomposition (atomization) of molecular species occurs giving a mixture of atoms, ions, electrons and various undissociated neutral and charged molecules. Some refractory elements yield difficultly dissociable, stable oxides; other elements give a good supply of atoms; whereas still others, easily thermally ionized, suffer depletion of ground state atoms. These atoms and ions are now available for excitation and spectrometric measurement.

The mechanism and extent to which the processes described above occur depend on many factors including the nature of the atomization/excitation cell, flame temperature, chemical environment and the nature of the aspirated solution. Common premixed flames used in absorption spectrometry are provided by the oxidant/fuel combinations air/acetylene, nitrous oxide/acetylene, air/propane, and air/hydrogen or entrained air/hydrogen/argon. The first two are by far the most popular, together capable of determining over 90% of the approximately 70 elements measurable by atomic absorption spectrometry. Flame temperatures are dependent on the nature of the oxidant and fuel, flame stoichiometry, burner type, and nature and amount of solvent nebulized. Maximum temperatures observed for stoichiometric gas mixtures are approximately 2400°, 2800°, 1800° and 2000°C for the air/acetylene, nitrous oxide/acetylene, air/propane, and air/hydrogen flames, respectively. High temperature chemical flames are efficient vehicles for atom formation for absorption spectrometry and thermal excitation for atomic emission. Electrothermal atomizers typically reach temperatures in the vicinity of 3000°C and serve to atomize and excite discrete drops of sample solution although some of the processes occurring there are different than in flames.

Chemical flames obtain their energy from the exothermic chemical reactions between fuel and oxidizer species; plasmas, on the other hand, obtain their energy from collisions induced by electrical currents. An inductively coupled plasma unit consists of a quartz plasma torch surrounded by an induction coil through which radiofrequency power is supplied. Seed electrons, introduced into the argon stream in the torch, interact with the intense high frequency (27.14 MHz) magnetic field and ionize the gas stream by collision. A high-temperature (typically 8000°C) plasma forms above the torch serving to atomize, ionize, and excite the sample introduced into the discharge. Atomic emission spectra are then monitored by the spectrometer.

Whereas in emission spectrometry, the flame or plasma serves the dual role of atomizing and exciting the sample, an external source of atom excitation is necessary for atomic absorption. Commonly this source is a hollow cathode lamp emitting light of the element to be determined in the form of sharp intense spectral lines characteristic of the analyte. The source spectral lines impinging on the gaseous atomized sample are attenuated by varying degrees. The resonance line is responsible for excitation of atoms of the element from the ground to the first excited state; its isolation and amplification by the spectrometer is the underlying principle of atomic absorption. In addition to single or multielement hollow cathode lamps as the most common excitation sources, applications have also been found for vapor discharge and electrodeless discharge lamps for elements such as the alkali metals, mercury, arsenic, and selenium.

Interferences

Optimistic statements regarding lack of interferences in flame atomic absorption have been tempered by experimental realities. A body of literature exists demonstrating that although spectrometric techniques are eminently suitable, regard must be given to parameters and conditions influencing the analytical signals in order to ensure good performance. The numerous phenomena interfering with the spectrometric signal to varying degrees, arise from spectral, physical, and chemical processes. Spectral interferences are defined as any radiation which overlaps that of the element being measured. In general, spectral interferences originate from the light source (AAS) and the atomization/excitation source. These interferences encompass, principally, atomic and ionic lines, bands emitted and absorbed by molecular compounds, and continuum emission from incandescent particles. In AAS, the situation in this regard is simpler than in emission with very few direct overlaps of spectral lines evident. Coincident lines observed in AAS are listed in Table 7.7 reflecting all of the overlaps observed thus far. Note that very few complications are to be expected with this highly selective technique. Nonatomic or background absorption, caused by light absorption or scattering at the wavelength of interest by the flame, vapor in EAAS, molecules or particles therein, is of more concern in AAS. Suitable background corrections can be made using hydrogen or deuterium continuum lamps. The case in emission, even with the high-temperature ICP source is different as exemplified by data in Table 7.8. In this example, dealing with only one analytical line of one element, numerous lines within ±0.1 nm arising from other elements can be observed. Seriousness of interference depends on the spectral bandpass used, intensities of the lines and the relative concentrations of the analyte, and interfering elements. It is clear that even with the narrow spectral bandpasses used in emission spectrometry, spectral interference potential exists and is in fact a common concern in emission work.

TABLE 7.7. SPECTRAL OVERLAP INTERFERENCES OBSERVED IN FLAME ATOMIC ABSORPTION SPECTROMETRY[a]

Analyte	Wavelength, nm	Interferent	Wavelength, nm	Line separation, nm
Al	308.215	V	308.211	0.004
Cd	228.802	As	228.812	0.010
Ca	422.673	Ge	422.657	0.016
Co	252.136	In	251.137	0.001
Cu	324.754	Eu	324.753	0.001
Ga	403.298	Mn	403.307	0.009
Fe	271.903	Pt	271.904	0.001
Hg	253.652	Co	253.649	0.003
Mn	403.307	Ga	403.298	0.009
Sb	217.023	Pb	216.999	0.024
	231.147	Ni	231.097	0.050
Si	250.690	V	250.690	0.000
Zn	213.856	Fe	213.859	0.003

[a] All of the spectral interferences reported in the literature to 1975. Interferent concentrations 1–2500 times the concentration of the corresponding analyte are required to produce an equivalent absorption signal. Adopted from Robinson (1974), Lovett *et al.* (1975), and Parsons *et al.* (1975).

TABLE 7.8. SPECTRAL OVERLAP INTERFERENCES POSSIBLE FOR MEASUREMENT OF CHROMIUM AT 267.716 nm BY INDUCTIVELY COUPLED PLASMA ATOMIC EMISSION SPECTROMETRY[a]

Interferent	Wavelength, nm	Line separation from 267.716 nm	Interference potential at spectral bandpasses (nm) of 0.010	0.025	0.040
Au (I)	267.595	0.121	—	—	—
U (II)	267.641	0.075	—	—	*b*
Ta (II)	267.648	0.068	—	—	*b*
Hf (II)	267.663	0.053	—	—	*b*
U (II)	267.669	0.047	—	*b*	*b*
Ir (I)	267.683	0.033	—	*b*	*b*
Dy (II)	267.684	0.032	—	*b*	*b*
Re (I)	267.703	0.013	*b*	*b*	*b*
Tm(II)	267.712	0.004	*b*	*b*	*b*
Te (I)	267.713	0.003	*b*	*b*	*b*
Pt (I)	267.715	0.001	*b*	*b*	*b*
Lu (I)	267.725	0.009	*b*	*b*	*b*
W (I)	267.728	0.012	*b*	*b*	*b*
Dy (II)	267.734	0.018	*b*	*b*	*b*
Hf (II)	267.758	0.042	—	*b*	*b*
Nb (II)	267.766	0.050	—	*b*	*b*
Re (I)	267.776	0.060	—	—	*b*
V (II)	267.780	0.064	—	—	*b*
W (II)	267.785	0.069	—	—	*b*
Tm(II)	267.809	0.093	—	—	—
Eu (II)	267.829	0.113	—	—	—

[a] Interferent lines within ± 0.1 nm of the Cr 267.716-nm line are listed; adapted from Boumans (1980).
[b] Potential exists for spectral interference at the Cr 267.716-nm line; seriousness of interference depends on relative concentrations of analyte and interfering elements.

Physical interferences are related to physical properties of the sample solution and physical processes in the atomization/excitation cell. These factors influence the atom population thereby exerting effects on the emission/absorption signals. Solution properties of importance are viscosity, surface tension, temperature, bulk matrix composition, and nature of the solvent. Pertinent physical processes are those related to solution aspiration and nebulization, solution transport to the flame, solvent evaporation, and solute vaporization. Chemical interferences, which occur from reactions in both the condensed (liquid and solid) and vapor phases, arise from and are affected by a variety of factors. Important considerations are the following: flame temperature, nature of the fuel/oxidant and the nebulizer/burner system, region of observation, nature of the solvent, use of releasing and protective chelating agents, ionization, excitation, dissociation, and nature of compounds formed. All of these factors can have influences, at times severe, on the absorption or emission signals and must be appropriately controlled.

Comparison of Atomic Absorption and Emission Techniques

A comparison of some of the technical and operational features of the atomic absorption and ICP techniques are summarized in Table 7.9. ICPAES is the only one of the three capable of true multielement performance with commercial instrumentation available for a range of elements (from a total of about 70) including several of the important nonmet als such as B, P, and S. Another useful feature of this technique is its very large linear calibration range over about five orders of magnitude contributing to its capability of measuring over a wide range of concentration with a single calibration line. Both EAAS and ICPAES have more complex operational and theoretical requirements, and the need for an operator with high technical and theoretical competence is noted.

A comparison of these techniques with respect to detection capabilities of other techniques is indicated in Fig. 7.6.

TABLE 7.9. COMPARISON OF FLAME ATOMIC ABSORPTION (FAAS), ELECTROTHERMAL ATOMIZATION ATOMIC ABSORPTION (EAAS), AND INDUCTIVELY COUPLED PLASMA ATOMIC EMISSION (ICPAES) TECHNIQUES OF ANALYSIS

	FAAS	EAAS	ICPAES
Elements measurable	Range of metals, +As, Hg, Se	Range of metals, +As, P, Si	Range of metals, +B, P, S
Multielement capability	No	No	Yes
Determination rate (per hour)	200	20	400
Detection limit (ng/mL)	0.04–30,000, typically 5	0.002–50, typically 0.5	0.05–100, typically 5
Precision	0.5%	2%	0.5%
Accuracy	Good	Good	Good
Spectral interferences	Low	Low	Moderate
Chemical interferences	Low–moderate	Moderate–high	Low
Linear calibration range	10	10	10^5
Operational/ developmental expertise	Moderate	High	High
Data handling and reporting	Manual/ automatic	Manual/ automatic	Automatic
Purchase cost, K$	10–50	20–70	50–150

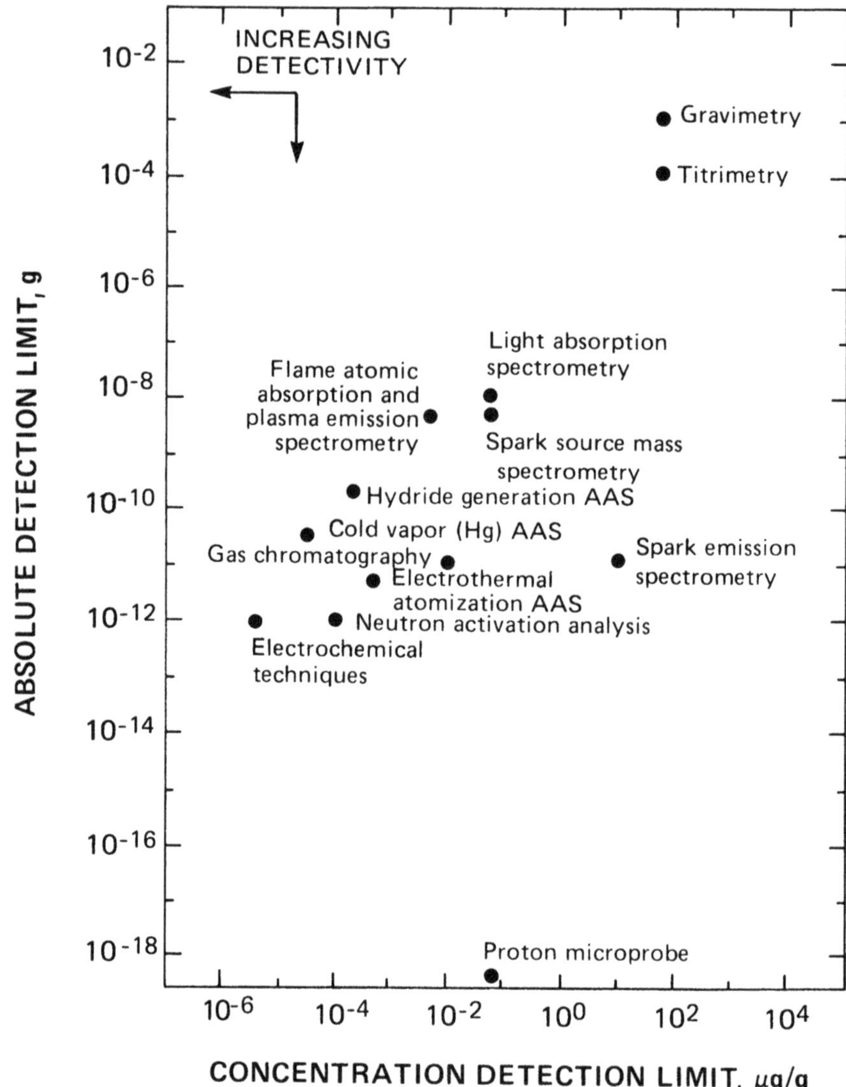

FIG. 7.6. Comparison of analyte detection capabilities of several analytical techniques. Absolute and concentration detection limits are for the form of the sample (original, ashed, in solution) usually required by the technique, averaged over the elements of interest in this report.

ANALYTICAL PROCEDURES FOR FOOD ANALYSIS

Preparation and Decomposition of the Analytical Sample

The initial bulk sample usually must be reduced in size prior to analysis. This may be accomplished by the usual methods of analysis employing various grinders and blenders exercising care to end up with a representative sample. When dealing with trace analysis, due regard must be paid to contamination of the bulk and analytical samples with extraneous sources of the elements of interest. Grinding of hard materials may lead to contamination with Co, Cr, Fe, Mn, Mo, and Ni from steel blades and Al, B, Ca, K, and Na from ceramic materials. Further contamination can occur with storage containers and vessels and reagents used for manipulations during analysis.

For each of the three spectrometric techniques, the preferred form of the sample is a liquid. Solid samples must therefore be brought into solution. This is accomplished by destruction of the organic matrix by acid digestion or dry ashing followed by acid dissolution of the residue resulting in almost complete dissolution of the sample (Hanson 1973; Pinta 1975, Chapter 9; Dunlop 1961; Gorsuch 1970; Bock 1979). Widely used acid digestion mixtures yielding clear digests and applicable to a wide range of foodstuffs are nitric/perchloric and nitric/perchloric/sulfuric acids with digestions performed in borosilicate glass vessels. Dry ashing is carried out in silica or platinum crucibles with an ashing aid such as sulfuric acid or magnesium nitrate if required. The resulting ash is dissolved in dilute hydrochloric acid. Although both acid digestion and dry ashing are satisfactory for many of the elements and commodities, one or the other is preferred in specific applications for which the reader is referred to the foregoing references and specific manuals and reports. Liquid food samples at times need only to be diluted with water or a suitable solvent before analysis.

Elements such as As, B, Cd, Co, Cr, Hg, Mo, Ni, Pb, Se, and V often occur in foods at concentrations too low to be measured by direct aspiration flame spectrometry. Spectrometric detectivity for such elements can be enhanced by chelation-solvent extraction. Besides enhancing detectivity, solvent extraction has the added advantage of leading to reduced interferences by virtue of reducing the matrix burden in the final solution. This is particularly advantageous for EAAS where matrix interelement interferences are more severe. Occasionally one resorts to other concentration/matrix-reduction techniques such as ion exchange and selective precipitation to prepare clean solutions for spectrometric measurement. In EAAS, various recipes for matrix alteration within the graphite furnace have been devised in order to reduce interferences. These procedures often call for the introduction of matrix-modifier compounds. The usual approach for elements such as As, Se, and Hg is to liberate them from the solution as vapors which are then introduced into the spectrometer.

Spectrometric Measurement

Spectrometer operating parameters such as wavelength, flame and sheathing gases, flame composition, observation height, light source current or wattage, and slit widths are selected and optimized using standard solutions. Standards are prepared in the same solvent (aqueous or organic) as the sample solutions and contain the major matrix elements of the sample if required for high accuracy. Spectral, physical, and chemical interferences are minimized by selection of suitable instrument operating parameters using manufacturer's instructions and published operating conditions as guides. Spectrometer response is calibrated with standard solutions, and samples are introduced into the spectrometer interspersed, as required, by standards and blanks. Sample responses corrected for appropriate reagent blanks are converted to element concentrations via comparison with the standard calibration curve using manual or computer computation.

Methods currently used for introduction of samples for flame and plasma spectrometry are listed in Table 7.10. Nebulization of solutions is by far the most common mode eliciting much research into nebulizer design. In order to save analytical time, nebulization of untreated or incompletely digested samples in the form of emulsions, suspensions, or slurries is possible with success dependent on the element/matrix combination. Discrete sample introduction can be by way of nebulizing small fixed volumes of solution, using flow injection, vapor generation, or volatilization of microsamples in cups or loops brought into the flame or externally using electrically heated chambers to produce vapor subsequently directed into the flame. Effluents from gas and liquid chromatographic columns have also been introduced into the spectrometer. In these instances, it is not the chromatographs that are looked upon as sample introduction techniques but it is the spectrometer which is considered to be the detector of elements

TABLE 7.10. METHODS OF SAMPLE INTRODUCTION FOR FLAME ATOMIC ABSORPTION AND INDUCTIVELY COUPLED PLASMA ATOMIC EMISSION

Flame Atomic Absorption	
· Nebulization of solutions	
· Nebulization of emulsions, suspensions and slurries	
· Discrete sample nebulization (microsampling, flow injections)	
· Volatilization—microsampling (Delves) cup, platinum loop, tungsten filament, arc, electrothermal, electrically heated chamber, capsule-in-flame	
· Vapor generation (hydride and mercury cold vapor)	
· Chromatograhy (GC, HPLC)	
Plasma Atomic Emission	
· Nebulization of solutions	· Electrothermal vaporization
· Nebulization of slurries	· Vapor generation
· Spark erosion	· Chromatography

eluting from the chromatograph used for speciation studies. While solid samples are at times introduced directly via the volatilization techniques, liquid forms of samples are by far the most common.

Evaluation of Method Performance

All of the steps comprising an analytical procedure, viz, sampling, sample treatment, analyte separation and manipulation, and analyte measurement, contribute their own inherent errors to the analysis. Of prime interest to the analyst is the resultant overall error in the final analytical result. Performance of the method with respect to accuracy may be estimated by testing the method using different sample masses or by measuring recoveries of elements added to the sample being analyzed then taken through the entire scheme. Satisfaction of these criteria, although necessary, is not sufficient for the complete demonstration of method validity. The best procedure for ultimate testing of method accuracy is its application to a reference material, compositionally similar to the analytical sample and certified for concentrations of interest. Commercially available biological reference materials pertinent to food analysis and certified for elemental content are listed in Table 7.11 and discussed in detail by Alvarez in Chapter 5 of this volume. It is evident that available reference materials cover neither the range of food matrices nor the range of elemental concentrations encountered in foods (Parr 1980; Wolf and Ihnat 1982) and that the shortfall in existing materials must be rectified by the development of additional materials. Steps are being taken in that direction (Muntau 1979; Ihnat *et al.* 1982).

TABLE 7.11. CURRENTLY AVAILABLE CERTIFIED BIOLOGICAL REFERENCE MATERIALS[a]

National Bureau of Standards, Washington, DC:			
SRM1566	Oyster tissue	SRM1572	Citrus leaves
SRM1567	Wheat flour	SRM1573	Tomato leaves
SRM1568	Rice flour	SRM1575	Pine needles
SRM1569	Brewers' yeast	SRM1577a	Bovine liver
International Atomic Energy Agency, Vienna:			
H-4	Muscle	H-8	Horse kidney[b]
A-11	Milk powder[b]	MA-A-1	Dried copepoda[b]
A-12	Animal bone[b]	MA-A-2	Homogenized fish flesh[b]
V-8	Rye flour[b]	A-13	Freeze-dried animal blood[b]
V-9	Cotton cellulose[b]		
Prof. H.J.M. Bowen, Reading University, UK:			
Kale[b]			

[a] Preparation of other food-related materials which may eventually be issued as reference or certified reference materials is in progress at the Commission of the European Communities, Ispra, Italy (Muntau 1979; Community Bureau of Reference 1982) and U.S. Department of Agriculture/Agriculture Canada (Ihnat *et al.* 1982).

[b] Defined as reference materials (RM) instead of certified reference materials (CRM).

Examples of Analytical Procedures

Brief descriptions of analytical procedures are presented exemplifying determinations using the three spectrometric techniques: FAAS, SEFAAS, and EAAS.

FAAS Determination. For the FAAS determination of Na and K in marine products, a 5-g sample is ashed in a platinum crucible at 500°C for 2 hr. The residue is dissolved in $(1+4)$ nitric acid, the solution filtered and brought to volume in a 100-mL volumetric flask. Aliquots are further diluted by 100- and 25-fold for sodium and potassium determinations, respectively. Solutions are aspirated into an air–acetylene flame for measurement of sodium at 589.00 nm and potassium at 766.49 nm.

SEFAAS Determination. Cobalt in plant tissues is determined by SEFAAS. The sample (2 g) is digested in a 100-mL Kjeldahl flask with nitric–perchloric acids. After dilution to about 40 mL, sodium citrate and an acid–base indicator are added and the pH is adjusted to 5.3–5.7. Ferric ion is reduced with hydrogen peroxide and the solution is transferred to a separatory funnel. Cobalt is chelated with 2-nitroso-1-naphthol and extracted with chloroform which is then returned to the rinsed funnel. Excess 2-nitroso-1-naphthol is removed with sodium hydroxide, the organic phase is evaporated to dryness, and the residue is dissolved in methylisobutylketone (MIBK). The MIBK solution is aspirated into an air–acetylene flame previously optimized on cobalt standard solutions in MIBK, and the element is determined at 240.73 nm.

EAAS Determination. For determination of cadmium by EAAS, 1–15 g (dry basis) of foodstuffs is digested, in a Vycor beaker, with nitric–perchloric acids. Upon completion of digestion, the solution with water rinses, heated to 90°C, is quantitatively transferred to a separatory funnel. Adjustment of pH is made with ammonium hydroxide and ammonium pyrollidine carbodithioate (APDC) is added to complex the Cd. The chelate is partitioned into MIBK. Stripping solution [3% and 8% mass/mass (m/m) HNO_3 and H_2O_2, respectively] is added to the separated MIBK phase. A 2-mL aliquot of the resulting aqueous phase is drained into a polypropylene bottle containing 6 mL of modification solution (0.33% and 0.096% m/m with respect to ammonium dihydrogen phosphate and ammonium hydroxide, respectively). Small volumes of solution are introduced into the graphite electrothermal atomizer and taken through an established drying, pyrolysis, and atomization program to measure Cd at 228.80 nm.

The preceding summaries are based on more complete descriptions for the bench analyst appearing in Ihnat (1982) based on work by the AOAC (Horwitz 1980), Subcommittee on Procedures of the Chemistry Task Force of the National Shellfish Sanitation Program (1975), Simmons (1973, 1975), and Dabeka (1979).

NEWER DEVELOPMENTS AND FUTURE DIRECTIONS IN ATOMIC SPECTROMETRY

Automation and the current explosive developments in microcomputers and their applications have left their imprints on atomic spectrometry. Sample changers have not only relieved the analyst of operational tedium but have significantly improved the precision of EAAS. They are beginning to be viewed as indispensable for that technique. New top-of-the-line atomic absorption spectrometers come equipped with built-in and/or external microcomputers for data handling and for instrumental control. Once a population of parameters is selected, in excess of 50 sample solutions can be run automatically in sequence for a large number of elements resulting in high sample throughput for the analytical laboratory. In emission spectrometry, computers have been, and continue to be, indispensable for instrument control and calculations. Cathode ray display tubes not only provide convenient data displays but depict, in real time, useful absorbance-time profiles in EAAS. It is expected that adoption of ICPAES by analytical laboratories in all fields will continue at an accelerating pace.

Research continues into various methods of sample introduction. For solutions, flow injection analysis whereby a small discrete volume of solution is injected into a continuously flowing stream of solvent is being investigated for ICPAES and FAAS. As sample dissolution is usually the slowest step in the analytical procedure, much interest abounds into means of introducing solid samples directly into absorption and emission spectrometers. Studies have dealt with sample atomization in the flame, atomization from electrically heated chambers, cathodic sputtering laser evaporation, with electrothermal atomization currently the preferred method for solid sample analysis. Introduction of the analyte into the flame or plasma as a hydride vapor has proved convenient and detective; it is applicable to the elements As, Bi, Ge, Pb, Sb, Se, Sn, and Te and many reports have appeared.

Although considered a single-element technique, recent work has demonstrated that atomic absorption can be endowed with simultaneous multielement detection capability. Salin and Ingle (1978) developed a flame spectrometer for the simultaneous determination of four elements using sequentially pulsed single-element hollow cathode lamps, and a multiple exit slit monochromator with a single photomultiplier tube. The approach of Harnly *et al.* (1979) was a multielement atomic absorption spectrometer with a high intensity continuum source and an echelle polychromator modified for wavelength modulation. With double beam operation and background correction on all channels, up to 16 elements could be measured simultaneously with flame or electrothermal atomization. An exciting application of the Zeeman effect to atomic absorption has been reported (Miller and Koizumi 1978). One important advantage is that nonatomic background absorption can be simultaneously and accurately corrected for at exactly the same wavelength as the atomic absorption line yielding more reliable data.

The combination of spectrometric techniques with other separation or measurement devices is giving rise to hybrid methodologies. An increase in interest in the actual organometallic forms of elements in biological samples has spawned combinations of gas and liquid chromatographic techniques with atomic spectrometry as element specific detectors (Van Loon 1981). Developments have centered on methods of interfacing the chromatographs geared to the investigation of As, Hg, and Pb speciation. A recent development of an ICP/mass spectrometer combination has been reported (Douglas 1983). In this hybrid, the ICP is simply a source of ions separated and analyzed by the mass spectrometer for elemental and isotopic abundance.

Finally, there continues to be slow but steady progress in the development of reference analytical methodology for atomic spectrometry. This facet of analytical science is an important complement to the new and exciting developments in spectrometric instrumentation. It is hoped that analytical methodology development will continue at a pace commensurate with the requirements of analysts in order to provide them with reliable, off-the-shelf protocols for routine and research applications.

BIBLIOGRAPHY

ALKEMADE, C.T.J., and MILATZ, J.M.W. 1955. Double-beam method of spectral selection with flames. J. Opt. Soc. Am. *45,* 583–584.

ALLEN, S.E., GRIMSHAW, H.M., PARKINSON, J.A., and QUARMBY, C. 1974. Chemical Analysis of Ecological Materials. Blackwell Scientific Publications, Oxford, UK.

APPROVED METHODS COMMITTEE, W.C. SCHAEFER (Chairman), 1976. Approved Methods of the American Association of Cereal Chemists. Am. Assoc. of Cereal Chemists Inc., St. Paul, MN.

BOCK, R. 1979. A Handbook of Decomposition Methods in Analytical Chemistry. John Wiley, New York. (English translation by I.L. Marr of 1972 German edition).

BOUMANS, P.W.J.M. 1980. Line Coincidence Tables for Inductively Coupled Plasma Atomic Emission Spectrometry. Vols. I and II. Pergamon Press, Toronto, Ont.

BOUMANS, P.W.J.M., and BOSVELD, M. 1979. A tentative listing of the sensitivities and detection limits of the most sensitive ICP lines as derived from the fitting of experimental data for an argon ICP to the intensities tabulated for the NBS copper arc. Spectrochim. Acta *34B,* 59–72.

BOUMANS, P.W.J.M., and de BOER, F.J. 1972. Studies of flame and plasma torch emission for simultaneous multi-element analysis—I. Preliminary investigations. Spectrochim. Acta *27B,* 391–414.

CANTLE, J.E. (Editor). 1982. Atomic Absorption Spectrometry. Elsevier Scientific Publishing Co., Amsterdam, Netherlands.

CHRISTIAN, G.D., and FELDMAN, F.J. 1970. Atomic Absorption Spectroscopy-Applications in Agriculture, Biology and Medicine. Wiley (Interscience), New York.

COMMUNITY BUREAU OF REFERENCE. 1982. Catalogue of BCR Reference Materials, Brussels.

CROSBY, N.T. 1977. Determination of metals in foods. A review. Analyst *102,* 225–268.

DABEKA, R.W. 1979. Graphite furnace atomic absorption spectrometric determination of lead and cadmium in foods after solvent extraction and stripping. Anal. Chem. *51,* 902–907.

DEAN, J.A., and RAINS, T.C. (Editors). Flame Emission and Atomic Absorption Spectrometry. 1969. Vol. 1, Theory; 1971. Vol. 2, Components and Techniques; 1975. Vol. 3, Elements and Matrices. Marcel Dekker, New York.

DICKINSON, G.W., and FASSEL, V.A. 1969. Emission spectrometric detection of the elements at the nanogram per milliliter level using induction-coupled plasma excitation. Anal. Chem. *41,* 1021–1024.

DOUGLAS, D. 1983. ICP/MS Technologies marry to produce better analysis. Can. Res. April (1983) 55–60.

DUNLOP, E.C. 1961. Decomposition and dissolution of samples: organic. *In* Treatise on Analytical Chemistry Part I, Vol. 2, Chapter 25. I.M. Kolthoff and P.J. Elving (Editors). Wiley (Interscience), New York.

DVOŘAK, J., RUBEŠKA, I., and REZAČ, Z. Flame Photometry—Laboratory Practice (English translation edited by R.E. Hester). CRC Press, Chem. Rubber Co., Cleveland, OH.

ELWELL, W.T., and GIDLEY, J.A.F. 1966. Atomic Absorption Spectrophotometry, 2nd Rev. ed. Pergamon Press, Oxford, UK.

FASSEL, V.A. 1979. Simultaneous or sequential determination of the elements at all concentration levels—the renaissance of an old approach. Anal. Chem. *51,* 1290A–1308A.

FRAUNHOFER, J. 1817. Determination of refractive and dispersive power of various glasses in relation to the production of achromatic telescopes. Ann. Physik (Gilbert's Ann.) *26,* 264–313 (German).

FULLER, C.W. 1977. Electrothermal Atomization for Atomic Absorption Spectrometry. The Chemical Society, London, UK.

GILBERT, P.T. JR. 1959. Analytical flame photometry: new developments. *In* Symposium on Spectroscopy, ASTM STP *No. 269,* pp. 73–156. Am. Soc. Testing Mats., Philadelphia, PA.

GORSUCH, T.T. 1970. The Destruction of Organic Matter. Pergamon, Toronto, Ont.

GREENFIELD, S., JONES, I. LL., and BERRY, C.T. 1964. High-pressure plasmas as spectroscopic emission sources. Analyst *89,* 713–720.

HANSON, N.W. (Editor). 1973. Official, Standardized and Recommended Methods of Analysis, 2nd ed. The Society for Analytical Chemistry, London, UK.

HARNLY, J.M., O'HAVER, T.C., GOLDEN, B., and WOLF, W.R. 1979. Background-corrected simultaneous multielement atomic absorption spectrometer. Anal. Chem. *51,* 2007–2014.

HART, F.L., and FISHER, H.J. 1971. Modern Food Analysis. Springer-Verlag, New York.

HERSCHEL, J.F.W. 1823. On the absorption of light by coloured media, and on the colours of the prismatic spectrum exhibited by certain flames; with an account of a ready mode of determining the absolute dispersive power of any medium, by direct experiment. Trans. Roy. Soc (Edinburgh) *9,* 445–460.

HORWITZ, W. (Editor). 1980. Official Methods of Analysis of the Association of Offical Analytical Chemists, 13th ed. Association of Official Analytical Chemists, Washington, DC.

IHNAT, M. 1976. Selenium in foods: evaluation of atomic absorption spectrometric techniques involving hydrogen selenide generation and carbon furnace atomization. J. Assoc. Offic. Anal. Chem. *59,* 911–922.

IHNAT, M. 1982. Application of atomic absorption spectrometry to the analysis of foodstuffs. *In* Atomic Absorption Spectrometry, Chapter 4d, pp. 139–220. J.E. Cantle (Editor). Elsevier Scientific Publishing Co., Amsterdam, Netherlands.

IHNAT, M., and MILLER, H.J. 1977. Analysis of foods for arsenic and selenium by acid digestion, hydride evolution atomic absorption spectrophotometry. J. Assoc. Offic. Anal. Chem. *60,* 813–825.

IHNAT, M., CLOUTIER, R.A., and WOLF, W.R. 1982. Preparation of a corn agricultural biological reference material. Abstracts 9th Annu. Meet. of the Fed. of Anal. Chem. and Spectroscopy Societies, Philadelphia, PA, Sept. 19–24, 1982. Abstract *No. 459.*

THE INSTITUTE OF BREWING ANALYSIS COMMITTEE, P.A. MARTIN (Chairman). 1973. Determination of iron and copper in beer by atomic absorption spectroscopy. J. Inst. Brew. *79,* 289–293.

THE INSTITUTE OF BREWING ANALYSIS COMMITTEE, J. WEINER (Chairman). 1974. Determination of zinc in beer by atomic absorption spectroscopy. J. Inst. Brew. *80,* 486–488.

THE INSTITUTE OF BREWING ANALYSIS COMMITTEE, J. WEINER (Chairman). 1977. Determination of lead in beer by atomic absorption spectroscopy. J. Inst. Brew. *83,* 82–84.

JOSLYN, M.A. (Editor). 1970. Methods in Food Analysis, Physical, Chemical and Instrumental Methods of Analysis. 2nd ed. Academic Press, New York.

KEIRS, C.D., and VICKERS, T.J. 1977. DC plasma arcs for elemental analysis. Appl. Spect. *31,* 273–283.

KIRCHHOFF, G., and BUNSEN, R. 1860. Chemical analysis by spectrum observations. Phil. Mag. *20,* 89–109.

KIRCHHOFF, G., and BUNSEN, R. 1861. Chemical analysis founded on observations of spectra. Ann. Chim. Phys. *62,* 452–486 (French).

KIRKBRIGHT, G.F., and SARGENT, M. 1974. Atomic Absorption and Fluorescence Spectroscopy. Academic Press, New York.

KOIRTYOHANN, S.R., and PICKETT, E.E. 1975. Agronomic applications (Chapter 15) and food analysis (Chapter 17). *In* Flame Emission and Atomic Absorption Spectrometry, Vol. 3, Elements and Matrices. J.A. Dean and T.C. Rains (Editors). Marcel Dekker, New York.

LEES, R. 1975. Food Analysis: Analytical and Quality Control Methods for the Food Manufacturer and Buyer. 3rd ed. Leonard Hill Books, London, UK.

LINK, W.E. (Editor). 1977. Official and Tentative Methods of the American Oil Chemists' Society, 3rd ed. Am. Oil Chemists' Society, Champaign, IL.

LOVETT, R.J., WELCH, D.L., and PARSONS, M.L. 1975. On the importance of spectral interferences in atomic absorption spectroscopy. Appl. Spect. *29,* 470–477.

LUNDEGÅRDH, H. 1934. The Quantitative Spectral Analysis of the Elements, II. Gustav Fischer, Jena.

MAC LEOD, A.J. 1973. Instrumental Methods of Food Analysis. Elek Science, London, UK.

MARGOSHES, M., and SCRIBNER, B.F. 1959. The plasma jet as a spectroscopic source. Spectrochim. Acta *15,* 138–145.

MASSMANN, H. 1967. Determination of arsenic by atomic absorption. Fres. Zeit. Analyt. Chem. *225,* 203–213 (German).

MASSMANN, H. 1968. Comparison of atomic absorption and atomic fluorescence in a graphite cuvette. Spectrochim. Acta *23B,* 215–226 (German).

MAVRODINEANU, R. (Editor). 1970. Analytical Flame Spectroscopy-Selected Topics. Springer-Verlag, New York.

McQUAKER, N.R., KLUCKNER, P.D., and CHANG, G.N. 1979. Calibration of an inductively coupled plasma-atomic emission spectrometer for the analysis of environmental materials. Anal. Chem. *51,* 888–895.

MEGGERS, W.F., CORLISS, C.H., and SCRIBNER, B.F. 1975. Tables of Spectral Line Intensities, Part I—Arranged by Elements, Part II—Arranged by Wavelength. NBS Monograph 145, 2nd ed. National Bureau of Standards, Washington, DC.

MELVILL, T. 1756. Observations on light and colours. Essays and Observations, Physical and Literary, Edinburgh *2,* 12–90. (From J.A. Dean and T.C. Rains 1969.)

MERCURY ANALYSIS WORKING PARTY OF THE BUREAU INTERNATIONAL TECHNIQUE DU CHLORE. 1976. Standardization of methods for the determination of traces of mercury. Part I. Determination of total mercury in materials containing organic matter. Anal. Chim. Acta *84,* 231–257.

MILLER, J.D., and KOIZUMI, H. 1978. Analytical applications of polarized Zeeman AA. Am. Lab. Nov. (1978) 35–51.

MINISTRY OF AGRICULTURE, FISHERIES AND FOOD. 1973. Technical Bull. 27. The Analysis of Agricultural Materials, A Manual of the Analytical Methods Used by the Agricultural Development and Advisory Service. Her Majesty's Stationery Office, London, UK.

MORRE, J. 1974. Pollution of milk by metals. Determination by atomic absorption spectrophotometry. Lait *54,* 139–152. (French). [Chem. Abstr. *81* (1974), 103299f.]

MUNTAU, H. 1979. Five years of environmental candidate reference material production at the Joint Research Centre Ispra. *In* Production and Use of Reference Materials, Proc. of the Int. Symp. held at the Bundesanstalt fur Materialprufung (BAM) Nov. 13–16, 1979.

PARR, R.M. 1980. The reliability of trace element analysis as revealed by analytical reference materials. *In* Trace Element Analytical Chemistry in Medicine and Biology. P. Bratter and P. Schramel (Editors). Walter de Gruyter and Co., Berlin.

PARSONS, M.L., SMITH, B.W., and BENTLEY, G.E. 1975. Handbook of Flame Spectroscopy. Plenum Press, New York.

PEARSON, D. 1976. The Chemical Analysis of Foods. 7th Edition. Churchill Livingstone, Edinburgh, UK.

PINTA, M. (Editor). 1975. Atomic Absorption Spectrometry. (Translated by K.M. Greenland and F. Lawson). Halstead Press, New York.

POMERANZ, Y., and MELOAN, C.E. 1978. Food Analysis: Theory and Practice. Rev. ed. Avi Publishing Co. Inc., Westport, CT.

PRICE, W.J. 1974. Analytical Atomic Absorption Spectrometry. Heyden and Son Ltd. London, UK.

PRICE, W.J. 1979. Spectrochemical Analysis by Atomic Absorption. Heyden and Son Ltd. London, UK.

RAMIREZ-MUÑOZ, J. 1968. Atomic Absorption Spectroscopy and Analysis by Atomic Absorption Flame Photometry. Elsevier Scientific Publishing Co., Amsterdam, Netherlands.

REYNOLDS, R.J. and ALDOUS, K. 1970. Atomic Absorption Spectroscopy—A Practical Guide. C. Griffin and Co., London, UK.

ROBINSON, J.W. 1966. Atomic Absorption Spectroscopy. Marcel Dekker, New York.

ROBINSON, J.W. (Editor). 1974. Handbook of Spectroscopy, Vol. I, Section D. CRC Press, Chem. Rubber Co., Cleveland, OH.

RUBESKA, I., and MOLDAN, B. 1969. Atomic Absorption Spectrophotometry. Iliffe Books Ltd., London, UK.

SAARLOOS, C.C. 1972. Atomic absorption spectrometry. III. Atomic absorption spectrometry applied to the food industry. Chem. Tech. (Amsterdam) *27* (8), 205–209 (Neth.). [Chem. Abst. *77* (1972), 60115r.]

SAKAI, K. 1975. Determination of heavy metals in foods by atomic absorption spectroscopy. Shokuhin Kogyo *18,* 71–78. (Japan). [Chem. Abstr. *83* (1975) 204864h.]

SALIN, E.D., and INGLE, J.D., JR. 1978. Design and construction of a time multiplex multiple slit multielement flame atomic absorption spectrometer. Anal. Chem. *50,* 1737–1744.

SCHRENK, W.G. 1975. Analytical Atomic Spectroscopy, Plenum Press, New York.

SCHULLER, P.L., and COLES, L.E. 1979. The determination of copper in foodstuffs (International Union of Pure and Applied Chemistry, Applied Chemistry Division, Commission on Food Contaminants). Pure and Appl. Chem. *51,* 385–392.

SCHULLER, P.L., and EGAN, H. 1976. Cadmium, Lead, Mercury and Methylmercury Compounds, A Review of Methods of Trace Analysis and Sampling with Special Reference to Food. FAO, Rome, Italy.

SCOTT, R.H., FASSEL, V.A., KNISELEY, R.N., and NIXON, D.E. 1974. Inductively coupled plasma-optical emission analytical spectrometry, a compact facility for trace analysis of solutions. Anal. Chem. *46,* 75–80.

SEILER, H. 1972. Methods for determination of metal ions in foods. Mitt. Geb. Lebensmitt Hyg. *63,* 180–187 (German). [Chem. Abstr. *78* (1973) 2787w.]

SLAVIN, W. 1968. Atomic Absorption Spectroscopy. Wiley Interscience Publishers, New York.

SIMMONS, W.J. 1973. Determination of low concentrations of cobalt in plant material by atomic absorption spectrophotometry. Anal. Chem. *45,* 1947–1949.

SIMMONS, W.J. 1975. Determination of low concentrations of cobalt in small samples of plant material by flameless atomic absorption spectrophotometry. Anal. Chem. *47,* 2015–2018.

SUBCOMMITTEE ON PROCEDURES OF THE CHEMISTRY TASK FORCE OF THE NATIONAL SHELLFISH SANITATION PROGRAM. 1975. Collection, Preparation and Analysis of Trace Metals in Shellfish, Publ. No. (FDA) *76-2006,* USDHEW, PHS, FDA.

TALBOT, H.F. 1826. Some experiments on coloured flames. Edinburgh J. Sci. (Brewster's) *5,* 77–81.

TEPER, I. 1977. Use of atomic absorption spectrophotometry in the feed industry. Krmivarstvi Sluzby *13* (3), 63 (Czech.). [Chem. Abstr. *87* (1977) 100691r.]

VAN LOON, J.C. 1981. Review of methods for elemental speciation using atomic spectrometry detectors for chromatography. Can. J. Spect. *26,* 22A–32A.

VARIAN TECHTRON. 1972A. Hollow Cathode Lamp Data. Varian Techtron, Springvale, Australia.

VARIAN TECHTRON. 1972B. Analytical Methods for Flame Spectroscopy. Varian Techtron, Springvale, Australia.

VARJU, M. 1971. Atomic absorption spectrophotometry and its application in investigations in food chemistry. Elelmiszervizsgalati kozlemenyek *17* (1–2), 64–71 (Hung). [Chem. Abstr. *76* (1972) 71059f.]

WALSH, A. 1955. The application of atomic absorption spectra to chemical analysis. Spectrochim. Acta *7,* 108–117.

WENDT, R.H. and FASSEL, V.A. 1965. Induction-coupled plasma spectrometric excitation source. Anal. Chem. *37,* 920–922.

WELZ, B. 1976. Atomic Absorption Spectroscopy. Verlag Chemie, New York.

WINGE, R.K., PETERSON, V.J. and FASSEL, V.A. 1979. Inductively coupled plasma-atomic emission spectroscopy:prominent lines. Appl. Spect. *33,* 206–219.

WOLF, W.R. and IHNAT, M. 1982. Evaluation of available reference materials for potential use in analysis of biological materials. Abstr. 9th Annu. Meet. of the Fed. of Anal. Chem. and Spectroscopy Societies, Phil. PA, Sept. 19–24, 1982. Abstract No. 108.

WOLLASTON, W.H. 1802. A method of examining refractive and dispersive powers by prismatic reflection. Phil. Trans. *92,* 365–380.

WOODRIFF, R. and RAMELOW, G. 1968. Atomic absorption spectroscopy with a high temperature furnace. Spectrochim. Acta *23B,* 665–671.

WOODRIFF, R., STONE, R.W., and HELD, A.M. 1968. Electrothermal atomization for atomic absorption analysis. Appl. Spect. *22,* 408–411.

WOODSON, T.T. 1939. A new mercury vapor detector. Rev. Sci. Instr. *10,* 308–311.

8

Reflectance Spectroscopy

Karl H. Norris[1]

INTRODUCTION

Reflectance spectrophotometry is widely used to measure the appearance of food products. Reflectance instruments developed for color measurement have been used to predict visual pigment concentrations of foods, but it should be recognized that such instruments do not represent the optimum design for constituent analyses. However, reflectance spectroscopy can be applied to constituent analyses as demonstrated by the developments of near infrared reflectance techniques for analyzing the composition of grains and oilseeds. This chapter is limited to the discussion of procedures for constituent analyses by reflectance even though there may be a strong relationship between the composition and the appearance of a sample. In addition, this chapter will discuss only those procedures that require little or no sample preparation.

OPTICAL GEOMETRY FOR DIFFUSE REFLECTANCE

Radiation striking a food sample will be reflected, absorbed, or transmitted depending on the scattering and absorption properties of the sample. The radiant energy reflected from the surface (specular component) contains no information about the composition of the sample, but the radiation which is transmitted into the sample and scattered and reflected back out of the sample contains information

[1] Instrumentation Research Laboratory, BARC, ARS, USDA, Beltsville, MD 20705

about the composition. This radiation, which is generally referred to as diffuse reflectance, has been called "body reflectance" by Birth (1983) to distinguish it from the surface reflectance from a rough surface which is also called diffuse reflectance. This "body reflectance," *R*, is the factor to measure to predict the composition of a sample, and it is necessary to minimize the contribution from the specular component.

Three different geometries are widely used in commercial instruments for compositional analyses by near infrared spectroscopy. These may be classed as to the method of collecting the reflectance as: (1) integrating sphere (Fig. 8.1), (2) large solid-angle detector (Fig. 8.2), and (3) small detector (Fig. 8.3). Each of these arrangements has advantages and disadvantages. The integrating sphere offers the most uniform collection of reflectance from the sample. It also offers an easy method to obtain dual-beam operation to minimize stability problems of the lamp and detector. The large solid-angle detector offers good collection efficiency, simplicity of construction, and minimum interfer-

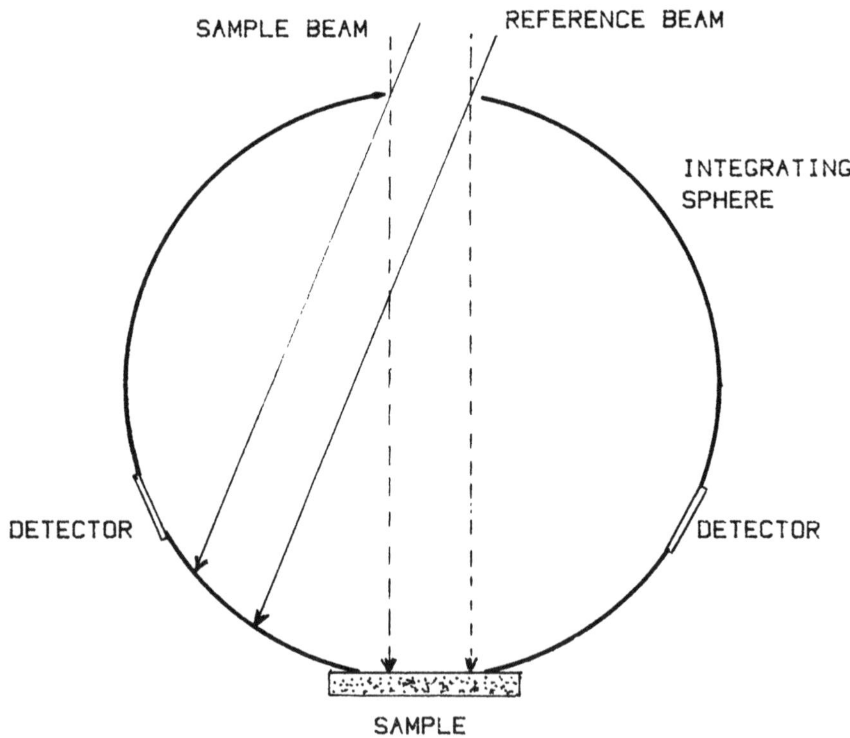

FIG. 8.1. Integrating sphere, reflectance geometry.

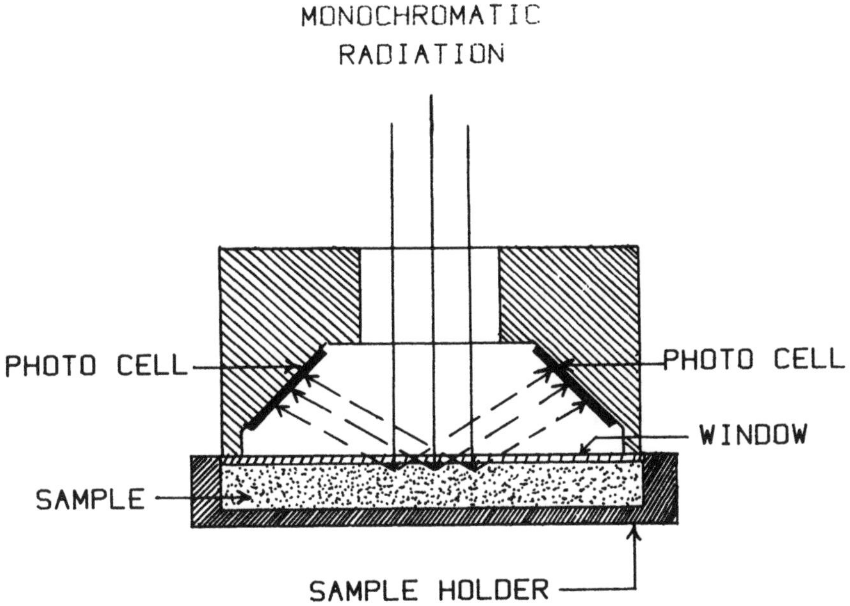

MONOCHROMATIC
RADIATION
PHOTO CELL
PHOTO CELL
WINDOW
SAMPLE
SAMPLE HOLDER

FIG. 8.2. Large, solid-angle detector geometry for reflectance.

WHITE LIGHT
FILTER
DETECTOR
SAMPLE

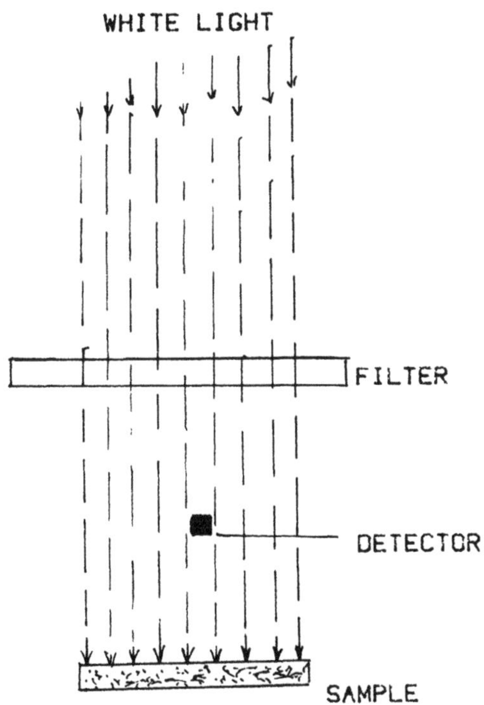

FIG. 8.3. Small detector, reflectance geometry.

ence from specular reflectance. The small detector design offers simplicity of construction, lowest detector noise, and lowest cost. All three geometries are used on filter instruments, but the large solid-angle detector and the integrating sphere units are also used with grating monochromators to provide full wavelength-scanning capabilities.

The data presented in this chapter were collected with an in-house-designed spectrophotometer using the grating-prism monochromator from a Cary 14 spectrophotometer, but with a large solid-angle detector in place of the normal dual-beam sampling optics. The instrument is coupled to a digital computer to control the scanning and collect and process the spectral data.

REFLECTANCE STANDARDS

Procedures have been developed for measurement of absolute reflectance, but these procedures are not practical for measurement of food products. Therefore, all reflectance measurements are made with reference to a standard. The ideal standard would have a body reflectance of 100% at all wavelengths of interest. Freshly smoked magnesium oxide has been widely used as a reflectance standard, but it is far from an ideal material. It provides a fragile surface which must be prepared with great care, and its reflectivity falls off in the ultraviolet and near infrared region. Purified barium sulfate powder and paint are widely used in the visible and ultraviolet region, but its reflectivity is low in the near-infrared region. The National Bureau of Standards now recommends the use of a commercial product, Halon, as a reference standard for reflectance in the ultraviolet, visible, and near infrared region (Anon. 1979). This powdered material is relatively easy to pack into a reference sample and is very stable as long as it is protected from contamination. Halon has a uniform, high reflectance over the ultraviolet, visible, and near infrared region, although it has a slight dip in reflectance in the 2100-nm region.

Powdered sulfur has been proposed as a reference standard for the near infrared region. Sulfur has a higher infrared reflectance than Halon, particularly in the 2100-nm region, but Halon standards are easier to prepare.

Reflectance stability and ruggedness are the most important characteristics for working standards. The reflectivity should be the same at all wavelengths of interest, but the reflectivity does not need to approach 100%. Teflon and opal glass provide suitable working standards for the visible and a special ceramic has been adopted for the near infrared region. These working standards can be cleaned with water and soap so that they provide very stable working standards. The thickness of all reflectance standards should be great enough such that changing the backing material has no effect. This may require standards having thicknesses greater than 1 cm. If thin standards are used, the backing material becomes a part of the standard, and if this backing material changes, the standard changes. This can create a stability problem.

FACTORS AFFECTING REFLECTANCE DATA

Sample parameters which affect reflectance data include the following: concentration and absorption coefficients of each constituent; scatter coefficient which is affected by particle size, shape, and packing density as well as by refractive index of particle; homogeneity of sample; and level of fluorescence. The reflectance varies exponentially with the concentration and absorption coefficient such that $\log(1/R)$ is approximately linear with concentration. Increasing concentration increases $\log(1/R)$ and decreases the reflectance. The reflectance of food products can be most easily changed by changing the particle size. Reflectance increases as particle size decreases. The difference between the refractive index of the particles and the fluid surrounding the particles determines the magnitude of the reflectance. Adjusting the refractive index of this fluid can have a major effect on the reflectance. Adding a material such as oil to cellulose powder causes the reflectance to decrease because the oil has a refractive index approaching that of the cellulose.

Fluorescence does not affect the reflectivity of a sample, but in many cases, it is impossible to separate the reflectance signal from the fluorescence signal, so that an apparent effect occurs. Fluorescence presents a serious problem to the use of reflectance spectroscopy in the ultraviolet and short wavelength visible portion of the spectrum, but causes no difficulty in the near infrared region.

Sample temperature can also affect reflectance spectra because some absorption bands are temperature sensitive. This is particularly true for water in the near infrared region with the absorption band shifting both in magnitude and wavelength of maximum absorption.

Changes in scattering properties affect reflectance at all wavelengths, but the effect is greatest in regions where absorption coefficients are large.

The change in reflectance from changes in particle size is illus-

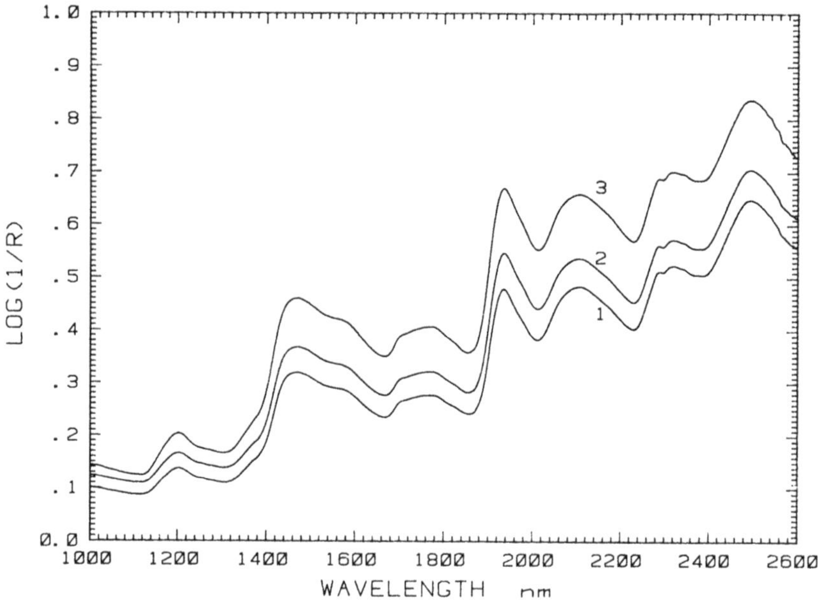

FIG. 8.4. Spectra for one sample of wheat ground by three different grinders to give a range in particle size. #1—160 μm, #2—200 μm, #3—330 μm.

trated in Fig. 8.4 with wheat ground in different grinders to produce samples having the same composition, but different particle size and shape. These curves show a greater effect in the longer wavelength region from particle size differences, but this is because the absorbers are stronger in the long wavelength region.

The change in reflectance from temperature differences is illustrated in Fig. 8.5 by the difference curves between a sample at 30°C and the same sample as a reference at 10°C. The reflectance changes are caused by changes in the water and water-related absorption characteristics.

The reflectance changes from changes in concentration of an absorber are illustrated with ground wheat at different moisture levels in Fig. 8.6. The greatest change is in the 1940-nm region where water has a strong absorption band. Smaller changes occur in the 1460-nm and the 2600-nm region where water also has significant absorption.

Instrumental factors that affect reflectance data include the following: spectral bandpass, wavelength accuracy and precision, photometric accuracy, photometric noise, specular component, stray light, and geometry. The technology developed for quantitative analysis by di-

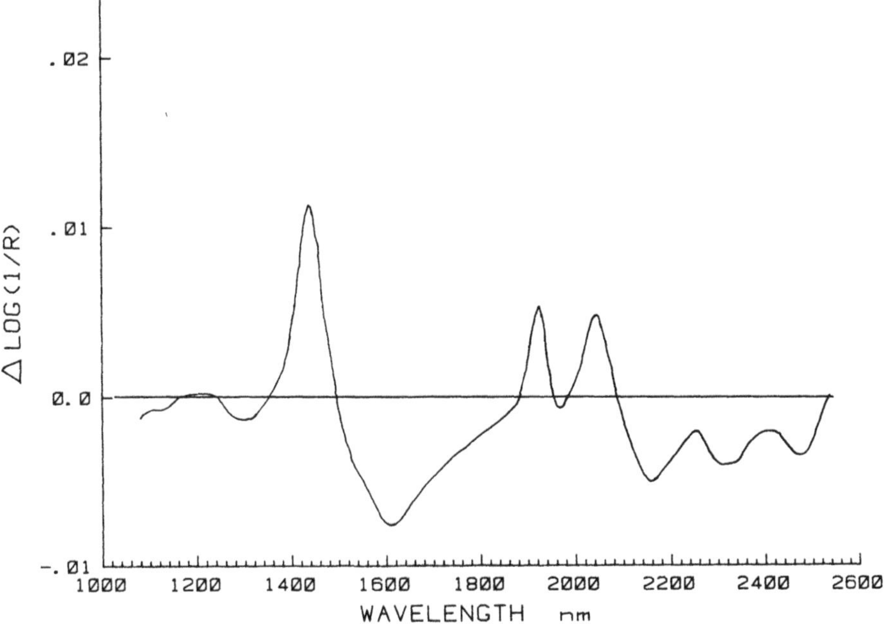

FIG. 8.5. Effect of sample temperature on wheat spectra. Difference curve between sample at 30°C and the same sample as a reference at 10°C.

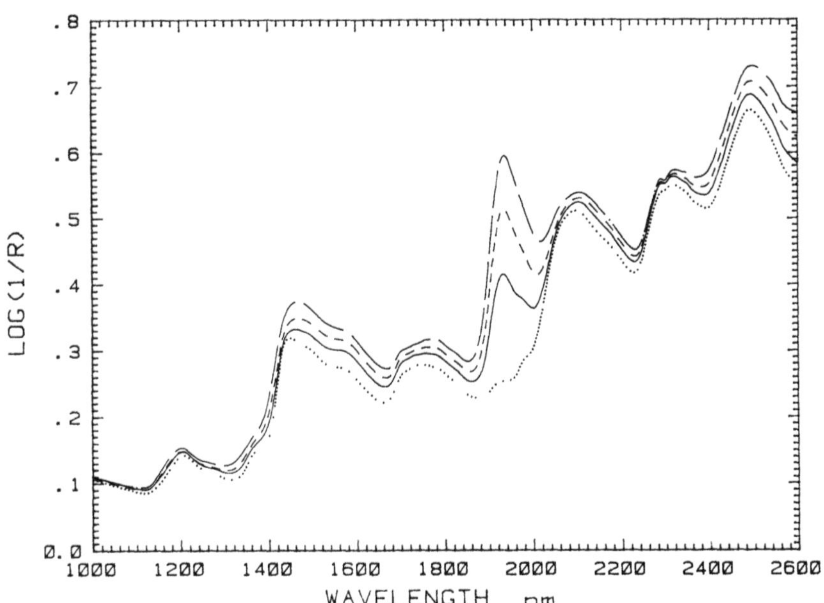

FIG. 8.6. Wheat spectra at different moisture levels: 0.2%, ···, 4.0%, —, 8.0%, – – –, 12.5%, – · – .

rect reflectance spectroscopy involves extensive data manipulation. As a result of this extensive data manipulation, it is imperative that high-quality data be available for processing. The two instrument parameters of photometric noise and wavelength precision are the most important because they determine the reproducibility of the analytical measurement. The photometric noise expressed as μ log(1/*R*) should be less than 100 at any wavelength, and the wavelength should be repeatable within ±0.1 nm on repeat scans. Instrument manufacturers are now exceeding these requirements for reflectance instruments in the near infrared region.

Spectral bandpass affects reflectance spectra, particularly if sharp absorption bands are present in the sample. Such sharp bands are not resolved if the instrument bandpass approaches that of the absorption bands. However, the actual bandpass is not as important as bandpass stability in an analytical instrument. A bandpass of 10–15 nm has proven adequate in the near infrared region for measuring the oil, protein, and moisture content of grains and oilseeds.

Photometric accuracy and wavelength accuracy are important in transferring a calibration from one instrument to another, but within a single instrument these are not of great consequence. Reflectance measurements do not put as great a burden as transmittance measurements on stray light requirements, but the stray light should not be greater than 0.1%. Sensitivity to the specular component can limit measurements when the measured reflectance is low. It is difficult without resorting to crossed polarizers to get the specular component below 0.5% or a value of 2.30 in log(1/*R*); therefore, samples giving a measured reflectance signal of 1.0% [log(1/*R*) = 2.0] will have a serious error from the specular component. The specular component is relatively constant within an instrument, and corrections for this constant can be applied to improve performance on low reflectance samples.

Geometrical stability between the sample and instrument is also an important parameter. A very slight displacement away from the integrating sphere can cause a big decrease in measured sample reflectance because of the vignetting from the sample port of the reflected radiation from deep within the sample. Therefore, if more than one sample cell is used, the precision of the positioning of these cells is important. The sample position is also important for the other types of geometries, but it is not as critical as for the sphere. In most measuring procedures, the sample is covered with a window through which the measured radiation must make two passes. This window, whether it is considered as part of the sample or part of the instrument, is very important. Again, stability of the window's properties is the most important feature. If the window has no absorption bands in the wavelength of interest, it will be less subject to variation from temperature. Infrared transmitting quartz has proven best for the near infrared region for wavelengths out to 2700 nm.

DATA TREATMENTS

The Kubelka–Munk theory (Kubelka 1948) is generally the one accepted for relating the transmittance and reflectance properties to the scatter coefficients and absorption coefficient of scattering samples. The Kubelka–Munk function is expressed as:

$$K/S = (1 - R)^2/2R$$

where K is the absorption coefficient and S is the scatter coefficient.

This function is widely used in the paint and paper industries to evaluate pigment concentration and scattering effects. The Kubelka–Munk function should vary linearly with concentration of an absorber since the absorption coefficient is the concentration multiplied by the absorptivity constant for the constituent. Therefore, a measurement of the reflectance R and a computation of K/S should give a linear measure of the concentration of a constituent. Unfortunately, this is true only if the measured R is the true absolute reflectance of the sample. It is not true if the measured R includes a significant specular component or if the reflectance is not corrected to the proper standard.

Converting the reflectance to $\log(1/R)$ gives a function that is linear with concentration changes over a wide range of concentrations, although it is not supported by theory. $\text{Log}(1/R)$ does not require that R be in absolute reflectance units, but it is sensitive to the specular component in the same way as the Kubelka–Munk function. In tests with near infrared spectral data, $\log(1/R)$ and the Kubelka–Munk function performed equally well in predicting protein, moisture, and oil content of grains. The $\log(1/R)$ function is easier to use, and as a result, it has become the accepted data treatment for quantitative analysis.

The near infrared reflectance spectra of cornstarch are shown in Fig. 8.7 expressed as K/S, $\log(1/R)$, and R. The $\log(1/R)$ spectra of starch, protein, water, and oil in Fig. 8.8 show the degree of overlap which occurs for just these four common constituents of food products. These spectra also identify choices of wavelengths to be used to measure each

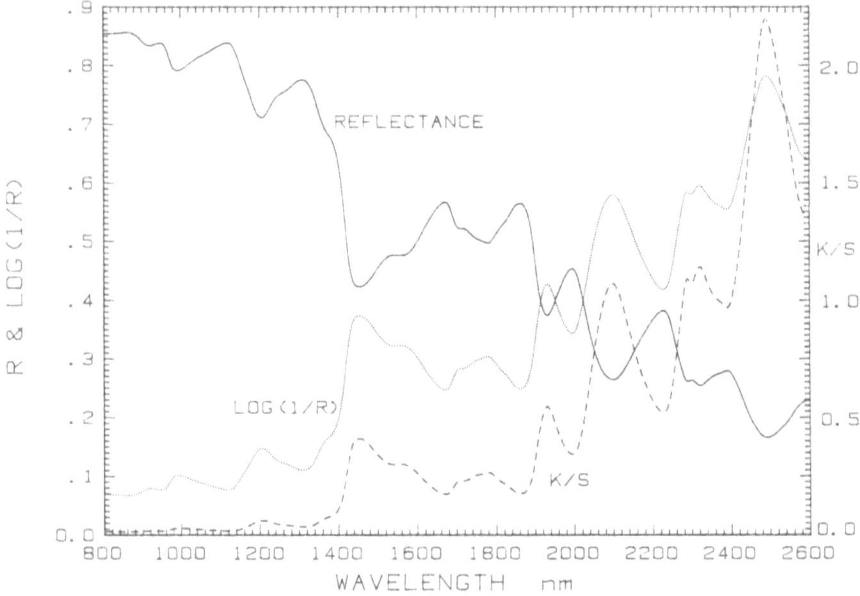

FIG. 8.7. Cornstarch spectra with different transformation.

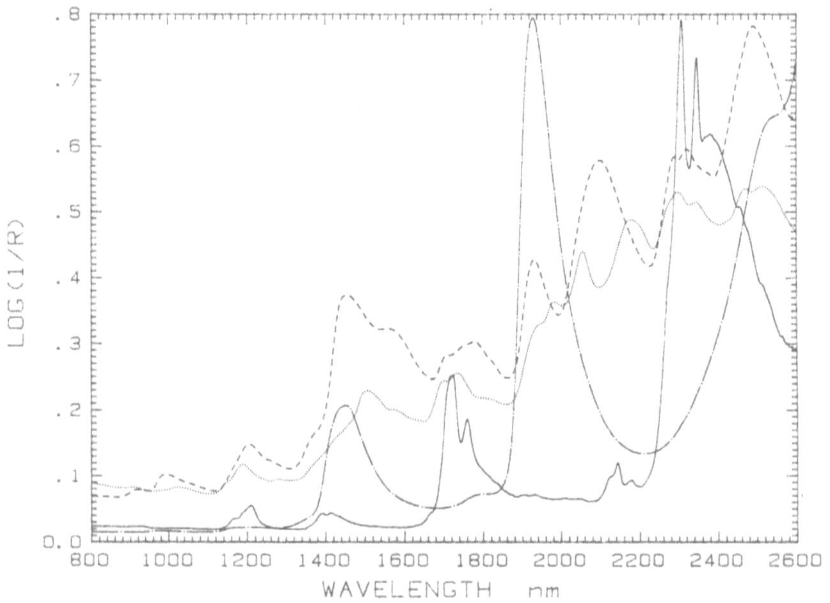

FIG. 8.8. Spectra of major constituents: Water, – · –; Starch, – – –; Protein, ···; Oil —.

of the constituents. Possible choices would be 1940 nm for water, 2100 nm for starch, 2070 or 2180 nm for protein, and 2310 nm for oil.

Reflectance spectra of food products expressed as *K/S*, log(1/*R*), or *R* do not show much character, and the value at any one wavelength is very closely correlated to the values at nearby wavelengths. To break up this correlation and provide more character in the spectra, derivative functions are widely used. Both first and second derivative treatments are used, with the second derivative spectra being easier to interpret because the sharp minimum of the second derivative of the log(1/*R*) curve coincides with the wavelength of maximum absorption for a constituent absorber. Derivatives may be applied to reflectance data expressed in *K/S*, log(1/*R*), or *R* to accomplish the same goal of providing spectra with more character, so that overlapping absorption bands can be evaluated.

REGRESSION TECHNIQUES FOR CALIBRATION

Quantitative analysis with reflectance spectroscopy must rely on a complex calibration procedure because we do not have a simple direct relationship between concentration and reflectance, and measurements are being made on complex mixtures of overlapping absorbers. The simplest calibration procedure involves using reflectance spectra of a large number of samples of known composition with the samples chosen to represent all the variability to be expected in the population under study. To calibrate for protein, a linear multiple regression is run between the protein content and the log(1/*R*) values at 1940 nm, 2100 nm, 2180 nm, and 2310 nm for each sample. The resulting regression equation becomes the calibration equation for predicting the protein content of unknown samples. The same regression procedure provides calibration equations for starch, moisture, and oil. The log(1/*R*) data at additional wavelengths can be used in the regression procedure, but care must be taken to avoid overfitting of data from the use of too many regression terms. Other choices from the four wavelengths listed above could be used, and computer programs are available for choosing the best wavelength points out of a selection of several hundred available wavelengths.

The data could be transformed to *K/S* instead of log(1/*R*), and the same regression procedure applied with equal performance, because both transformations provide a linear relationship to concentration. Data expressed as reflectance (*R*) can be used if the range of reflectance is low, because over small changes in reflectance, the value is approximately linear with concentration.

Data transformed to first or second derivative can also be used for the linear multiple-correlation procedure (Norris *et al.* 1976), and this data treatment is widely used for forages and animal feeds. The log(1/*R*) data are generally used for this derivative transformation, but results should be equally good with data expressed as *K/S* or *R*.

A linear single-term regression procedure using a ratio of two derivatives has also been developed (Norris 1983B). In this procedure, the numerator derivative is chosen to give the best relationship to the constituent being measured, and the denominator is chosen to best fit the sum of all of the constituents present in the samples. The use of a derivative ratio provides the means for minimizing interfering effects such as the effect of particle size. Multiterm regression procedures can also be used with derivative-ratio data to provide even more flexibility in obtaining the optimum calibration.

The above calibration procedures use only a few data points from a curve consisting of several hundred data points. Procedures using more of the data points have also been explored. Hruschka and Norris (1982) described a curve-fitting procedure which uses all of the spectrum or a selected continuous region of the spectrum for calibration. This procedure has the advantage of being less sensitive to photometric noise because the use of many data points provides a signal-averaging effect.

Validation of a calibration is a very important part of the procedure when such complex data treatments are applied to obtain the calibration. The validation is done by predicting the composition of another set of the same population of samples, but not including the samples used in the calibration. Thus, for a set of 100 samples with known composition, from 50 to 70 might be selected for calibration and the remaining 30–50 could be used for validation. Ideally, both the calibration set and the validation set of samples should be representative of the total population of samples.

It should be noted that reliable compositional data are essential to validate a calibration procedure, since the composition predicted by the reflectance data is compared with the composition determined by some accepted standard procedure to evaluate the performance of the measurement. Random errors in the composition data of the calibration samples have little effect, but random errors in the composition data of the validation samples may mask the true error of the reflectance procedure. The accuracy of the reflectance procedure is, of course, dependent on the accuracy of the test method used in determining the composition of the calibration samples.

APPLICATIONS

Instrumentation applying near infrared reflectance spectroscopy is widely used to predict the oil, moisture, and protein content of grains and oilseeds. The type of performance which can be attained with this technique is illustrated in Table 8.1. In this experiment, 100 samples of Hard Red Spring Wheat were ground with a Udy Cyclone grinder using the 1.0-nm screen. Sixteen subsamples were drawn from each of the 100 samples for duplicate Kjeldahl analyses by the AACC method 46-15. Moisture on the 100 samples was determined using AACC method 44-15A. Four subsamples were drawn from each of the 100 ground samples for spectral scans from 1000 to 2600 nm. The four spectral scans were averaged to obtain one curve for each of the 100 samples. A random selection of 50 of these 100 spectra were used for calibration against the moisture and average Kjeldahl values. The other 50 samples formed a prediction set for validation. The ratio of two second derivatives was used for calibration and the resulting calibration equations were used to predict the protein and moisture of the prediction set as shown in Table 8.1

The use of average chemical data reduces the error in the reference data so that the resulting standard error of prediction represents the error in the reflectance procedure. These results were obtained using the single-term, derivative-ratio procedure using a second derivative at 2187 nm divided by the second derivative at 2328 nm for protein. A first derivative at 1931 nm divided by the first derivative at 2146 nm was used for the moisture calibration.

Raw and cooked beef and pork samples represent an additional problem for near infrared analyses because the water absorption band

TABLE 8.1. PREDICTION ERRORS FOR DRY BASIS PROTEIN AND MOISTURE IN HRS WHEAT

	SEP[a]	Bias[b]	Precision[c]
Protein	0.16	−0.04	0.13
Moisture	0.11	−0.02	0.09

[a] SEP is the standard error of prediction computed as $\sqrt{[\Sigma(\text{diff.})^2]/(n-1)}$.
[b] Bias is the difference between the mean of the predicted and the mean of the chemical data.
[c] Precision is the standard deviation of the four replicate predictions.

at 1940 nm is too strong for a measurement. The log(1/*R*) curves for two raw pork samples (Fig. 8.9) show that the low-fat, high-moisture pork sample exceeded the range of the spectrophotometer in the 1940-nm region. This does not preclude useful measurements, although the data from about 1850 to 2050 nm are of no value. Similar curves were obtained for cooked pork although water absorption is not quite as strong (Fig. 8.10). Raw and cooked beef have very similar curves to those for pork. The raw beef curves (Fig. 8.11) also show very strong water absorption bands which limit the use of some wavelengths for linear correlations. The fat absorption bands are clearly evident in these spectra at 1210, 1722, 1760, 2310, and 2345 nm. The meat samples were homogenized in a food processor to provide a uniform paste for measurements.

In a preliminary experiment, 30 raw pork samples ranging from 5 to 38% fat were analyzed using the single-term, derivative-ratio procedure. The second derivative at 1208 nm divided by the second derivative at 1126 nm was used for calibration for fat. The calibration results shown in Fig. 8.12 provide a correlation coefficient of 0.9997 and a standard error of calibration of 0.17% fat. Three different numerator wavelengths: 1208, 1722, and 1764 nm gave essentially equal

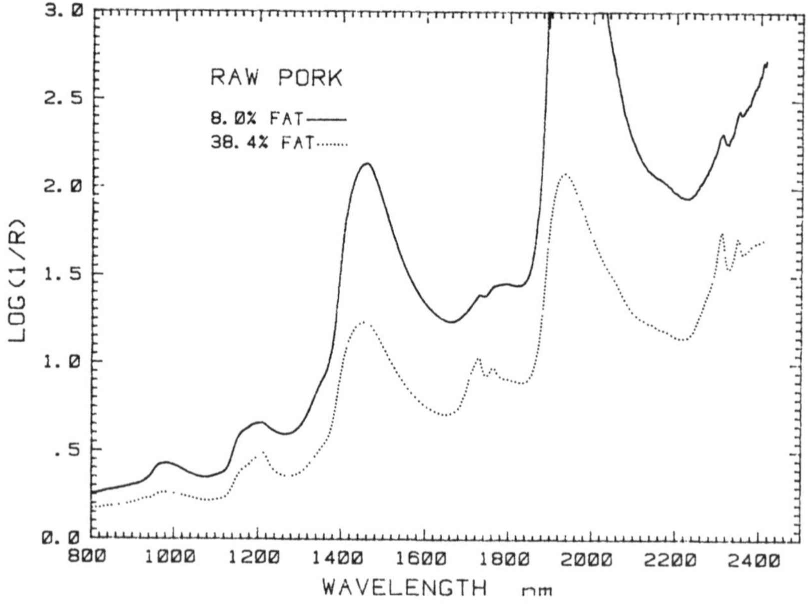

FIG. 8.9. Spectra of homogenized raw pork at two levels of fat.

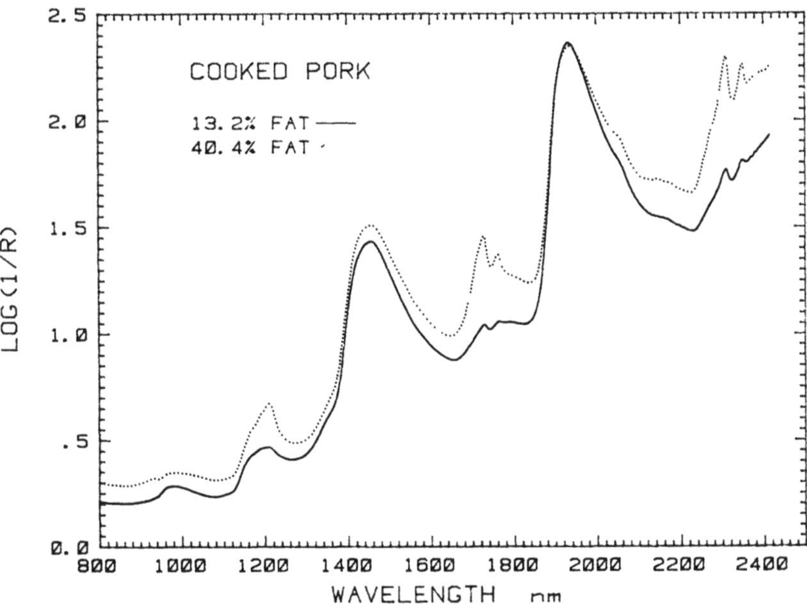

FIG. 8.10. Spectra of homogenized cooked pork at two levels of fat.

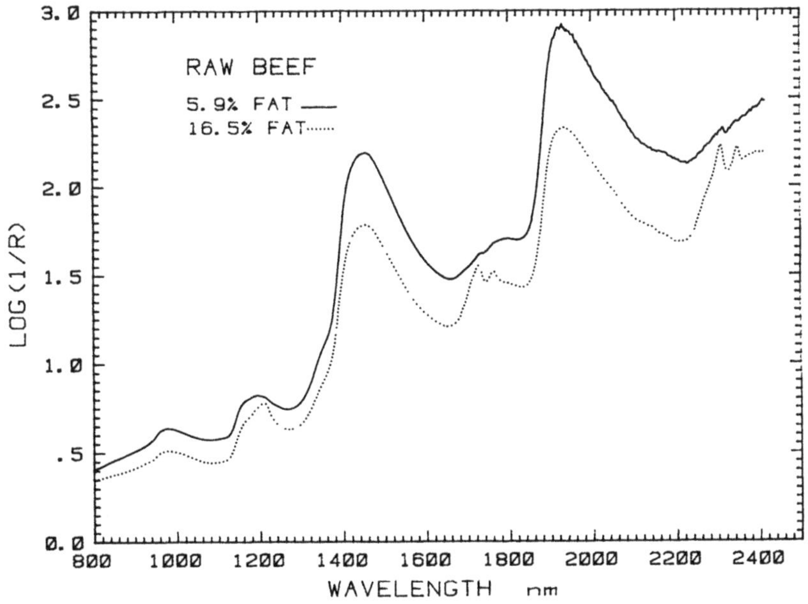

FIG. 8.11. Spectra of homogenized raw beef at two levels of fat.

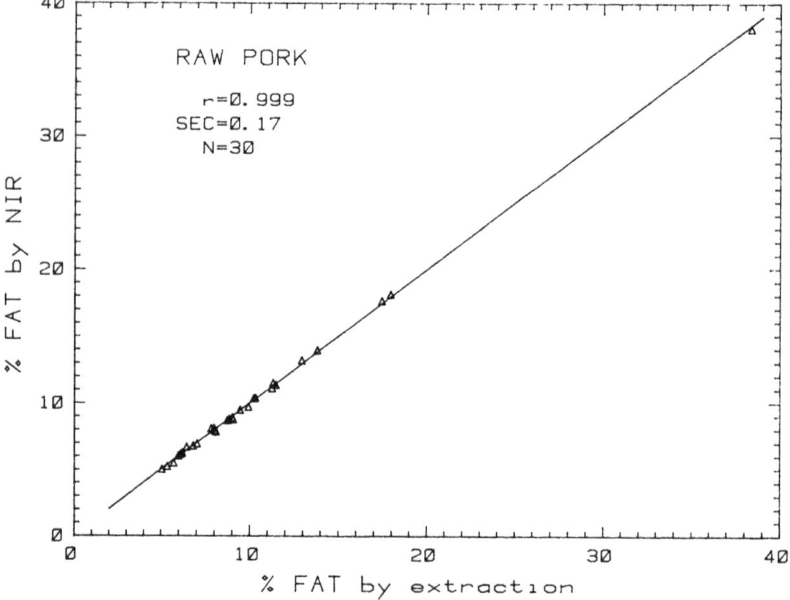

FIG. 8.12. Calibration plot for near infrared measurement of fat content of raw pork.

performance. These are all fat absorption bands, but the bands at 2310 and 2345 nm did not provide a high correlation. The high absorption of water at these two wavelengths introduces nonlinearity into the $\log(1/R)$ values.

The calibration results for water, Fig. 8.13, are not as impressive, with a correlation coefficient of 0.987 and a standard error of 0.98% water. A second derivative at 1702 nm divided by a second derivative at 1011 nm was used for the water calibration. Two samples had errors of greater than 2.3%, but they were not excluded from the data even though the reflectance data suggest that the oven moisture data on these two samples are in error. A validation test was not performed on these pork samples because so few samples were available, but the high correlations indicate that reflectance spectrocopy can be used for analysis of fat and water.

A similar study was conducted with homogenized raw beef with samples ranging from 6.7 to 60% fat. Two replicates were drawn from each of 12 homogenized samples, and reflectance measurements and chemical analyses were done on both replicates. The second derivative at 1722 nm divided by the second derivative at 1360 nm provided

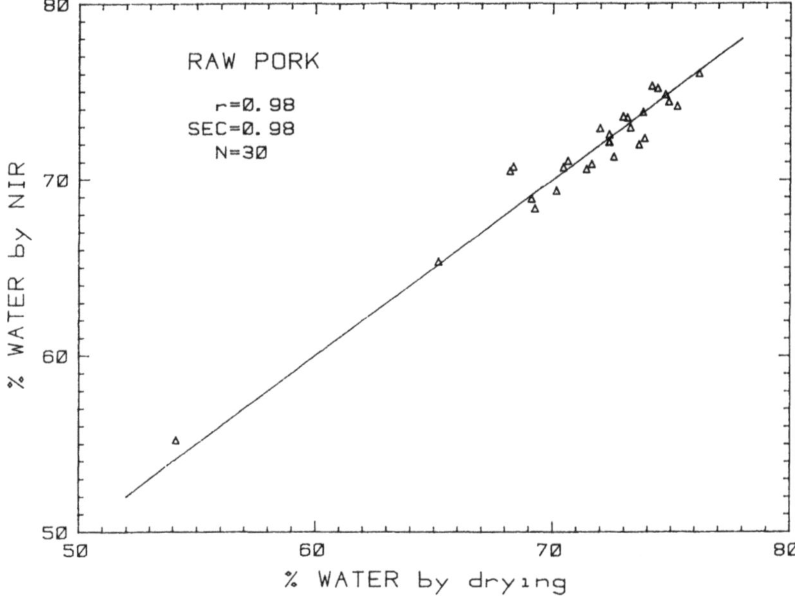

FIG. 8.13. Calibration plot for near infrared measurement of water content of raw pork.

the best calibration for fat; see Fig. 8.14. The correlation coefficient of 0.98 is very high, but it would be even higher except for one sample which has an apparent error in the chemical data because one of the replicate samples fitted the infrared data and the other did not.

The calibration results for water in the beef samples, Fig. 8.15, show a correlation coefficient of 0.998 and a standard error of 0.61% water. This was obtained by using a second derivative at 1795 nm divided by a second derivative at 1608 nm. The reproducibility of the reflectance technique can be evaluated from these data. The computed standard deviations of repeats were 0.42% for fat and 0.40% for water. These errors include both the instrument error and the sampling error.

The protein content was not available on these beef and pork samples, but the reflectance technique should predict protein content as well as the fat and water content. These results on meat are similar to the results reported by Lanza (1983) using a more complex treatment of the reflectance data. It should be noted that each of the meat components has several absorption bands; therefore, many choices of wavelengths are possible for predicting each of the components. For

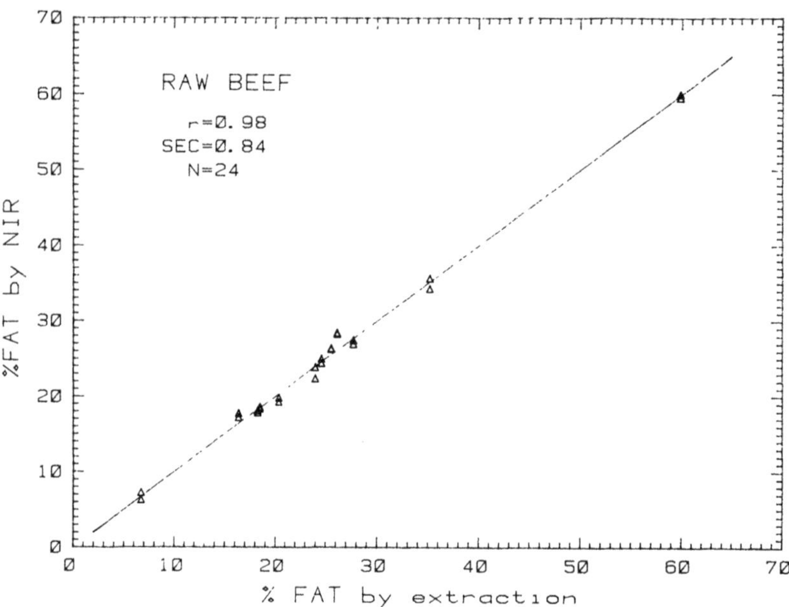

FIG. 8.14. Calibration plot for near infrared measurement of fat content of raw beef.

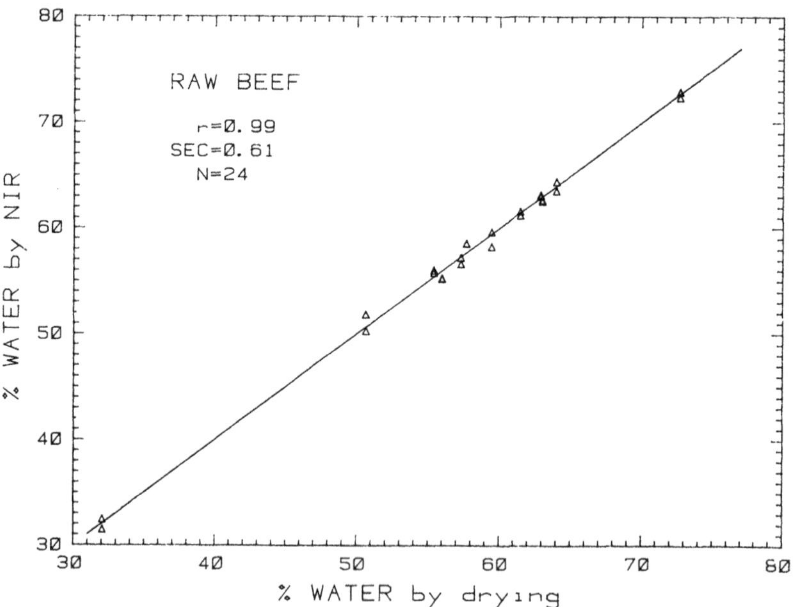

FIG. 8.15. Calibration plot for near infrared measurement of water content of raw beef.

example, the wavelengths selected for fat in pork were different from those for beef, but the difference in performance using either set of wavelengths is very small, and either set can be used for both pork and beef.

The possibility of measuring the fiber content of breakfast cereals with near infrared reflectance was explored by Baker *et al.* (1979). They used a multiterm data treatment with two derivative ratios to correlate with the fiber content expressed as neutral detergent fibers. They reported a correlation coefficient of 0.985 and a standard error of 1.36% when using log(1/*R*) data from the computerized spectrophotometer and essentially the same performance with a commercial filter instrument designed to analyze grains and oilseeds.

CONCLUSION

Near infrared reflectance instruments are now widely used to analyze grains and oilseeds, with instruments being manufactured by three companies in the United States and two companies in Europe. The technique is beginning to be applied to a more general class of food products, and many more applications can be anticipated in the future.

The data treatments and regression procedures used in the near infrared region have not been applied to pigment analyses in the visible region, but they should be useful in any spectral region. The same type of data treatments have been applied to diffuse transmittance data in the visible and near infrared (Norris 1983A; Watada *et al.* 1976).

BIBLIOGRAPHY

AMER. ASSOC. CEREAL CHEMISTS. 1969. Approved Methods of the AACC. Method 44-15A, Method 46-15. AACC, Inc., St. Paul, MN.

ANON. 1979. Optical properties of pressed Halon coatings. Information Sheet from Radiometrics Physics Division, NBS, 10 pp.

BAKER, D., NORRIS, K.H., and LI, B.W. 1979. Food fiber analysis: Advances in methodology. *In* Dietary Fibers: Chemistry and Nutrition, G.E. Inglett and I. Falkehag (Editors), pp. 67–77. Academic Press, New York.

BIRTH, G.S. 1983. Personal communication. Research Engineer, USDA, Russell Research Center, Athens, GA.

HRUSCHKA, W.R., and NORRIS, K.H. 1982. Least-squares curve fitting of near infrared spectra predicts protein and moisture content of ground wheat. Appl. Spectrosc. *36,* 261–265.

KUBELKA, P. 1948. New contributions to the optics of intensely lightscattering materials. Part I. J. Opt. Soc. Am. *38,* 448–452.

LANZA, E. 1983. Determination of moisture, protein, fat, and calories in raw pork and beef by near infrared spectroscopy. J. Food Sci. *48,* 471–474.

NORRIS, K.H. 1983A. Extracting information from spectrophotometric curves. Predicting chemical composition from visible and near infrared spectra. *In* Proc, IUFOST Symp. on Food Research and Data Analysis, Oslo, Norway, pp. 95–113. Applied Science Pub. Ltd., London, England.

NORRIS, K.H. 1983B. Instrumental techniques for measuring quality of agricultural crops. *In* Proc. NATO Advanced Study Institute, Post Harvest Physiology and Crop Preservation, Sounion, Greece, pp. 471–484. Plenum Pub. Corp., New York.

NORRIS, K.H., BARNES, R.F., MOORE, J.E., and SHENK, J.S. 1976. Predicting forage quality by infrared reflectance spectroscopy. J. Animal Sci. *43,* 889–897.

WATADA, A.E., NORRIS, K.H., WORTHINGTON, J.T., and MASSIE, D.R. 1976. Estimation of chlorophyll and carotenoid contents of whole tomatoes by light absorbance technique. J. Food Sci. *41,* 329–332.

9

Biological and Biochemical Assays in Food Analysis

John R. Whitaker[1]

Several of the chapters presented in this volume on *Modern Methods of Food Analysis* use living systems to determine the presence or concentration of a desired constituent. Chapter 10 is Microbiological Assays by Dr. J.M. Jay and Chapter 11 is Sensory Analysis as an Analytical Laboratory Tool in Food Research by Professor R.M. Pangborn. Both of these chapters depend upon a living system for the analysis. In this chapter, the use of animals and single cell systems will be described. In addition, the use of certain types of molecules that have the ability to selectively bind and in some cases subsequently to catalyze a reaction will be discussed.

The advantages of biological analyses are several fold. In many cases, it is essential to determine how a compound affects a living system, with all its multifaceted functions. In animals, there is an opportunity to measure intake, digestion via the gastrointestinal system, absorption of the digested material, and its resynthesis into compounds to meet cellular and/or energetic needs. By the use of radioactive tracer techniques, the incorporation of the material into specific compounds can be determined. In the living animal, use of the compound for growth, repair and maintenance, and its excretion can be measured. The toxicological aspects can be determined. In animals, as well as the analytical techniques using antibodies and enzymes, the effect of different isomeric forms of a compound can be tested.

There are disadvantages in the use of animals for analytical tech-

[1] Department of Food Science and Technology, University of California, Davis, CA 95616

niques. The responses are quite variable from one animal to another requiring several replications. The need to reduce this variability leads to use of animals bred specifically for experimentation, having identical genetic composition, age, sex, etc. This increases the expense of the experiments. The response time is rather slow, requiring several weeks to complete. It is costly to maintain experimental animals, especially the larger ones. Diseases can wipe out a colony of animals. Animals have an advantage over other analytical techniques in that they more closely resemble the human. But how comparable are the results?

With all these limitations in the use of animals for experimentation, it is reasonable to look at other biological systems, such as cells, antibodies, and enzymes for use in analysis.

THE WHOLE ANIMAL IN ANALYSIS

The advantages of use of whole animals for analyses have been outlined above. To that list should be added the advantages of being able to evaluate a complex mixture of compounds such as foods and feeds. This leads to an overall response but it is thought that the response should have some resemblance to that elicited in the human. If this is to be, careful selection of the experimental animals becomes very important. The human is the ideal animal but only limited testing is permitted. Primates (monkeys, chimpanzees, and baboons) are the second type of animals of choice. Unfortunately, the cost of maintaining large colonies of these animals for widespread experiments is prohibitive. Fortunately, the U.S. Federal Government has foreseen the need to maintain some primates for this purpose in Primate Centers located around the United States.

The results obtained by use of sheep, goats, and cows do not readily translate into human requirements because they are ruminants. The pig has been judged to have a metabolic system resembling the human and a breed of small pigs has been developed for experimental purposes. Chicks and chickens are also used, but generally not directly for experiments to be applicable to humans. Guinea pigs have an absolute requirement for ascorbic acid as does the human but the guinea pig has many disadvantages as an experimental animal. The rat has become the animal of experimental choice and much of the data and responses have been correlated with limited experiments on humans. Attention must be given to sex, age, breed and prior feeding experience of the rat. The male 21-day-old Sprague-Dawley rat has become the standard of reference for most nutritional studies.

Levels of Measurement

There are several levels of measurements that can be made using animals.

Acceptance. Acceptance can best be made with human subjects because they can also provide verbal feedback on the level of acceptance. Yet animals, such as the dog and rat, are selective in the food they eat not only as to kind but also as to composition. In our animal feeding studies, we noted there was appreciably more intake of casein, as the sole source of protein in the diet, when it was supplemented with methionine to increase its nutritional quality. This increased acceptability occurred whether the methionine was added as free L-methionine to the ration or was covalently bound as *N*-acetyl-L-methionine to the ϵ-amino group of lysine residues. We found that the protein intake was considerably less when carbohydrates (glucose, fructose, or lactose) were covalently attached via Schiff base to the ϵ-amino group of the lysyl residues followed by reduction with $NaBH_4$. Apparently, the animals have a feedback mechanism that will permit them to discriminate against these less nutritious foods.

Digestion. Conversion of a food or feed to amino acids, monosaccharides, free fatty acids, and purine and pyrimidine bases by *in vitro* methods (acid or base) and determination of the composition may have little correlation with digestibility as determined on the whole animal.

The quality of proteins is generally of more interest to nutritionists. By measuring the total protein nitrogen of the food ingested and correcting it for the total converted nitrogen excreted in the feces, a true digestibility can be measured. The correction needed is due to the excretion of sloughed intestinal membrane cells, of enzymes, and of microorganisms in the digestive tract. This is shown in Eq. (1).

$$\text{True digestibility coefficient} = \frac{\text{Food protein N} - (\text{Fecal N} - \text{Metabolic fecal N})}{\text{Food protein N}} \qquad (1)$$

Nutrient Content. There are a number of determinations that are designed to measure the nutrient content of a food or feed. These can be grouped into two categories for ease of discussion.

The first group of measurements is made directly on the animal itself and usually involves only determining the weight of food ingested and the weight gain. These methods include the protein efficiency ratio (PER; see Osborne *et al.* 1919), the net protein ratio (NPR; see Bender and Doell 1957) and modifications of these such as the slope-ratio assay, involving feeding multiple levels of the nutrient to be determined.

$$\text{PER} = \frac{\text{Gain in body weight}}{\text{Protein consumed}}, \tag{2}$$

$$\text{NPR} = \frac{W_t - W_0}{P}, \tag{3}$$

where W_t = weight gain of animals fed a diet containing the test protein, W_0 = weight gain of animals fed a protein-free diet, and P = amount of protein consumed during the period by the animals fed on the protein-containing diet.

The second group of measurements involves determining both the intake and output of nutrients by the animal. These methods include the biological value (BV), net protein utilization (NPU), net energy (NE), and nitrogen balance.

The biological value (Allison 1955) of a protein (for example) is defined as the percentage of absorbed nitrogen retained in the body of the animal as given in Eq. (4).

$$\text{“Apparent” BV} = \frac{\text{N intake} - \text{Fecal N} - \text{Urinary N}}{\text{N intake} - \text{Fecal N}} \times 100. \tag{4}$$

However, not all the fecal nitrogen comes from the food. It comes from microorganisms, enzyme secretions and cell sloughing. Therefore, the “true” BV is given by Eq. (5).

$$\text{“True” BV} = \text{N intake} - \frac{(\text{Fecal N} - \text{MFN}) - (\text{Urinary N} - \text{EUN})}{\text{N intake} - (\text{Fecal N} - \text{MFN})} \times 100, \tag{5}$$

where MFN = metabolic fecal nitrogen and EUN = endogeneous urinary nitrogen.

The net protein utilization is designed to correct for the extent of digestibility of the protein. The NPU is defined as the difference in final body nitrogen content between a group of animals fed a diet with a test protein (test animals) and a group fed the same diet without the test protein (control animals), expressed as a percentage of the nitrogen intake of the test group as shown in Eq. (6).

$$\text{NPU} = \frac{\begin{array}{c}\text{(Dry body wt of test animals)(\%N)} - \\ \text{(Dry body wt of control animals)(\%N)}\end{array}}{\begin{array}{c}\text{Nitrogen content of food consumed} \\ \text{in test period, dry wt basis}\end{array}} \times 100. \tag{6}$$

The net energy (NE) is given by Eq. (7).

$$\text{NE} = \text{Metabolizable energy} - \text{Heat increment}, \tag{7}$$

where the metabolizable energy is the total available from the food, the heat increment is that required for digestion and assimilation of the food, and NE is the net energy available for maintenance, growth, work and production of milk, etc. All metabolizable compounds contribute to this value.

The nitrogen balance is calculated by use of Eq. (8).

$$\text{Nitrogen balance} = \text{N intake} - \text{Fecal N} - \text{Urine N}. \tag{8}$$

Other Parameters. Other measures of the quality of a food can be determined by measuring the free amino acid content of the plasma after the consumption of a test food (Longnecker and Hause 1959), or measuring the plasma proteins or the plasma enzymes. Xanthine oxidase levels of the liver of rats have been correlated with body weight gain (Litwack *et al.* 1952).

Toxicological Analyses

The potential toxicity of new compounds intended for human consumption must be determined using animals. The usual studies are done with 100–1000 times the amount of the compound anticipated for consumption by humans. This permits the use of smaller numbers of animals.

Two types of tests are run. The first test uses lethality as an index and is designed to determine the dose required to kill 50% of the test animals, the LD_{50} value (Trevan 1927). This is the acute toxicity test. The LD_{50} may need to be determined for several levels of the compound and to be supplemented by extensive histological and biochemical studies at the cellular and enzyme level, respectively.

The second test is designed to determine the chronic effect of the compound and involves feeding the animals sublethal doses over an extended time, involving a minimum of three months but can extend over several years. Associated with these tests may be determinations of mutagenicity, carcinogenicity, reproduction, and teratology.

Analyses using animals are indeed expensive and of long duration. Therefore, other biological assays have been developed as described below. However, these methods need to be tested for correlation with responses in the whole animal.

WHOLE CELLS AS ANALYTICAL TOOLS

Microorganisms have been used for many years as analytical tools to determine the concentration of vitamins, amino acids, and minerals, among other compounds. They have also been used effectively to determine the presence of mutagenic compounds in air, food, and feeds. These have been thoroughly covered by Dr. J.M. Jay in Chapter 10 in this volume entitled Microbiological Assays.

It is now possible to grow dispersed plant and animal cells in culture over several generations. Hybridoma cells formed by the fusion of mice or rat spleen cells with mice or rat myeloma cells can be kept alive indefinitely (Yelton and Scharff 1981; Galfré and Milstein 1981). Thus, animal and plant cells are now available as analytical tools.

The use of cell cultures has several advantages over the use of whole animals or plants in that (a) a single-cell type can be used, (b) the results will be statistically more significant in that several hundred thousand cells are used in each assay, (c) the assay times are a few hours compared to days for animals, (d) the cost of maintaining cell cultures is far cheaper than for animals, and (e) for the first time normal human cells can be used as the analytical test material. Whole cells have advantages over the use of antibodies and enzymes as analytical tools in that (a) uptake of compounds by the cell can be measured, and (b) the effect can be measured using a living system.

With whole cells, the test parameter measured may be the (a) uptake (absorption) of a compound through the cell membrane (or wall), (b) the lysis of the cell membrane (or wall), (c) cytotoxic effects, (d) cytostatic effects, and (e) cytogenetic effects.

Cytogenetic Effects

One of the most used cell culture techniques involving microorganisms is that for measuring the potential mutagenicity of compounds of the air (CO, NO, NO_2, O_3, SO_3), naturally occurring compounds in foods and feeds such as aflatoxins, compounds added to food materials during growth and storage, and those formed during the processing of foods such as the nitrosamines. The Ames test (Ames *et al.* 1975), using *Salmonella typhimurium* is now used extensively and routinely to assist in evaluating the potential mutagenicity and carcinogenicity of large numbers of chemicals. This is a rapid and inexpensive assay which can be used to screen large numbers of compounds. Those shown to be mutagenic can then be tested further on animals and by epidemiological surveys. The basis for the assay is the following. The histidine autotroph mutants of *Salmonella typhimurium* are unable to grow in the absence of histidine. Through their effect on the DNA of the microorganism, some compounds can cause reversion to the prototrophic forms which can grow in the absence of histidine. Other microorganisms can also be used. The reader is referred to Chapter 10 on Microbiological Assays for a more complete discussion of the Ames test.

Human lymphoid cells have been used to study the cytogenetic and cytokinetic effect of some 14 organophosphorus insecticides (Robti *et al.* 1982). The cytotoxic effect was dose related and often led to extensive cell kill. Incubation of cells with 20 μg/mL of the various organophosphorus compounds produced 6–18% (Azodrin) M_1 metaphases compared to no M_1 metaphases in control cultures. Eleven of the 14 compounds tested significantly increased the sister chromatid exchange (SCE) frequency. Of the nine organophosphorus compounds tested following metabolic activation by liver microsomal S9 preparation, diazinon, dimethoate, Dursban, and Phosdrin caused significant increases in sister chromatid exchange.

Gruenwedel and his collaborators have used cultured human cervix HeLa S3 carcinoma cells to determine the effects of methylmercury (II), sodium selenite, and early Maillard products on the viability (measured with the trypan blue dye exclusion test) and intracellular metabolic activity (incoporation of ^{3}H-labeled precursors into DNA, RNA, and protein). It was found that methylmercury concentrations above 1–10 μM (30×10^{-16} mol/cell intracellular amounts) caused an abrupt effect on both viability and intracellular metabolic activities (Gruenwedel and Cruikshank 1979A; Gruenwedel *et al.* 1979; Gruenwedel 1981; Gruenwedel *et al.* 1981). Sodium selenite intracellular concentrations above 6.63 μM (0.5 ppm) also caused somewhat similar effects (Gruenwedel and Cruikshank 1979B). The presence of both methylmercury and sodium selenite, at equimolar concentrations, had less effect on the HeLa S3 cells than did either compound alone (Gruenwedel 1983). The Amadori compound 1-(*N*-L-tryptophan)-1-deoxy-D-fructose (Trp-Fru), its nitrosated product, NO-Trp-Fru, and sodium nitrite had no effect on the viability of the cells or the intracellular synthesis of RNA and protein (Lynch *et al.* 1983). However, there was a marked enhancement of DNA synthesis by NO-Trp-Fru at concentrations even lower than 1 μM.

Whole Cell Lysis

The lysis of cells is a sensitive analytical tool for the presence of certain microbial enterotoxins and for antibody and antigen determinations.

Enterotoxin A of *Clostridium perfringens* can be assayed by the guinea pig skin test (Stark and Duncan 1972), by the inverted gut method (Duncan and Strong 1969), or by rocket immunoelectrophoresis (Duncan and Somers 1972). However, these methods lack precision or sensitivity or require several hours for the test. The enterotoxin A, at nanogram levels, has been shown to cause the lysis of rat liver and Vero cells (Skjelkvåle *et al.* 1980; McClane and McDonel 1980). The degree of lysis, directly dependent on the concentration of active enterotoxin, can be measured by counting the number of lysed cells or better by determining the amount of lactate dehydrogenase released from the lysed cells, measured by adding lactate and NAD^+ and following the increase in absorbance at 340 nm spectrophotometrically.

Red blood cells containing attached endogenous antigens, or exogenous antigens which have been artifically attached, lyze when the specific antibodies of the antigen and complement are added. The extent of lysis can be measured by counting the number of lysed cells (best done in an agar plate so the cells are immobilized), by measuring the amount of hemoglobin released or by measuring the release of $^{51}Cr^{3+}$ incorporated into target cells carrying the desired antigen (Galfré and Milstein 1981; Goodman 1961; Pearson *et al.* 1977; Jerne and Nordin 1963). The procedure can also be carried out by attaching the antibody to the red blood cells and mixing with the specific antigen and complement (reverse hemolytic assay; Molinaro and Dray 1974). When the agar plate technique is used, there is a linear relationship between the hemolytic areas and the log of the antigen concentrations. When a 0.5% concentration of indicator cells is used, the sensitivity is 0.1 μg/mL of antigen with a precision of about 10%.

Absorption, Hydrolysis, and Utilization

Use of microorganisms to determine the concentration of selected amino acids is usually done with protein hydrolyzates. Sometimes it is important to determine if the cell can utilize a compound in its original state.

Tetrahymena pyriformis, a single-cell protozoan, can be cultured in

TABLE 9.1. GROWTH OF *TETRAHYMENA THERMOPHILI* ON CASEIN AND CHEMICALLY PHOSPHORYLATED CASEIN[a]

Protein[b]	Growth rate[c], $b^{-1} \times 10^2$	Relative growth rate
Defined media	4.8	100.0
Hammarsten casein	3.8	79.2
Phosphorylated casein	3.6	75.0

[a] Matheis *et al.* (1983).
[b] 0.08% protein in growth medium, 30°C.
[c] Slope of plot of log absorbance vs. time.

the same way as microorganisms. Changing the nutrient broth weekly, along with dilution of the cells, permits *Tetrahymena pyriformis* cultures to be maintained indefinitely. The protozoan has the same essential amino acid requirements as the human. It can hydrolyze proteins, preferably in insoluble suspension, to amino acids. The rate of growth of the organism can be measured readily by determining the increase in turbidity of the cell culture. We have successfully used this technique to screen the effect of chemical modification of proteins on their digestibility/nutritional quality (Table 9.1; Matheis *et al.* 1983). The results correlated well with those obtained by rat feeding experiments. The advantages are savings in cost, time, and the amount of material needed.

The ε-*N*-acetyl-L-methionyl-L-lysine, covalently bound into casein, is nutritionally available to rats. Liver, spleen, and intestinal cell cultures were used successfully to show that [^{14}C] ε-*N*-acetyl-L-methionyl-L-lysine is taken up by the cells and is hydrolyzed to acetate, methionine, and lysine, and the amino acids are used by the cells for growth (Puigserver *et al.* 1982).

Effect of Salinity, Hormones, Herbicides

Plant cells can now be readily grown in cell suspension. These cell cultures can be used to test the effect of hormones, herbicides, and salinity on growth. They can also be used to select for salt-tolerant mutants, herbicide resistant mutants, disease resistant mutants, and effect of other environmental factors. Because of the many hundreds of thousands of cells which can be grown in a single Erlenmeyer flask, the selection process can be greatly speeded up over that of conventional field selection techniques.

TABLE 9.2. ANALYTICAL USES OF IMMOBILIZED CELLS

Use	Cells	Comments	Reference
BOD sensors	Various microorganisms from soils and activated sludges *Clostridium butyricum*	Time about 40 min compared to 5 days; 10% reproducibility; stable for more than 30 days	Karube *et al.* (1976); Karube *et al.* (1977A,B)
Acetic acid sensor	*Trichospora brassicae*	Correlation coefficient of 1.04 with GC; stable for 1500 assays; 15-min assays.	Hikuma *et al.* (1979A)
Alcohol sensor	*Trichospora brassicae*	Correlation coefficient of 0.98; stable over 2100 assays.	Hikuma *et al.* (1979B)
Nystatin sensor	"Yeast" cells	0.5 unit/mol sensitivity	Karube *et al.* (1977C)
Cephalosporin sensor	*Citrobacter freundi*	10 min per assay; 125 μg/mL sensitivity	Matsumoto *et al.* (1979)
Nicotinic acid sensor	*Lactobacillus arabinosus*	Range 5×10^{-8} to 5×10^{-6} g/mL; useful for at least a month	Matsunaga *et al.* (1978A)
Vitamin B_1	*Lactobacillus fermentii*	$>2.5 \times 10^{-8}$ g/mL; 15-min assay	Matsunaga *et al.* (1978B)

Immobilized Cells as Analytical Tools

Immobilized enzymes have proved to be of great importance in many industrial processes as well as in analytical techniques. The advantages of immobilization can be carried over to intact cells (Suzuki and Karube 1981). Some of these uses are shown in Table 9.2.

IMMUNOASSAY TECHNIQUES

Presently, one of the most used biological assay techniques is that involving a specific antibody–specific antigen complexation technique. This technique is so widely used because of its great specificity, great sensitivity, and the relative ease of running many samples repetitively and continuously once the reagents and equipment are available.

The basic technique depends upon the complexation of an antigen with a specific antibody that has been elicited against the antigen. The antibody is a large protein of 150,000 MW and belongs to the immunoglobulin fraction of animal sera. As shown in Fig. 9.1, immunoglobulin G is comprised of four polypeptide chains, two heavy chains (H),

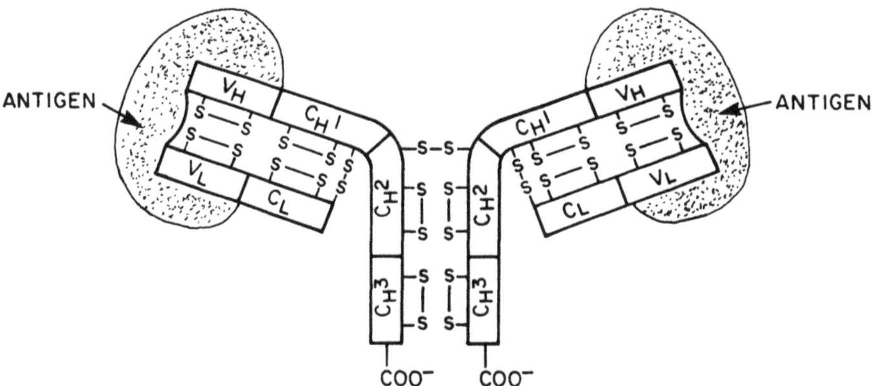

FIG. 9.1. Schematic structure of immunoglobulin G (antibody) and the specific binding of antigen. Immunoglobulin G is composed of two identical light (L) chains and two identical heavy (H) chains. Each has constant amino acid regions (C_H3, C_H2, C_H1 and C_L1) and variable amino acid regions (V_H and V_L). The antigen recognition site involves the variable regions.

and two light chains (L). Binding of the antigen occurs at the V_H and V_L regions and involves specific recognition sites for antigen on both the L and H chains. Normally produced antibodies are divalent; i.e., they can bind two molecules of antigen per molecule of antibody. Antibodies specific against a given antigen are produced by the repeated injection of the purified antigen into an animal such as a rabbit, rat, cow, or sheep. Most biochemical reagent companies sell a variety of antibodies because of the popularity of this assay method.

The antigen can be a protein, a carbohydrate, a nucleic acid, a lipid, or a small molecule (known as a hapten which when attached to a macromolecule will elicit specific antibodies)—such as a drug, toxin, or nicotine. It may be a cell-surface antigen, virus, or bacteria. Because of the very large range of compounds, cells, etc., to which antibodies can be prepared (some 10^6 different antibody producing cell types exist per animal), the popularity of this method will undoubtedly continue to grow. It is one that the food scientist has used very little to date.

Fundamental Principles

The specific binding of antigen with antibody is a result of the specific interaction of the antigen with a complimentary portion of the surface of the antibody (see Fig. 9.1). This binding requires careful alignment of both electrostatic groups and hydrophobic groups on the two molecules. The strength of the complex formed between the two molecules is a function of the type and number of interacting groups ($K_d = 10^{-4}$–10^{-10} M). Most of the assays to detect antigen–antibody complexation depend upon an equilibrium between the complex and the free molecules [Eq. (9)] as we shall see.

$$\text{Antibody} + \text{Antigen} \overset{K_1}{\rightleftharpoons} \text{Antibody} \cdot \text{Antigen} \xrightleftharpoons{K_2,\ \text{Antigen}} \text{Antibody} \cdot (\text{Antigen})_2. \quad (9)$$

The K_1 is considerably larger than K_2 (expressed as dissociation constants) and the complex of antibody·antigen is much more soluble than is the antibody·(antigen)$_2$ complex where both antigenic sites on the antibody are occupied. The first quantitive use of the antibody·antigen complexation took advantage of this to determine the antibody titer of blood sera. Figure 9.2 shows a titration of antibody with antigen; precipitation occurs at the antibody·antigen ratio of 1:2.

The desired qualities in an antibody preparation are (1) high binding affinity, (2) uniformity of binding affinity, (3) high specificity for the compound (antigen) to be determined, (4) stablility during storage, (5) relatively low cost, and (6) easy assay method. Fortunately for cost of the method, it is generally not necessary to separate the immunoglobulins prior to use in the assay. Treatment of blood serum with 28% saturated ammonium sulfate (at pH 8.1–8.2) at room temperature precipitates the immunoglobulins while albumin and most other serum proteins remain soluble. The precipitate is removed by centrifugation and dissolved in 0.15 M NaCl containing 1 mM NaN_3. The precipitation step is usually repeated once. If necessary, antibod-

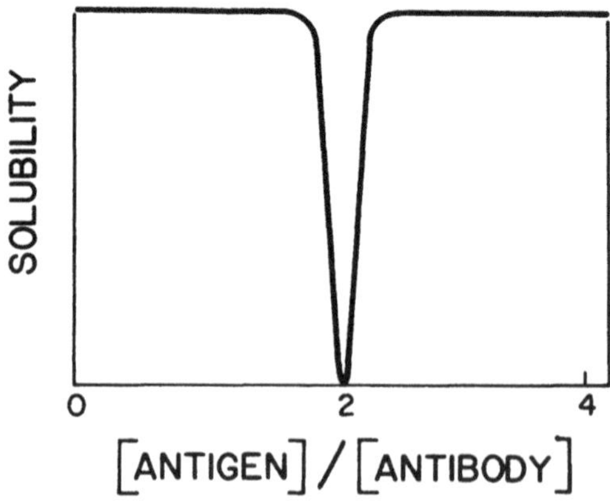

FIG. 9.2. Effect of ratio of antibody to antigen on solubility of the complex. Free antibody and free antigen are soluble. At ratios of <2 and >2, the complex is soluble. At the specific ratio of 2 antigen molecules to 1 antibody molecule the complex is quite insoluble.

ies specific against the desired antigen can be isolated by affinity chromatography (Sepharose-specific antigen complex) or can be enriched from the start by use of monoclonal antibody preparation techniques (Galfré and Milstein 1981; Yelton and Scharff 1981).

Analytical Techniques

A number of analytical techniques are available for the quantitative determination of an antigen in a biological system. It is our purpose here to describe the principles of these techniques and not the details of their application to the myriad compounds.

Single Radial Immunodiffusion (SRID) (Vaerman 1981). This procedure is one of the simplest of the techniques to perform. The antibody is mixed uniformly with the melted agarose (in 0.02 *M* borate buffer at pH 7.5–8.5 and ionic strength of 0.1 with sodium chloride) and poured onto a plate to a thickness of 3–4 mm. Petri dishes with flat bottom surfaces are statisfactory. After the agarose has solidified, small identical size reservoirs (wells) are cut out of the agarose. The wells are filled with a fixed volume of the antigen-containing solution to be tested, covered, and incubated for several hours (time depends on size of antigen) until the precipitin rings around the well do not migrate further. The antigen migrates into the antibody-containing agarose, complexing with the antibody as it goes. A stable and permanent precipitin band (visible) will form when the stoichiometry of the complex is 1:2 (antibody:antigen). Typical results are shown in Fig. 9.3.

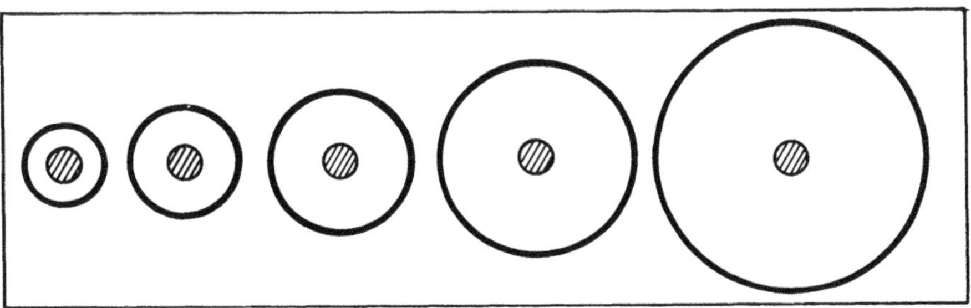

FIG. 9.3. Schematic representation of the single radial immunodiffusion (SRID) technique. Antibody is uniformly distributed in the agar gel plate. Small wells (striped circle in center) are filled with the specific antigen solutions. The antigen diffuses from the well, contacting the antibody. When the ratio of 2:1 of antigen:antibody is reached a precipitin line is formed (the outer circle).

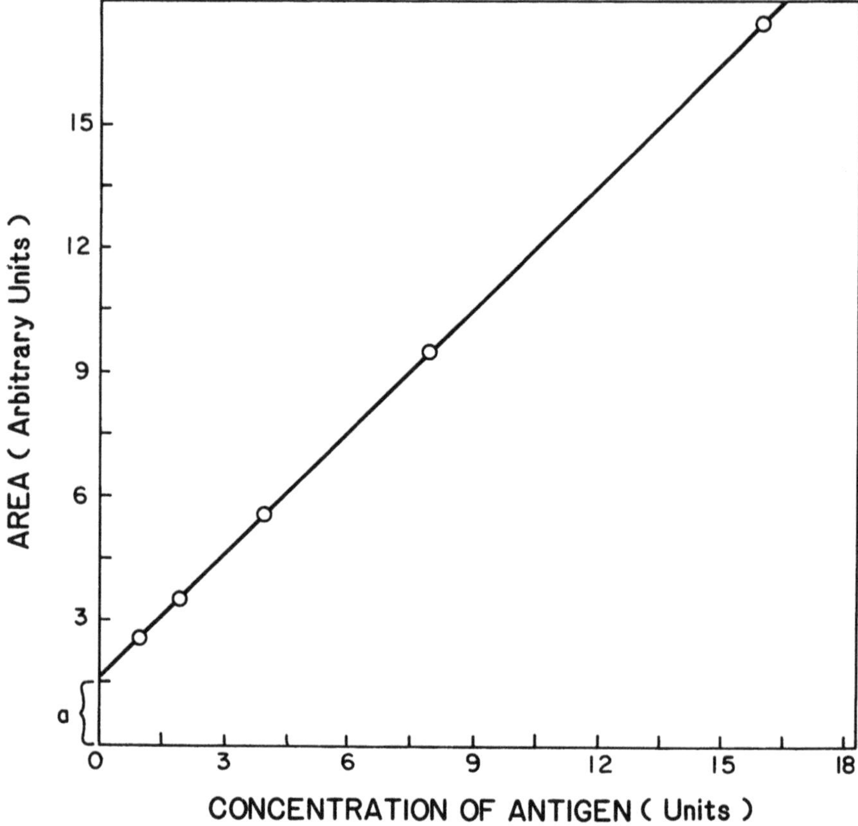

FIG. 9.4. Relation between antigen concentration placed in the center well (Fig. 9.3) and the area inside of the precipitin line. The data fit the equation $A = a + bC_{Ag}$; a is the area of the small well.

There is a linear relation between the amount of antigen placed in the well and the area inside the precipitin ring surrounding the well (Fig. 9.4). The area is inversely proportional to the concentration of antibody in the agarose which is kept constant within a set of experiments.

The relationship between concentration of antigen (C_{Ag}) and the area inside the precipitin ring A) is given by Eq. (10) where b is the slope of

$$A = a + b \cdot C_{Ag} \tag{10}$$

the line (determined by antibody concentration) and a is the intercept on the y-axis (the intercept term should be equal to the area of the well).

A more sophisticated variation of this technique is the two-cross immunodiffusion method (Pokric and Pucar 1981) which permits the calculation of antigen conentrations and diffusion coefficients without the use of a standard curve.

Electroimmunoassays (EIA). The principle of using the specific stoichiometric interaction between antigen and antibody to determine antigen concentration (or purity) is the same for EIA as described above. However, an additional specificity is added to the system, that of differential migration of proteins in an electrical field.

There are three techniques used: (1) immunoelectrophoresis, (2) crossed immunoelectrophoresis, and (3) rocket immunoelectrophoresis.

In immunoelectrophoresis (Laurell 1966; Laurell and McKay 1981), the antigen(s) is (are) applied at a single spot on a slab of agarose gel and electrophoresis performed in the usual way (Fig. 9.5). The electrophoresis is terminated before the proteins reach the ends of the slab. A slot parallel with direction of migration of the proteins and extending the length of migration is cut into the agarose gel and filled with antibody(ies) prepared against the antigen(s). The antigen(s) and antibody(ies) diffuse toward each other and form a precipitin line at the point of stoichiometry (1:2, antibody:antigen). The white precipitin curves can be made more visible by staining the protein(s) with Coomassie Brilliant Blue. This method is semiquantitative at best, being used primarily to determine purity and/or identity of sample (such as proteins of meat, fish, cereals, etc.).

Crossed immunoelectrophoresis involves electrophoresis in two directions (Laurell 1965; Axelsen 1976). A buffered solution of agarose (1% in Tris-barbital buffer, pH 8.6, ionic strength 0.02) is poured as a 1-cm strip on a glass plate (10 × 10 cm) and allowed to solidify. A well, 2.5 mm in diameter, is cut out of the gel (+ in Fig. 9.6) and 5 μL of the antigen solution added. Electrophoresis is performed for the desired time. A solution of agarose (in above buffer) but containing 1.25 μL of antiserum (antibodies) against the antigen is poured along side the original strip and allowed to solidify. Electrophoresis is then performed at a right angle to the original direction so as to move the antigen(s) into the antiserum-containing agarose to form precipitin bands. After electrophoresis the gel is washed, dried gently with filter paper,

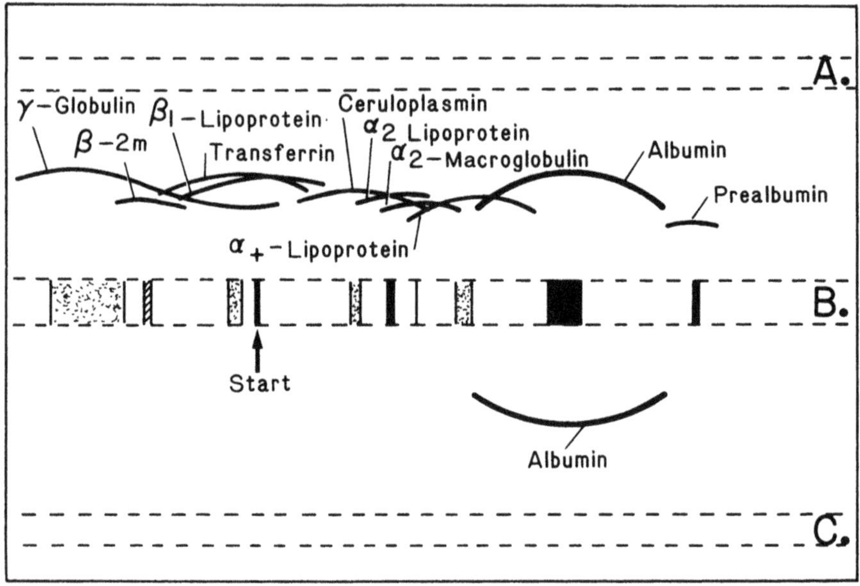

FIG. 9.5. Schematic representation of immunoelectrophoresis. The antigen-containing sample (blood serum here) was placed into an agar gel at arrow labeled Start. Electrophoresis was performed in the horizontal direction to give separation of proteins in track labeled B. Following electrophoresis, horizontal grooves, A and C, were cut out of the agar gel and antibody-containing solutions were placed in them. In groove A, the solution contained antibodies against all the blood serum proteins. Groove C contained antibodies against blood serum albumin only. The agar gel plate was incubated overnight to permit lateral diffusion of the antigens and antibodies. Specific antigen·antibody interaction (ratio 2:1) led to the precipitin arcs.

and stained with Coomassie Brilliant Blue R-250. Results obtained with human serum are shown in Fig. 9.6.

In rocket immunoelectrophoresis (Laurell and McKay 1981) the antibody(ies) is (are) incorporated into the agarose-buffer solution (pH, 8.0–8.5) and poured as a thin layer onto a 10×10-cm glass surface. After solidifying, 2.5-mm diameter sections of the agarose gel are removed (see Fig. 9.7) and about 5 μL of antigen solution is added. Electrophoresis is performed in the indicated direction. The antigen(s) move in the electrophoretic field until each reaches a point where there is a stoichiometric complex (1:2, antibody:antigen) formed at which the precipitin line becomes stationary. Because of lateral diffusion the precipitin line looks like a rocket. The precipitin lines are generally

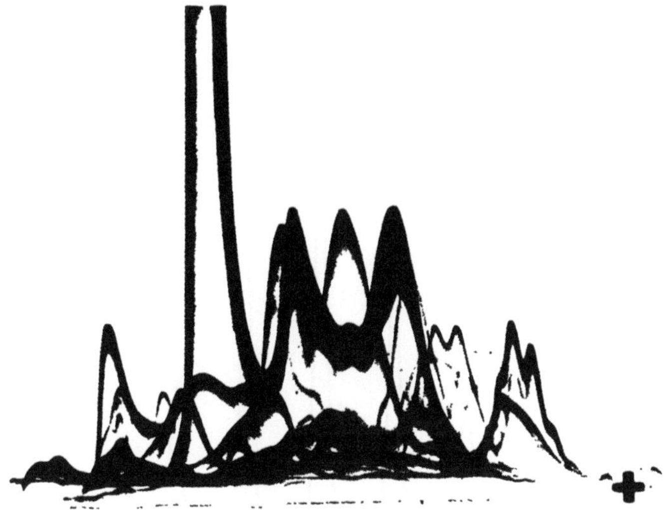

FIG. 9.6. Representation of the crossed immunoelectrophoresis technique. A sample of human blood serum was applied to an agar gel strip at the point marked + and the first electrophoresis was performed in the horizontal direction. The proteins moved from right to left. Additional agar gel, containing antiserum (antibodies against human blood serum produced in rabbit, for example) was poured alongside the original strip (above as shown in figure). Electrophoresis was then performed in the vertical direction so as to cause the human blood serum proteins (antigens) to migrate into the antibody-containing agar gel. The antigen·antibody precipitin lines were stained with Coomassie Brilliant Blue dye. (Reproduced by permission of Pharmacia.)

stained with Coomassie Brilliant Blue. The area under the rocket is proportional to the amount of antigen present (for a fixed concentration of antibody). More frequently, the height of the rocket, also proportional to antigen concentration, is used for quantitation.

Laurell and McKay (1981) have shown that the shape of the precipitin loops, density of the precipitates, sharpness in demarcation on both sides of the line, and formation of adjacent peaks (for a single antigen) provide clues as to physicochemical differences between samples from different sources. A detailed theory for rocket and crossed immunoelectrophoresis has been developed by Cann (1975).

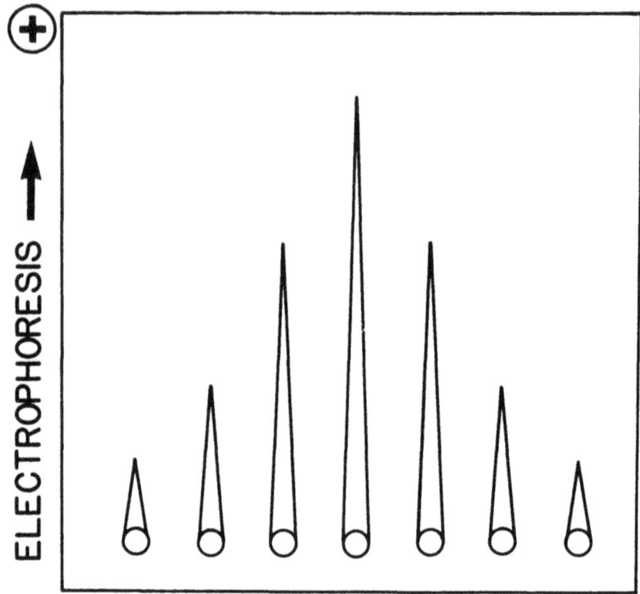

FIG. 9.7. Schematic representation of rocket immunoelectrophoresis. An agar solution, containing antibody against the antigen and buffered at pH 8–8.5, is poured on a flat surface and allowed to solidify. Antigen-containing solution is placed in small wells cut into the agar gel. Electrophoresis is performed in the direction indicated (vertical in figure) to move the antigen into the antibody-containing agar gel. Precipitin lines are formed at antigen/antibody ratios of 2/1. The precipitin lines have been stained with Coomassie Brilliant Blue dye.

Use of Insolubilized (or Immobilized) Antibody. Under this topic are described (1) the radioimmunoassay (RIA), (2) the fluorescence immunoassay (FIA), and (3) the enzyme-linked immunoadsorbent assay (ELISA). There are numerous modifications of these methods but the principles are the same.

Two requirements of the antigen–antibody reaction for quantitative use are that one be able to separate the bound and free antigen (or antibody) and then be able to determine quantitatively the amount of bound material (or free material).

The purpose of immobilizing the antibody (usually) is to accomplish the first need. Separation of the antigen–antibody complex and free antigen and antibody can be accomplished by differential ammonium sulfate or ethylene glycol precipitation. However, these precipitation methods are tedious [because of the several steps involved (see below)] and do not lend themselves in a quantitative manner to microanalytical techniques. Techniques that have been used successfully are covalent coupling of the antibodies to solid particles [Sephadex or cellulose (Wide *et al.* 1967; Wide 1969)], or magnetic supports, such as iron oxide–cellulose particles, polycarbonate-coated ferromagnetic spheres, postmagnetized Sepharose, carbonyl iron–starch, and microspheres (Guesdon and Avranlas 1981); microencapsulation (Lim and Buehler 1981); and by use of Sepharose-bound protein A (MacSween and Eastwood 1981) all designed to permit easy removal of the immobilized antibody–antigen complex from the system at the end of reaction. The most successful technique is due to the observation of Catt and Tregear (1967) that plastic surfaces adsorb antibodies quite strongly without interfering with their antigen-binding abilities. Therefore, plastic tubes and plastic microtiter trays are used widely.

Redioimmunoassay (RIA). The second requirement is for a rapid, precise, and sensitive method of determining the amount of free (or bound) antigen (or antibody) left at end of complexation. The first and still most sensitive method is that using radioimmunoassay (Yalow and Berson 1960). The principle of the method is competition between radiolabeled antigen, added in known amounts, and the antigen in the unknown, in binding to the antibody as shown in Eq. (11).

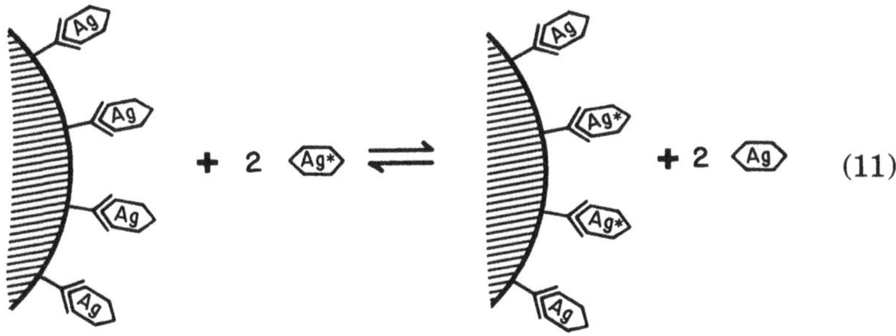

As shown in Fig. 9.8, the amount of Ag* bound to the immobilized antibody is inversely related to the Ag content of the unknown. The Ag and Ag* can be added to the immobilized antibody simultaneously or in consecutive order. If the principle of equilibrium shown in Eq. (9) applies, the same results should be attained either way.

While a number of radiolabels could be used, the most useful, in order, has proved to be ^{125}I-labeled antigen, prepared by use of Chlo-

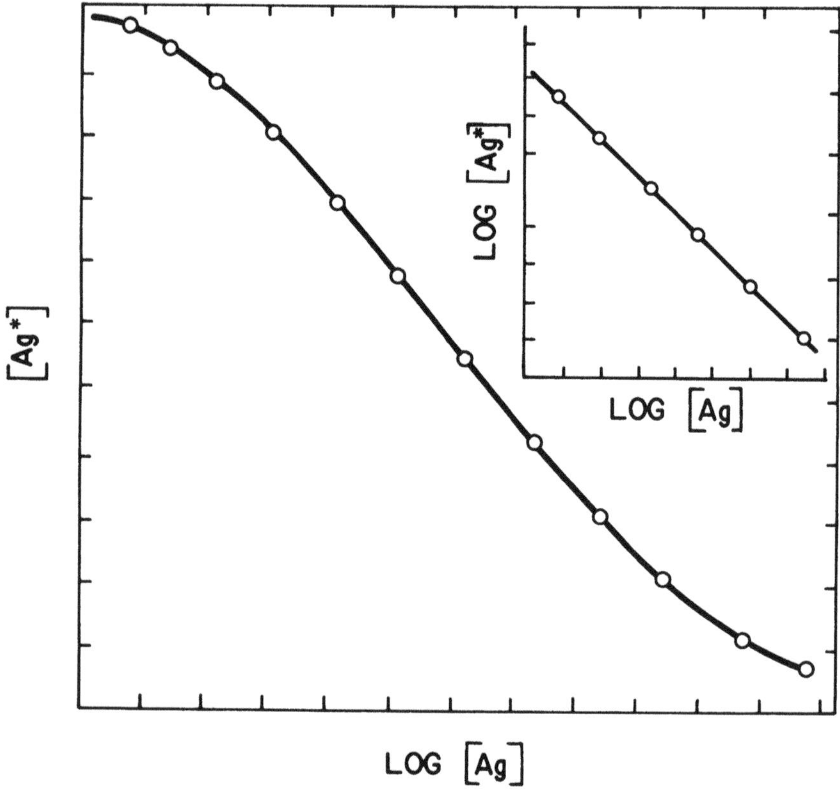

FIG. 9.8. Relationship between the concentration of radiolabeled antigen, Ag*, and antigen, Ag, from a sample whose concentration of antigen is to be determined. The insert shows that a plot of log[Ag*] vs. log[Ag] is linear and that the higher the [Ag], the less of the radiolabeled Ag* will be bound.

roamine-T (Corrie and Hunter 1981) and reductive alkylation using formaldehyde and titriated borohydride (Means and Feeney 1968; Tack and Wilder 1981).

The ^{125}I-labeled antidinitrophenyl (DNP) antibodies have been proposed as a universal tracer, eliminating the need to prepare ^{125}I-labeled specific antigens (Neurath 1981). The rationale of this method is shown in Fig. 9.9.

Fluorescent Immunoassays (FIA). Because of the hazards of working with radiotracers, especially ^{125}I-labeled compounds, and the specialized counting equipment needed, other methods of quantitating the

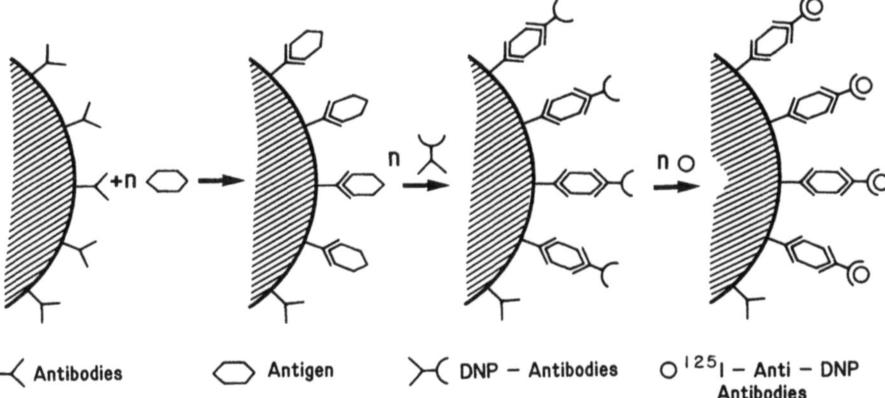

FIG. 9.9. Schematic illustration of the use of ^{125}I-anti-DNP antibodies to determine the concentration of antigen in an unknown sample. Antibodies against the antigen are immobilized. The antigen-containing solution is added, leading to formation of antigen·antibody complex (immobilized antibody in excess). The immobilized antibody·antigen complex is treated with excess DNP-antibodies, the excess DNP-antibodies removed and the immobilized antibody·antigen·DNP-antibody complex treated with excess ^{125}I-labeled anti-DNP antibodies. The amount of ^{125}I-anti-DNP antibodies bound (or conversely unbound), determined by scintillation counting, is proportional to the amount of antigen present in the unknown.

results have been developed. Several fluorescent methods have been developed in which the fluorescent group is attached to the antigen used in competitive binding to the immobilized antibody. These methods include fluorescent polarization techniques (Dandliker *et al.* 1981), fluorescence excitation transfer immunoassay techniques (Ullman and Khanna 1981), indirect quenching fluoroimmunoassay techniques (Nargessi and Landon 1981), and substrate-labeled fluorescent immunoassay techniques (Burd 1981). The advantage of the first three of these techniques over regular fluorescent assay techniques is that the complex has different fluorescent properties than does the free fluorescent labeled antigen, permitting the amount of binding of antigen to be determined without separating the antigen·antibody complex from the free reagents.

Enzyme-Linked Immunoadsorbent Assay (ELISA). The ELISA method has proved to be the most popular of all the techniques used at the present time, because of its great sensitivity especially when the enzyme is peroxidase. (Other enzymes such as alkaline phosphatase, β-galactosidase, and glucosidase can also be used.)

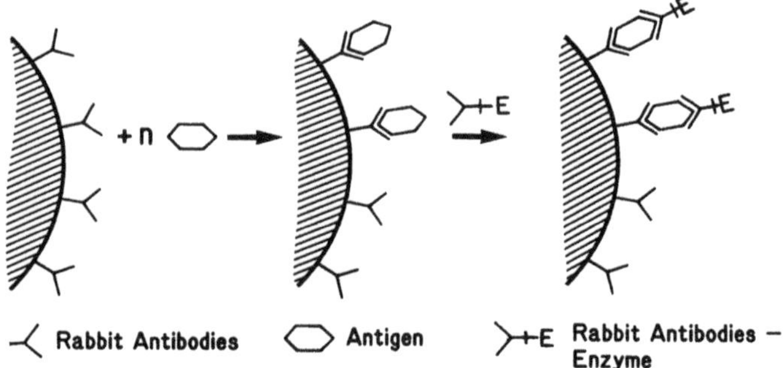

FIG. 9.10. Schematic illustration of the "sandwich" enzyme-linked immunoadsorbent assay (ELISA). See text, along with symbols used in the figure, for explanation.

There are two versions of the ELISA technique, the "sandwich" technique (Engvall and Perlmann 1971) and the "double-sandwich" technique (Notermans *et al.* 1982). The principle of each is shown in Figs. 9.10 and 9.11.

In the "sandwich" ELISA method, polystyrene tubes (or wells in microtiter trays) are coated with rabbit antibodies (used only to illustrate the principle of the technique) against the specific antigen to be determined. Unbound immunoglobulins (antibodies) are washed from the tubes and the sample containing the antigen to be measured is added and incubated to form the antibody·antigen complex. After

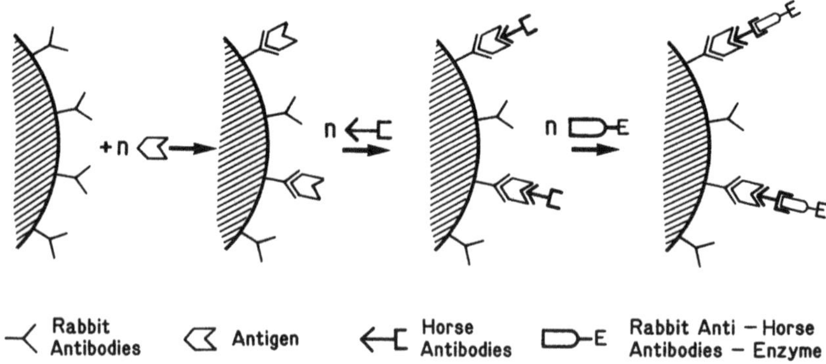

FIG. 9.11. Schematic illustration of the "double-sandwich" enzyme-linked immunoadsorbent assay (ELISA). See text, along with symbols used in the figure, for explanation.

washing, rabbit antibodies against the antigen, which were previously covalently attached to horseradish peroxidase, are added and incubated. After washing, buffered substrate solution (for example, S-amino salicylic acid and H_2O_2) for peroxidase is added, incubated for the desired time, and measured for the increase in absorbance of the solution. The increase in [product] absorbance measured is directly proportional to the amount of enzyme-labeled rabbit antibodies bound, which in turn is directly proportional to the amount of antigen present in the unknown sample (Fig. 9.12).

The "double-sandwich" ELISA method is performed in a similar fashion as the sandwich ELISA method. As shown in Fig. 9.11, the polystyrene-adsorbed rabbit antibodies are incubated with the sample containing the antigen to form the rabbit antibody·antigen complex. Following washing, the complex is incubated with horse antibodies (as an example) against the antigen to form the rabbit anti-

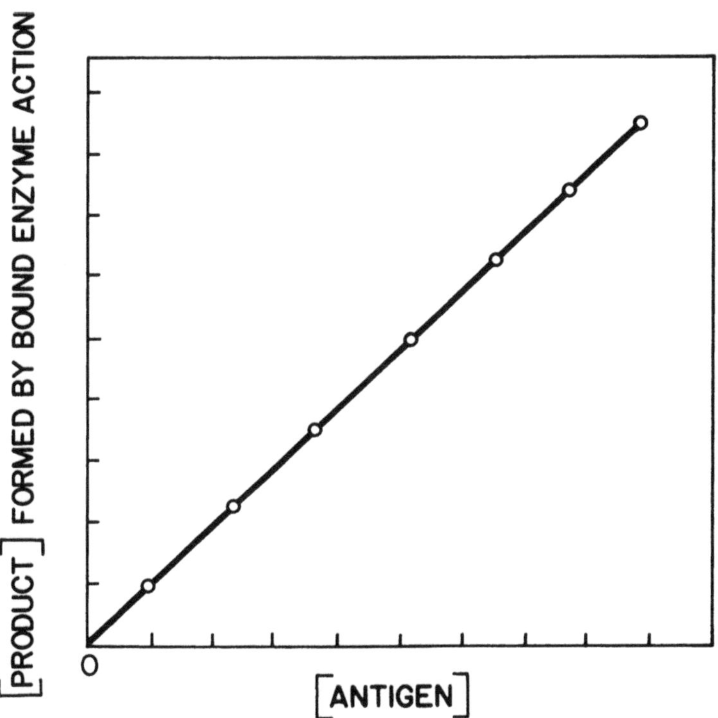

FIG. 9.12. Relationship between antigen concentration of sample and formation of product due to attached enzyme.

body·antigen·horse antibody complex. Following washing, the complex is then treated with peroxidase-coupled rabbit antihorse antibodies. The complex is washed and assayed for bound peroxidase activity as described above.

The advantage of the "double-sandwich" ELISA technique is that the same enzyme–antibody conjugate (against horse antibodies) can be used for the detection of all antigens while in the "sandwich" ELISA technique a different enzyme–antibody conjugate must be synthesized for each antigen. The "double-sandwich" ELISA technique does require antibodies against the antigen be produced in two different animals (rabbit and horse above), but this is presently far easier than production of the enzyme–antibody conjugate.

ENZYME-BASED ANALYTICAL ASSAYS

Because of their high sensitivity, great specificity and ease of measurement, enzyme-based analytical assays are often the method of choice. The cost of reagents, primarily that of the enzyme, is more than compensated for by savings in time and expense as a result of being able to perform the assay on crude biological extracts. Enzymatic reactions are usually performed between 25° and 35°C, near neutral pH, and within a few minutes thereby minimizing other changes in the compound during the assay. Side reactions are not a problem with enzyme assays when sufficiently pure enzymes are used.

Enzyme-based assays can be used to determine the level of enzymes in biological systems as a result of nutritional state and health, cultural practices, genetic manipulation, maturity, storage, and processing. They can also be used to determine the concentrations of compounds in biological systems based on the action of the compounds as substrates, activators, or inhibitors of specific enzyme systems. Furthermore they are premier analytical tools for determining the structure of complex molecules, such as proteins, nucleic acids, carbohydrates, and lipids.

While the first reports of the use of enzymes of analytical purposes date to 1846 (Osann 1846; use of peroxidase to determine H_2O_2), it has been only during the last 10–15 years that enzymes have become routine analytical tools. This is a result of the ready availability of enzymes in sufficient purity at reasonable cost and convenience (i.e., enzyme assay kits), suitable instrumentation, and methodology maximizing the sensitivity of the enzymes and sufficient knowledge of individual enzyme characteristics to ensure reproducibility within and between laboratories.

For extensive information on the principles and techniques of the analytical use of enzymes, the reader is referred to the four volume compendium on *Methods of Enzymatic Analysis* edited by Bergmeyer (1978) or *Handbook of Enzymatic Methods of Analysis* by Guilbault (1976). *Methods in Enzymology* (1955–1983; Academic Press), now up to some 100 volumes, provides information on the purification, properties and applications of enzymes. Whitaker (1974, 1983) has reviewed the principles and uses of enzymes for analytical purposes. Boehringer-Mannheim Corporation (Indianapolis, IN) and Worthington Biochemical Corporation (Freehold, NJ) in particular provide detailed information on the analytical uses of enzymes.

General Considerations

There are two basically different assay techniques that can be used with enzyme-based analytical assays. These are the total change method and the rate assay (kinetic) method.

Total Change Method. In the total change method, the reaction is allowed to go to completion and as a result of the total observed change (spectrophotometric, fluorometric, pH, or other) the concentration of a compound which serves as a substrate can be measured [Eq. (12)].

$$\text{Substrate} \xrightarrow{\text{Enzyme}} \text{Product}. \tag{12}$$

A typical result is shown in Fig. 9.13. From the total change, C_n, and comparison to a standard curve of response vs. concentration, the concentration of the compound can be determined.

An example is the determination of glucose in an extract of a biological material [Eqs. (13) and (14)].

$$\text{Glucose} + O_2 \xrightarrow{\text{Glucose oxidase}} \delta\text{-gluconolactone} + H_2O_2, \tag{13}$$

$$H_2O_2 + \text{Indicator compound} \xrightarrow{\text{Peroxidase}} \text{Product (colored or fluorescent)}. \tag{14}$$

The amount of colored or fluorescent product formed in the second reaction [Eq. (14)] gives a direct measure of the amount of glucose in the sample. This is a coupled enzyme assay, permitting greater sensitivity in determination of the compound. The H_2O_2 could be determined directly since it has ϵ_m of 100 $M^{-1}\text{cm}^{-1}$ at 234 nm. However,

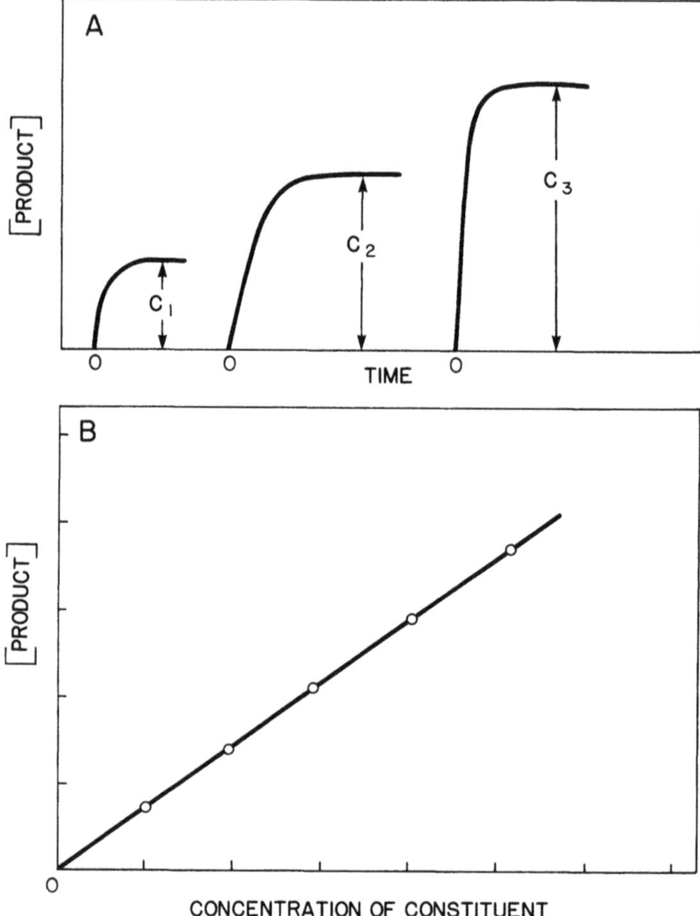

FIG. 9.13. Schematic illustration of the total change method for determination of concentration of compounds which serve as substrates of enzymes. (A) The enzyme concentration is high enough to convert all the substrate to product in a few minutes. The product concentration is measured spectrophotometrically, fluorometrically or electrometrically. (B) Relationship between concentration of product formed and concentration of constituent being measured.

the sensitivity is increased by a factor of 10^3–10^5 by use of the coupled assay.

Alternatively, the coupled assay using hexokinase and glucose-6-phosphate dehydrogenase can be used so the reaction can be determined on a continuous basis. The reactions involved are shown in Eqs. (15) and (16).

$$\text{Glucose} + \text{ATP} \xrightarrow{\text{Hexokinase}} \text{Glucose-6-phosphate} + \text{ADP}, \tag{15}$$

$$\text{Glucose-6-phosphate} + \text{NADP}^+ \xrightarrow[\text{dehydrogenase}]{\text{Glucose-6-phosphate}} \text{6-phosphoglucono-}\delta\text{-lactone} + \text{NADPH} + \text{H}^+. \tag{16}$$

The concentration of NADPH, determined spectrophotometrically (at 340 nm) or fluorimetrically (at 420 nm), is equal to the glucose concentration in the unknown.

The procedure can be easily modified to permit the determination of starch or sucrose present in a sample. For starch analysis, the sample is first treated with glucoamylase which converts all the starch to glucose which is then determined as described above. For sucrose analysis, the sample is first treated with invertase to hydrolyze all the sucrose to glucose and fructose. The glucose is then determined as described above.

Rate Assay Method. The total change method can only be used for determining the concentration (and amount) of a compound which serves as a substrate. The rate assay method must be used to determine the concentration of enzymes or of compounds which serve as inhibitors or activators for an enzyme system. It can also be used to determine compounds which serve as substrates, but the total change method is preferred for this case.

The principle of the rate assay method is shown in Fig. 9.14. Results are more easily interpreted when initial velocities v_0 are determined so that effects of stability of the enzyme during assay, changes in substrate concentration, product inhibition, or approach-to-equilibrium ($S \rightleftharpoons P$) in some cases are minimized. The initial velocity v_0 is determined from the experimental data by drawing a tangent to the initial progress curve as shown in Fig. 9.14. A plot of v_0 vs. the concentration of enzyme, substrate, activator, or inhibitor gives a predictable response which can be used to interpret the results from a sample containing an unknown concentration. The relationship between $[E_0]$ and v_0 is linear (Fig. 9.14B). For compounds which serve as substrate, the relationship will follow a hyperbolic relationship because of saturating effect (Fig. 9.15A). The results are best expressed in a linear relationship, by Eq. (17).

$$\frac{1}{v_0} = \frac{K_m}{V_{max}} \frac{1}{[S]_0} + \frac{1}{V_{max}}, \tag{17}$$

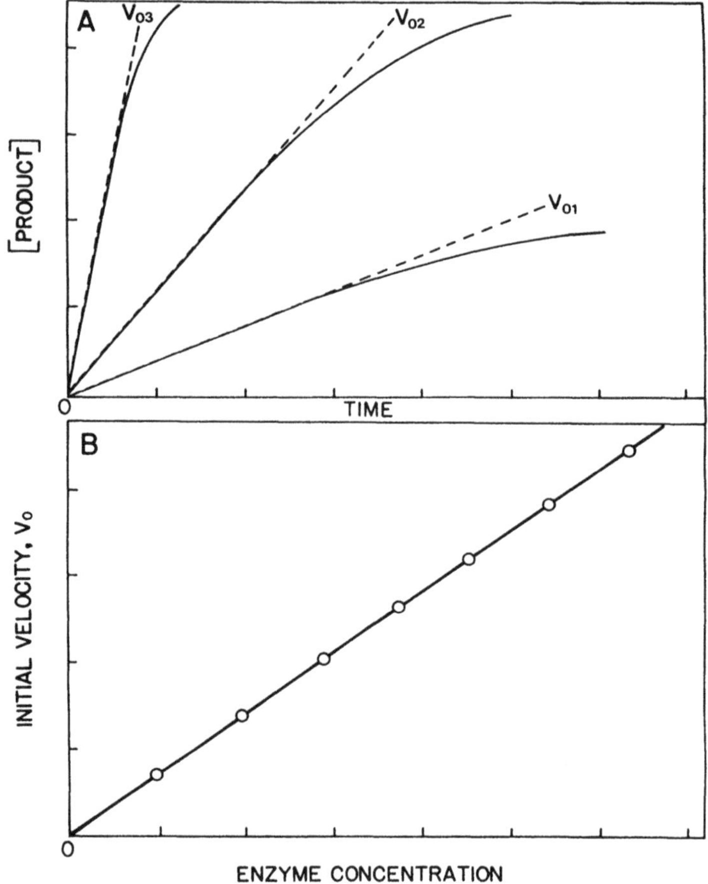

FIG. 9.14. Schematic illustration of use of the rate assay method to determine the concentration of enzyme in a sample. (A) The increase in product concentration as a function of time due to enzyme action is shown. The initial velocity, v_0, from the tangent drawn to the initial part of the curve, is determined. (B) Relationship between initial velocity v_0 and the enzyme concentration.

where V_{max} is the observed v_0 when $[S]_0 >> K_m$ and K_m is the Michaelis constant. A typical plot is shown in Fig. 9.15B.

With tight binding activators and inhibitors the relationship between v_0 and concentration is expected to be linear, while with loose binding activators and inhibitors ($K > 10^{-6}$ M) the relationship will be hyperbolic as for substrate and an analogous plot as shown in Eq. (17) is best used.

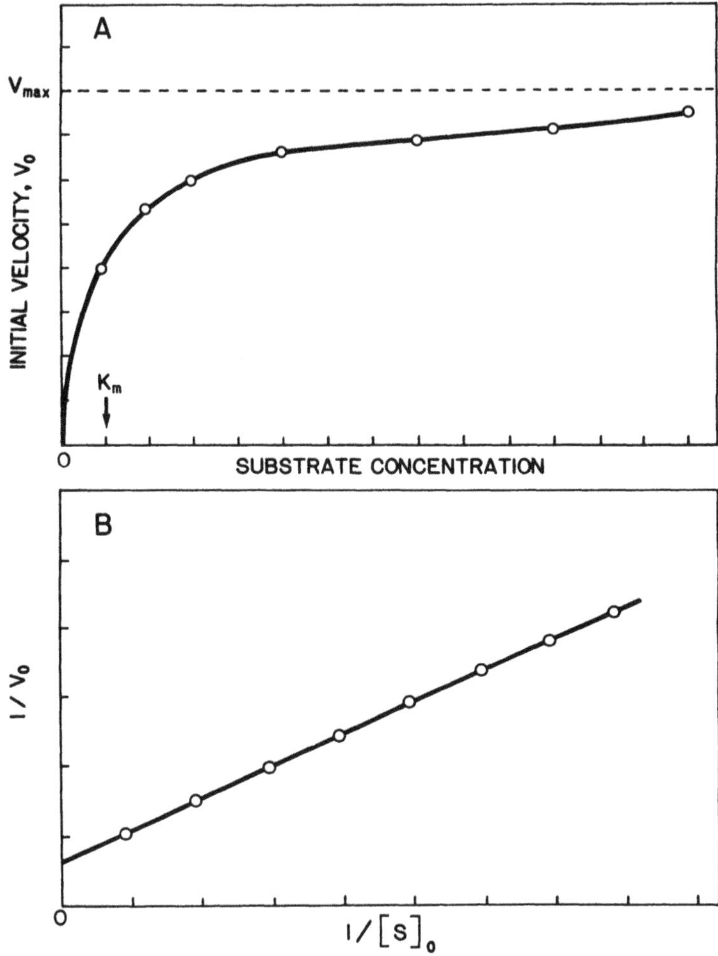

FIG. 9.15. Effect of substrate concentration on initial velocity v_0 for an enzyme-catalyzed reaction. (A) Hyperbolic relationship between substrate concentration and v_0, plotted relative to K_m for the reaction. (B) Plot of data as $1/v_0$ vs. $1/[S]_0$ according to Eq. (17) of text.

Measurement of Reaction Rates

The analytical use of enzymes requires suitable methods for following the conversion of substrate to product. The method selected is required to detect changes occurring as a result of the conversion of substrate to product. It is preferable to have a method that follows the conversion of substrate to product in a continuous fashion with as high a sensitivity as needed by the specific procedure. In this regard, spectrophotometric, fluorometric, polarimetric, electrometric, and thermal analysis methods are best. Examples of spectrophotometric and fluorometric methods are given above. These usually can follow reactions where the change in concentration is in the range of 10^{-5} M (spectrophotometric) to 10^{-8} M (fluorometric).

The polarimetric method, depending upon changes in optical rotation which are easy and convenient to measure, is usually less sensitive than the spectrophotometric and fluorometric assays and would require changes of 10^{-5}–10^{-4} M of the compound being measured.

The electrometric method for measuring ionic and molecular concentration changes is convenient and sensitive. A hydrogen ion electrode is used when changes in H^+ are to be monitored in the reaction. The sensitivity to change is a function of the pH of the reaction, being in the range of 10^{-4}–10^{-7} mM at pH 6–9. Other types of electrodes are commercially available; some actually incorporate an enzyme. Electrodes are available to specifically measure oxygen, ammonia, amino acids, glucose, phosphate, sulfate, nitrate, nitrite, etc. An example of an oxygen–enzyme electrode used to determine the concentration of glucose in a sample is shown in Fig. 9.16.

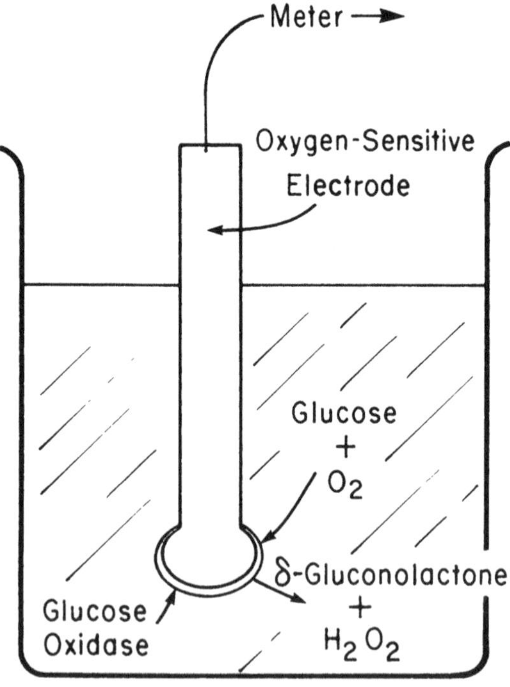

FIG. 9.16. Schematic diagram of an oxygen–electrode/glucose oxidase system for measuring glucose concentrations of solutions. The glucose oxidase is immobilized to the surface of the glass where changes in O_2 concentration are measured.

The most universal detector system for an enzyme-catalyzed reaction is the thermister. In the conversion of substrate to product, heat will be either released or used in all reactions. This change in heat content of the system can be detected with great sensitivity (<0.001 calories) provided the system can be isolated from its surroundings. The requirement for temperature control of 0.001°C or less puts a real limit on the sensitivity of the use of thermal detection systems unless a differential measuring system is used.

Recently, a competitive enzyme binding assay method has been proposed for quantitatively measuring compounds which are not turned over during an enzyme-catalyzed reaction (Myers *et al.* 1975; Arons *et al.* 1975). The binding constant needs to be large enough ($\geq 10^7\ M^{-1}$) so that the compound does not dissociate during the course of the assay.

The competitive enzyme binding assay method will be demonstrated with the assay of methotrexate, a folic acid analog with antimetabolic activity which is used in cancer chemotherapy (Erlichman *et al.* 1982).

The assay involves combining the enzyme tetrahydrofolate reductase with [^{3}H]methotrexate, with serial dilutions of the sample containing an unknown amount of unlabeled methotrexate and with NADPH. Equilibrium is reached immediately. The unbound [^{3}H]methotrexate and methotrexate are adsorbed to activated charcoal. An aliquot of the solution, containing [^{3}H]methotrexate and methotrexate adsorbed to the enzyme, is added to a vial containing Aquasol (New England Nuclear Corp.) and counted in a scintillation counter. The amount of bound [^{3}H]methotrexate will vary inversely with the amount of unlabeled methotrexate according to Eq. (18).

$$\text{Enzyme}\cdot[^3\text{H}]\text{Methotrexate} + \text{Methotrexate} \overset{K_b}{\rightleftharpoons} \text{Enzyme}\cdot\text{Methotrexate} + [^3\text{H}]\text{Methotrexate}. \quad (18)$$

The K_b for methotrexate and tetrahydrofolate reductase is $2.1 \times 10^8\ M^{-1}$. The type of data obtained, at a fixed concentration of [^{3}H]methotrexate and varying concentrations of methotrexate, is shown in Fig. 9.17. The sensitivity of the assay is 1.5 nM with a reproducibility of ~10%.

This method should receive increasing attention as it permits a sensitivity greater than that of most other techniques and requires only a few minutes to run. A similar technique has been used by Rothenberg *et al.* (1972) for determining the concentration of folates in samples. Assay kits are commercially available.

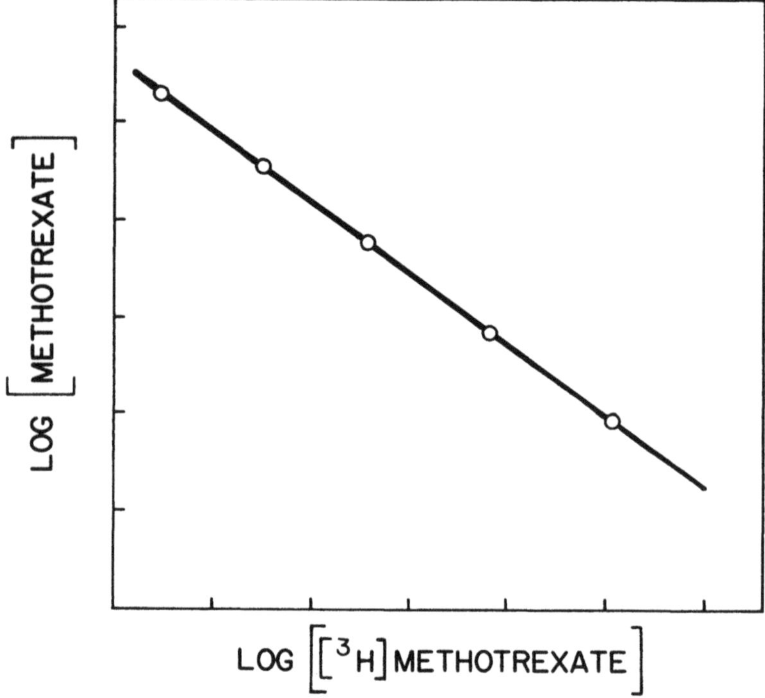

FIG. 9.17. Relationship between unbound methotrexate and [^{3}H]methotrexate concentrations. The two compounds compete for binding to the folate-binding site of the enzyme, tetrahydrofolate reductase.

Structural Analyses Using Enzymes

Because of the close fit required between the active site of an enzyme and its substrate, only one isomer generally serves as a substrate. This specificity can be used to distinguish between L- and D-isomers, α- and β-anomers and cis- and trans-isomers.

Enzymatic analyses are invaluable in distinguishing among the positional isomers of triglycerides after their separation by gas chromatography, by thin-layer chromatography, by high performance chromatography, and/or by gel permeation chromatography (Brockerhoff 1965; Litchfield 1972). Specific sequential enzymatic hydrolysis has been useful in determining the primary structure of pectins, gums, and other plant polysaccharides (Aspinall 1970), of the mucopolysaccharides from higher animals (Jeanloz 1970), of microbial polysaccharides (Troy and Koffler 1969; MacWilliam 1970; Phaff 1971), of amylopectin (Mercier 1973), and of glycogen (Larner *et al.* 1952).

The availability of highly purified specific nucleases and of efficient nucleotide separation methodology has made it relatively easy to elucidate the primary structures of RNA (~ 3000–4000 residues) as well as of DNA (10^4–10^6 residues). In general, the procedure involves limited treatment with a specific RNase (RNA) or specific DNase (DNA), separation of the large fragments by cellulose acetate electrophoresis, polyacrylamide gel electrophoresis, or homochromatography. Each separated fragment is then further digested by an enzyme with different specificity. This digestion, followed by separation, is repeated until the sequence is known. Identification of the fragments is helped by known effect of base composition and sequence on migration behavior. The complete base sequence of alanine transfer RNA, with 77 nucleotides (26,000 mol wt) was worked out in 1965 (Holley *et al.* 1965) using ribonuclease T_1 (specific for bonds involving guanidine) and ribonuclease A (specific for bonds involving pyrimidines). The same rationale was used for the successful elucidation of the primary sequence of DNAs (Wu 1978).

Use of enzymes for the cleavage of specific peptide bonds has made it relatively easy to determine the primary sequence of proteins. The classic work on insulin (Sanger 1956) provided the experimental strategy that has been used in the elucidation of the primary structure of many proteins and peptides (Dayhoff 1976). As shown in Fig. 9.18, the protein is first fragmented with an endoprotease with narrow specificity, such as trypsin, which hydrolyzes peptide bonds only at arginine and lysine residues. The large fragments are separated by electrophoresis or chromatography and each treated with a second endoprotease, such as chymotrypsin, with a specificity different from trypsin. The smaller fragments are again separated and the amino acid sequence of each is determined by use of the Edman degradation technique (often by use of a peptide sequenator) or by the use of exosplitting proteases such as carboxypeptidases A and B, cathepsin C, prolidase, or amino peptidase.

SUMMARY

Biologically based analytical techniques have the advantage over chemically and physically based techniques in that they routinely distinguish between or among the different isomeric forms of compounds and thereby more readily predict the effect of the compound in the

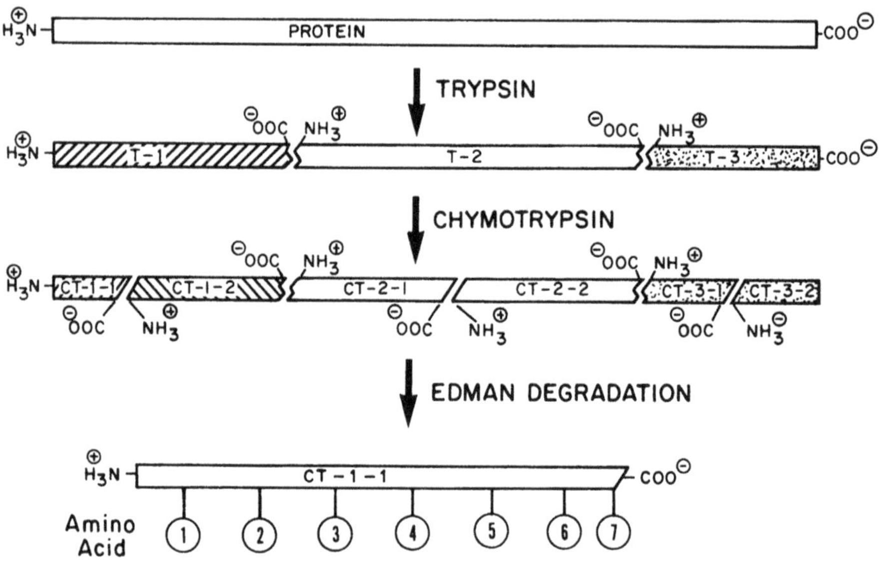

FIG. 9.18. Determination of the primary structure of a protein by use of the proteases, trypsin and chymotrypsin, and the Edman degradation method.

human. The ideal assay, from the human standpoint, is the testing of the compound on humans. However, initial testing on humans is confined to those experiments of discrimination in acceptance parameters of foods and beverages known to be nontoxic. Initial testing of any new compound must first involve animals to determine the possible acute and chronic toxicity or the proposed biological response in terms of nutrient or health enhancement. Such data require a long time to obtain, are expensive, and not always readily extrapolated to humans.

Scientists have looked for other methods that have a high degree of correlation and predictability with animal feeding experiments. Such methods involve the use of cells. Initially restricted to use of microorganisms, these tests can now be performed on animal cells, including human cells, from different organs and on plant cells. The use of normal human cells should greatly facilitate the extrapolation of results to humans. The disadvantages of such methods include lack of determination of effect of ingestion and digestion of the compounds and the effect on multicellular tissues, either of single-type or multitype cells.

Antibody-antigen reactions are highly specific and will become more so with the use of hybridoma cells to produce monoclonal antibodies. Major advances have been achieved in the analytical uses of the antibody antigen complexation reactions in medicine. These methods should certainly be extended into the area of food science to permit more rapid assay for the source of food ingredients, especially proteins, microbial and other toxins, vitamins, and products produced by food processing and modification methods.

Enzyme analyses are rapid, sensitive, and very specific. They can be used for analysis of compounds that serve as substrates, inhibitors, or activators of enzymes. Enzyme assays are also invaluable in determining changes in the enzyme content of food materials as a result of maturation, heat processing, or microbial contamination.

We encourage the food scientist, biochemist, and chemist to consider the advantages of enzymatic assays. Readily available enzyme kits and instrumentation make these analyses applicable to technician-level use.

ACKNOWLEDGMENT

The author is most appreciative of the help of Virginia DuBowy in checking references and typing the manuscript.

BIBLIOGRAPHY

ALLISON, J.B. 1955. Biological evaluation of proteins. Physiol. Rev. *35,* 664–700.

AMES, B.N., McCANN, J., and YAMASAKI, E. 1975. Methods for detecting carcinogens and mutagens with the *Salmonella*/mammalian-microsome mutagenicity test. Mutat. Res. *31,* 347–364.

ARONS, E., ROTHENBERG, S.P., DaCOSTA, M., FISCHER, C., and IQBAL, M.P. 1975. Direct ligand-binding radioassay for the measurement of methotrexate in tissues and biological fluids. Cancer Res. *35,* 2033–2038.

ASPINALL, G.O. 1970. Pectins, plant gums and other plant polysaccharides. *In* The Carbohydrates, Chemistry and Biochemistry, 2nd ed., Vol. IIB, pp. 515–536. W. Pigman, D. Horton, and A. Herp (Editors). Academic Press, New York.

AXELSEN, N.H. 1976. Analysis of human candida precipitins by quantitative immunoelectrophoresis. A model for analysis of complex microbial antigen–antibody systems. Scand. J. Immunol. *5,* 177–190.

BENDER, A.E., and DOELL, B.H. 1957. Biological evaluation of proteins: a new aspect. Brit. J. Nutr. *11,* 140–148.

BERGMEYER, H.U. (Editor). 1978. Methods of Enzymatic Analysis (4 vols.). Academic Press, New York.

BROCKERHOFF, H. 1965. A stereospecific analysis of triglycerides. J. Lipid Res. *6,* 10–15.

BURD, J.F. 1981. The homogeneous substrate-labeled fluorescent immunoassay. Methods Enzymol. *74,* 79–87.

CANN, J. 1975. Phenomenological theory of rocket immunoelectrophoresis. Biophys. Chem. *3,* 206–214.

CATT, K., and TREGEAR, G.W. 1967. Solid-phase radioimmunoassay in antibody-coated tubes. Science *158,* 1570–1572.

COLOWICK, S.P., and KAPLAN, N.O. (Editors-in-Chief). 1955–1983. Methods Enzymol. (1–100 vols.). Academic Press, New York.

CORRIE, J.E.T., and HUNTER, W.H. 1981. ^{125}Iodinated tracers for hapten-specific radioimmunoassays. Methods Enzymol. *73,* 79–112.

DANDLIKER, W.B., HSU, M.-L., LEVIN, J., and RAMANATH RAO, B. 1981. Equilibrium and kinetic inhibition assays based upon fluorescence polarization. Methods Enzymol. *74,* 3–28.

DAYHOFF, M.O. (Editor). 1976. Atlas of Protein Sequence and Structure, Vol. 5. Nat Biomedical Research Foundation, Silver Spring, MD.

DUNCAN, C.L., and SOMERS, E.B. 1972. Quantitation of *Clostridium perfringens* type A enterotoxin by electroimmunodiffusion. Appl. Microbiol. *24,* 801–804.

DUNCAN, C.L., and STRONG, D.H. 1969. Illeal loop fluid accumulation and production of diarrhea in rabbits by cell-free products of *Clostridium perfringens.* J. Bacteriol. *100,* 86–94.

ENGVALL, E., and PERLMANN, P. 1971. Enzyme-linked immunosorbent assay (ELISA). Quantitative assay of immunoglobulin G. Immunochemistry *8,* 871–874.

ERLICHMAN, C., DONEHOWER, R.C., and MYERS, C.E. 1982. Competitive protein binding assay of methotrexate. Methods Enzymol. *84,* 447–458.

GALFRÉ, G., and MILSTEIN, C. 1981. Preparation of monoclonal antibodies: strategies and procedures. Methods Enzymol. *73,* 3–46.

GOODMAN, H.S. 1961. Radiobiology. A general method for the quantitation of immune cytolysis. Nature (London) *190,* 269–270.

GRUENWEDEL, D.W. 1981. Effect of methylmercury(II) on the size of HeLa S3 carcinoma cells. Cell Pathol. *37,* 153–166.

GRUENWEDEL, D.W. 1984. Differential effects of sodium selenite and methylmercury(II) on membrane permeability and DNA replication in HeLa S3 carcinoma cells: A preliminary report regarding the modification of organomercurial toxicity by selenium compounds. Adv. Exp. Med. Biol. (in press).

GRUENWEDEL, D.W., and CRUIKSHANK, M.K. 1979A. Effect of methylmercury(II) on the synthesis of deoxyribonucleic acid, ribonculeic acid and protein in HeLa S3 cells. Biochem. Pharmacol. *28,* 651–655.

GRUENWEDEL, D.W., and CRUIKSHANK, M.K. 1979B. The influence of sodium selenite on the viability of intracellular synthetic activity (DNA, RNA, and protein synthesis) of HeLa S3 cells. Tox. Appl. Pharmacol. *50,* 1–7.

GRUENWEDEL, D.W., GLASER, J.F., and FALK, R.H. 1979. A scanning electron microscope study of the surface features of HeLa S3 suspension-culture cells treated with methylmercury(II). J. Ultrastructure Res. *68,* 296–307.

GRUENWEDEL, D.W., GLASER, J.F., and CRUIKSHANK, M.K. 1981. Binding of methylmercury(II) by HeLa S3 suspension-culture cells: intracellular methylmercury levels and their effect on DNA replication and protein synthesis. Chem.-Biol. Interaction *36,* 259–274.

GUESDON, J.-L., and AVRANLAS, S. 1981. Magnetic solid-phase enzyme immunoassay for the quantitation of antigens and antibodies: application to human immunoglobulin E. Methods Enzymol. *73,* 471–482.

GUILBAULT, G.G. 1976. Handbook of Enzymatic Methods of Analysis. Marcel Dekker, New York.

HIKUMA, M., KUBO, T., YASUDA, T., KARUBE, I., and SUZUKI, S. 1979A.

Amperometric determination of acetic acid with immobilized *Trichosporon brassicae*. Anal. Chim. Acta *109,* 33–38.

HIKUMA, M., KUBO, T., YASUDA, T., KARUBE, I., and SUZUKI, S. 1979B. Microbial electrode sensor for alcohols. Biotechnol. Bioeng. *21,* 1845–1853.

HOLLEY, R.M., APGAR, J., EVERETT, G.A., MADISON, J.T., MARQUISEE, M., MERRILL, S.H., PENSWICK, J.R., and ZAMIR, A. 1965. Structure of a ribonucleic acid. Science (Washington) *147,* 1462–1465.

JEANLOZ, R.W. 1970. Mucopolysaccharides of higher animals. *In* The Carbohydrates, Chemistry and Biochemistry, 2nd Edition, Vol. IIB, pp. 589–625. W. Pigman, D. Horton, and A. Herp (Editors). Academic Press, New York.

JERNE, N.K., and NORDIN, A.A. 1963. Plaque formation in agar by single antibody-producing cells. Science *140,* 405.

KARUBE, I. MATSUNAGA, T., TSURU, S., and SUZUKI, S. 1976. Continuous hydrogen production by immobilized whole cells of *Clostridium butyricum*. Biochim. Biophys. Acta *44,* 338–343.

KARUBE, I., MITSUDA, S., MATSUNAGA, T., and SUZUKI, S. 1977A. A rapid method for estimation of BOD by using immobilized microbial cells. J. Ferment. Technol. *55,* 243–248.

KARUBE, I., MATSUNAGA, T., MITSUDA, S., and SUZUKI, S. 1977B. Microbial electrode BOD sensors. Biotechnol. Bioeng. *19,* 1535–1547.

KARUBE, I., MATSUNAGA, T., TSURU, S., and SUZUKI, S. 1977C. Biochemical fuel cell utilizing immobilized cells of *Clostridium butyricum*. Biotechnol. Bioeng. *19,* 1727–1733.

LARNER, J., ILLINGWORTH, B., CORI, G.T., and CORI, C.F. 1952. Structure of glycogens and amylopections. II. Analysis by stepwise enzymatic degradation. J. Biol. Chem. *199,* 641–651.

LAURELL, C.-B. 1965. Antigen–antibody crossed electrophoresis. Anal. Biochem. *10,* 358–361.

LAURELL, C.-B. 1966. Quantitative estimation of proteins by electrophoresis in agarose gel containing antibodies. Anal. Biochem. *15,* 45–52.

LAURELL, C.-B., and McKAY, E.J. 1981. Electroimmunoassay. Methods Enzymol. *73,* 339–369.

LIM, F., and BUEHLER, R.J. 1981. Microencapsulation of antibody for use in radioimmunoassay. Methods Enzymol. *73,* 254–261.

LITCHFIELD, C. 1972. Analysis of Triglycerides. Academic Press, New York.

LITWACK, G., WILLIAMS, J.N., JR., CHEN, L., and ELVEHJEM, C.A. 1952. A study of the relationship of liver xanthine oxidase to quality of dietary protein. J. Nutr. *47,* 299–306.

LONGNECKER, J.B., and HAUSE, N.L. 1959. Relationship between plasma amino acids and composition of the ingested protein. Arch. Biochem. Biophys. *84,* 46–59.

LYNCH, S.C., GRUENWEDEL, D.W., and RUSSELL, G.F. 1983. On the mutagenic activity of a nitrosated early Maillard product: DNA-synthesis (DNA-repair) in HeLa S3 carcinoma cells induced by nitrosated 1-(*N*-L-tryptophan)-1-deoxy-D-fructose. Food Chem. Toxicol. *21,* 551–556.

MacSWEEN, J.M., and EASTWOOD, S.L. 1981. Recovery of antigen from Staphylococcal protein A-antibody adsorbents. Methods Enzymol. *73,* 459–471.

MacWILLIAM, I.C. 1970. The structure, synthesis and functions of the yeast cell wall—a review. J. Inst. Brew. *76,* 524–535.

MATHEIS, G., PENNER, M.H., FEENEY, R.E., and WHITAKER, J.R. 1983. Phosphorylation of casein and lysozyme by phoshorus oxychloride. J. Agric. Food Chem. *31,* 379–387.

MATSUMOTO, K., SEIJO, H., WATANABE, T., KARUBE, I. SATOH, I., and SUZUKI, S. 1979. Immobilized whole cell-based flow-type sensor for cephalosporins. Anal. Chim. Acta *105,* 429–432.

MATSUNAGA, T., KARUBE, I., and SUZUKI, S. 1978A. Rapid determination of nicotinic acid by immobilized *Lactobacillus arabinosus*. Anal. Chim. Acta *99,* 233–239.

MATSUNAGA, T., KARUBE, I., and SUZUKI, S. 1978B. Electrochemical microbioassay of vitamin B_1. Anal. Chim. Acta *98,* 25–30.

McCLANE, B.A., and McDONEL, J.L. 1980. Characterization of membrane permeability alterations induced in Vero cells by *Clostridium perfringens* enterotoxin. Biochim. Biophys. Acta *600,* 974–985.

MEANS, G.E., and FEENEY, R.E. 1968. Reductive alkylation of amino groups in proteins. Biochemistry *7,* 2192–2201.

MERCIER, C. 1973. The fine structure of corn starches of various amylose-percentage: normal and amylomaize. Starke *25,* 78–83.

MOLINARO, G.A., and DRAY S. 1974. Antibody coated erythrocytes as a manifold probe for antigens. Nature (London) *248,* 515–517.

MYERS, C.E., LIPPMAN, M.E., ELIOT, H.M., and CHABNER, B.A. 1975. Competitive protein binding assay for methotrexate. Proc. Natl. Acad. Sci. USA *72,* 3683–3686.

NARGESSI, R.D., and LANDON, J. 1981. Indirect quenching fluoroimmunoassay. Methods Enzymol. *74,* 60–79.

NEURATH, A.R. 1981. Use of ^{125}I-labeled anti-2,4-dinitrophenyl (DNP) antibodies as a general tracer in solid-phase radioimmunoassays. Methods Enzymol. *73,* 127–138.

NOTERMANS, S., HAGENAARS, A.M., and KOZAKI, S. 1982. The enzyme-linked immunosorbent assay (ELISA) for the detection and determination of *Clostridium botulinum* toxins A, B, and E. Methods Enzymol. *84,* 223–238.

OSANN, G. 1846. Guajakharz als Reagenz auf electrische Strome. Poggendorf's Ann. *67,* 372–374.

OSBORNE, T.B., MENDEL, L.B., and FERRY, E.L. 1919. Method of expressing numerically the growth-promoting value of proteins. J. Biol. Chem. *37,* 223–229.

PEARSON, T., GALFRÉ, G., ZIEGLER, A., and MILSTEIN, C. 1977. A myeloma hybrid producing antibody specific for an allotypic determination on "IgD-like" molecules of the mouse. Eur. J. Immunol. *7,* 684–690.

PHAFF, H.J. 1971. Structure and biosynthesis of the yeast call envelope. *In* The Yeasts, Vol. 2, pp. 135–210. H.H. Rose and J.S. Harrison (Editors). Academic Press, New York.

POKRIC, B., and PUCAR, Z. 1981. The two-cross immunodiffusion technique for determining diffusion coefficients and precipitating titers of antigen and antibody. Methods Enzymol. *73,* 306–319.

PUIGSERVER, A.J., GAERTNER, H.F., SEN, L.C., FEENEY, R.E., and WHITAKER, J.R. 1982. Covalent attachment of essential amino acids to proteins by chemical methods: Nutritional and functional significance. Adv. Chem. Ser. *198,* 149–167.

ROBTI, R.C., KRISHAN, A., and PFAFFENBERGER, C.D. 1982. Cytokinetic and cytogenetic effects of some agricultural chemicals in human lymphoid cells *in vitro:* organophosphates. Mutat. Res. *102,* 89–102.

ROTHENBERG, S.P., DaCOSTA, M., and ROSENBERG, Z. 1972. A radioassay for serum folate: use of a two-phase sequential-incubation, ligand-binding assay. New Eng. J. Med. *286,* 1335–1339.

SANGER, F. 1956. The structure of insulin. *In* Currents in Biochemical Research, pp. 434–459. D.E. Green (Editor). Wiley (Interscience), New York.

SKJELKVÅLE, R., TOLLESHAUG, H., and JARMUND, T. 1980. Binding of enterotoxin from *Clostridium perfringens* type A to liver cells *in vivo* and *in vitro*. Acta Path. Microbiol. Scand. Sect. B *88,* 95–102.

STARK, R.L., and DUNCAN, C.L. 1972. Transient increase in capillary permeability induced by *Clostridium perfringens* type A enterotoxin. Infect. Immun. *5,* 147–150.

SUZUKI, S., and KARUBE, I. 1981. Bioelectrochemical sensors based on immobilized enzymes, whole cells, and proteins. *In* Applied Biochem. Bioeng. 3. Analytical Applications of Immobilized Enzymes and Cells, pp. 145–174. L.B. Wingard, Jr., E. Katchalski-Katzir, and L. Goldstein (Editors). Academic Press, New York.

TACK, B.F., and WILDER, R.L. 1981. Tritiation of proteins to high specific activity: application to radioimmunoassay. Methods Enzymol. *73,* 138–147.

TREVAN, J.W. 1927. The error of determination of toxicity. Proc. Royal Soc. Lond. (Biol.) *101,* 483–514.

TROY, F.A., and KOFFLER, H. 1969. The chemistry and molecular architecture of the cell walls of Penicillium. J. Biol. Chem. *244,* 5563–5576.

ULLMAN, E.F., and KHANNA, P.L. 1981. Fluorescence excitation transfer immunoassay (FETI). Methods Enzymol. *74,* 28–60.

VAERMAN, J.-P. 1981. Single radial immunodiffusion. Methods Enzymol. *73,* 291–305.

WHITAKER, J.R. 1974. Analytical applications of enzymes. *In* Food Related Enzymes, pp. 31–78. J.R. Whitaker (Editor). Adv. Chem. Ser. *136,* 31–78.

WHITAKER, J.R. 1984. Analytical uses of enzymes. *In* Food Analysis: Principles and Techniques. D.W. Gruenwedel and J.R. Whitaker (Editors). Marcel Dekker, New York (in press).

WIDE, L. 1969. Radioimmunoassays employing immunosorbents. Acta Endocrinol. Suppl. *142,* 207–221.

WIDE, L., AXEN, R., and PORATH, J. 1967. Radioimmunosorbent assay for proteins. Chemical couplings of antibodies to insoluble dextran. Immunochemistry *4,* 381–386.

WU, R. 1978. DNA sequence analysis. Annu. Rev. Biochem. *47,* 607–634.

YALOW, R.S., and BERSON, S.A. 1960. Immunoassay of endogenous plasma insulin in man. J. Clin. Invest. *39,* 1157–1175.

YELTON, D.E., and SCHARFF, M.D. 1981. Monoclonal antibodies: a powerful new tool in biology and medicine. Annu. Rev. Biochem. *50,* 657–680.

10

Microbiological Assays

J..M. Jay[1]

The use of microorganisms to determine or assess a variety of biologically active substances either qualitatively or quantitatively is well established. Among the earliest applications of microbiological assays were those for vitamins and later for amino acids. During the early 1940s, microorganisms were employed to determine the presence and quantity of many antimicrobial agents, and more recently, to assess the mutagenicity of chemicals. Although other methods have largely supplanted microbiological assays for amino acids and to some extent antibiotics, the microbiological methods continue to be of value, for unlike chemical and physical methods, microbiological assays enable one to assess the biological activity of the agent or compound in question.

Detailed methods and procedures for conducting all assays covered in this chapter are available in one or more of several standard reference works, and the reader who wishes to carry out a microbiological assay is referred to one or more of the following.

1. AOAC Methods (13th ed., 1980). Official methods for foods and feeds are covered for amino acids, vitamins, and antibiotics.
2. United States Pharmacopeia (USP, 19th ed., 1975). The USP covers the microbiological assay of antibiotics and calcium pantothenate.
3. Analytical Microbiology, Vol. II (1972). This volume, as well as Volume I, covers the theory of antibiotic inhibition zones, plate assays, and detailed procedures for antibiotics, amino acids, vitamins, and some other compounds.

[1]Department of Biological Sciences, Wayne State University, Detroit, MI 48202

4. ASM Manual of Methods (Guirard and Snell 1981). Chapter 7 (on biochemical factors in growth of microorganisms) provides excellent background material on the growth requirements of many microorganisms including those used in the assay of B vitamins and amino acids.

5. Difco Supplemental Literature (1972). The composition and handling of assay media for amino acids and vitamins are provided in addition to the use of each for the respective assay.

6. ATCC Catalogue of Strains (13th ed., 1978). All catalogued strains used for the assay of vitamins, amino acids, antibiotics, and the like are listed in the appendix in addition to a list of some strains that may be used in mutagenesis assays. Literature references to many of the strains are listed in the main body.

7. Handbook of Microbiology, III (1973). Media, cultures, and methods of assay for B vitamins and selected antibiotics are covered.

AMINO ACIDS

Conventional microbiological assays for amino acids may be carried out in three ways: (1) by plate diffusion, (2) by turbidimetry, and (3) by acidimetry. Regardless of the method chosen, the overall basis is similar for each. An assay medium which contains all essential growth factors except the amino acid in question is selected. When increasing quantities of the amino acid are added to the medium followed by inoculation with an appropriate assay microorganism, growth or acid production occurs in a linear fashion over a given concentration of the test amino acid. Upon the construction of a standard curve by use of known quantities of the amino acid, one can determine the quantity of that amino acid in an unknown sample by reading from the standard curve. According to Guirard and Snall (1981), all 20 of the amino acids found in proteins can be assayed using the basal medium in Table 10.1 with *Pediococcus acidilactici* (ATCC 8042). The medium composition is presented here to illustrate the general growth requirements of this assay organism which is fairly typical of the lactic acid bacteria in general. For actual assay use, one should obtain fresh lots of the respective assay medium from a reputable manufacturer. Some other organisms that may be used to assay amino acids are presented in Table 10.2.

It may be noted that the amino acid requirement of some lactobacilli such as *L. casei* is influenced by the presence and quantity of pyridoxal or pyridoxamine while *Streptococcus faecium* (ATCC 9790) is

TABLE 10.1. COMPOSITION OF BASAL MEDIUM WHICH CAN BE USED TO ASSAY FOR EACH OF THE 20 AMINO ACIDS WITH *PEDIOCOCCUS ACIDILACTICI* (ATCC 8042)[a]

Ingredients	Quantity/L	Ingredients	Quantity/L
Glucose	25 g	Thiamin–HCl	0.5 mg
Sodium acetate	20 g	Pyridoxine–HCl	1.0 mg
NH_4Cl	3.0 g	Pyridoxal–HCl	0.3 mg
K_2HPO_4	0.6 g	Pyridoxamine–HCl	0.3 mg
KH_2PO_4	0.6 g	Ca–pantothenate	0.5 mg
NaCl	10 mg	Riboflavin	0.5 mg
$MnCl_2 \cdot 4H_2O$	20 mg	Nicotinic acid	1.0 mg
$MgSO_4 \cdot 7H_2O$	200 mg	*p*-Aminobenzoic acid	0.1 mg
$FeSO_4 \cdot 7H_2O$	10 mg	Biotin	0.001 mg
		Folic acid	0.01 mg
Amino acids (omit the one being assayed)			
DL-α-alanine[b]	200 mg	DL-leucine	250 mg
L-arginine–HCl	242 mg	L-lysine–HCl	250 mg
L-asparagine	400 mg	DL-methionine	100 mg
L-aspartic acid	100 mg	DL-phenylalanine	100 mg
L-cysteine	50 mg	L-proline	100 mg
L-glutamic acid	300 mg	DL-serine	50 mg
Glycine	100 mg	DL-threonine	200 mg
L-histidine–HCl	62 mg	DL-tryptophan	40 mg
DL-isoleucine	250 mg	L-tyrosine	100 mg
		DL-valine	250 mg

[a] Modified from Guirard and Snell (1981).
[b] To assay for alanine with this medium, folinic acid must be added.

TABLE 10.2. SOME MICROORGANISMS USED IN THE ASSAY OF AMINO ACIDS

Organisms	ATCC no.	Amino acids
Lactobacillus casei subsp. *rhamnosus*	7469	Arginine, glutamic acid
L. casei	9595	Aspartic acid, glutamic acid, serine
L. delbruckii	9649	Alanine
L. fermentum	9338	Alanine, histidine
L. leichmannii	7830	Cystine
L. plantarum	8014	Cystine, isoleucine, leucine, methionine, phenylalanine, tryptophan, valine, others
Neurospora crassa	12949, 14183	Arginine
Pediococcus acidilactici	8042	All 20 found in proteins
P. acidilactici	8081	Alanine, arginine, cystine, methionine, threonine, tyrosine
Proteus vulgaris	13315	Leucine, valine
Streptococcus faecium	8043	Arginine, histidine, lysine, isoleucine, leucine, methionine, threonine, tryptophan, tyrosine, valine
S. faecium	9790	Arginine, histidine, isoleucine, leucine, threonine, tryptophan, valine
Tetrahymena furgasoni (*T. pyriformis* W)	10542	Arginine, histidine, lysine, methionine

not. These vitamers substitute for lysine and threonine in some lactobacilli. Further, ornithine and citrulline can substitute for arginine in some *Neurospora* mutants while they cannot replace arginine for *S. faecium* 9790 (Stokes *et al.* 1945). A large number of amino acid antagonists is known for most amino acids and the presence of these must be avoided [see *Handbook of Microbiology* (1973) for list].

Turbidimetric Methods

When assaying amino acids by this as well as by other methods, the choice of assay organism is critical and different species and strains may be recommended for the same amino acid in different standard procedures. One should obtain the desired strain directly from the American Type Culture Collection (ATCC) in lyophilized form. Prior to use, the culture should be rehydrated and carried on the appropriately recommended medium such as micro assay or micro inoculum agar slants. Stock cultures should be maintained on slants rather than in broth and transferred at least monthly. The use of broth for stock maintenance may lead to changes in the culture relative to its sensitivity to the amino acid in question. For inoculum use, an overnight broth culture of the assay strain should be prepared in the appropriate broth followed by centrifugation and resuspension in physiologic saline and centrifugation for two to three times. The density of inoculum should be such that one drop or 0.1 mL when added to 10 mL of assay medium yields 10^4–10^5 cells/mL. For best results, the inoculum should be a logarithmic phase culture. Turbidity is normally read at 37°C between 540 and 660 nm after 16–20 hr incubation.

Although some references differ on the temperature of incubation for the same assays and strains, the incubator employed should be one that does not allow temperature variation of more than ± 0.5°C. A notable example of how temperature of incubation can affect an assay was presented by Braekkan (1960) who found that when pantothenic acid was assayed by turbidity employing *Lactobacillus plantarum* (ATCC 8014), a lag occurred consistently in the growth response curve with incubation at 37°C but not when incubation was at 30°C.

It is important that duplicate or even triplicate tubes be used to construct the standard curve and that a standard curve is run each time an unknown sample is assayed. This is especially true when samples from different products or sources are assayed. Only L-amino acids should be used since in general these are the biologically active forms. Only the straight-line portion of the reference curve may be used for determining the quantity of unknown amino acids. Nonlinear curves may be linearized by use of certain statistical devices (see below under Improved Methods). The standard curve may be plotted by use of semilog paper with amino acid concentrations on the log scale and percent transmission (relative degree of turbidity) on the arithmetic scale.

The care and maintenance of glassware and utensils for amino acids are critical. It is obvious that trace quantities of the test substances must not be in or on tubes or pipettes. While acid cleaning of glassware may be employed, it is very important that all traces of chromate or acid are removed by extensive rinsing in hot tap water followed by at least three rinses in distilled water. Deionized water should not be substituted for glass distilled water. As a further means of assuring cleanliness, the heating of cleaned and dry glassware in a dry-air oven at 180°C for 3 hr will pyrolyze any remaining organic matter. When sufficient organic matter is present, the glassware will appear darkened by this heat treatment in which case it should be rejected. When photometric tubes are inoculated for subsequent reading without transfer, it is important that the caps or covers not reflect light in such a way that the readings are affected.

Acidimetric Methods

By these methods, the quantity of acid produced by the inoculum is measured rather than turbidity. The same general precautions noted above must be observed. While with turbidimetry an incubation period of 16–20 hr is suitable, up to 72 hr are generally allowed for maximum acid production. The quantity of acid produced is measured by titrating with a weak base such as 0.02 *N* NaOH to a neutral endpoint. This may be achieved manually with burette or by use of an autotitrator. The standard curve is plotted in the same general way as noted above for the turbidimetric method. When properly done, results by acidimetry and by turbidimetry are comparable.

Large Plate Assays

Large glass plates such as flat-bottom baking dishes or sheets of glass 25.4 cm × 30.5 cm or even larger are employed. With the latter, metal borders may be clamped or sealed on so as to make an enclosed area for agar. In either case, the amount of agar necessary to give a layer of 0.64–0.95 cm is determined unless a thin-agar plate is desired (see below). Following sterilization, the covered plate or dish is placed on a flat surface. It is imperative that the surface be entirely flat and a leveling device should be used. The predetermined quantity of the appropriate assay medium is sterilized and allowed to cool to about 45°C. The inoculum, previously prepared and washed with saline as indicated above, is added to the cooled agar so as to give a concentration of 10^6–10^7 cells/mL and then is poured into the large plate. Upon hardening, the medium is ready for the amino acid standard and unknown samples. For this purpose, round holes of approximately 0.64 cm diameter may be made to about one-half the agar depth, or stainless steel cylinders (8-mm o.d., 6-mm i.d., 10-mm height) may be placed on the surface of the inoculated medium at least 5 cm apart. In either case, the standard curve is prepared by adding to the wells or cylinders increasing quantities of the amino acid standard in duplicate along with duplicates of the unknown sample. Following incubation at 37°C for up to 24 hr (3–4 hr may be sufficient), the diameter of zones of growth around the wells or cylinders is carefully measured by the use of vernier calipers or a Fisher–Lily zone reader.

The standard curve is constructed by averaging the duplicate readings and plotting diameter of zones of growth against the quantity of amino acid in the standard as noted above for turbidimetry. The quantity of amino acid in unknown samples is read from the standard by use of the averaged duplicate readings. An excellent review of large plate assay procedures for amino acids has been presented by Bolinder (1972).

Application of Amino Acid Assays

Although amino acid assays of hydrolyzed proteins are more likely to be done manually by chromatography or by automated analysis, microbiological procedures continue to be used where one is interested in the presence and quantity of one or more amino acids or in the presence of the 10 essential amino acids. This is especially true in the evaluation of protein quality by microbiological assay where either lysine, lysine and methionine, or the 10 essential amino acids are assayed.

According to Ford (1962), *S. faecalis* subsp. *zymogenes* (ATCC 23655, NCDO 592) has an absolute requirement for the following eight amino acids: arginine, histidine, isoleucine, leucine, methionine, glutamic acid, tryptophan, and valine. Ford then used this organism to assess the nutritional value of proteins by its ability to measure the available amino acids noted. The values for protein quality determined by *S. faecalis* were closely correlated with the rat assay for some proteins but correlation was poor for others. This study and others provided the basis for a collaborative study by 10 European researchers on the assessment of protein quality by use of amino acid assay by two microorganisms. In this study reported by Boyne *et al.* (1967), a provisional method was proposed to evaluate protein quality by an assay for available lysine and methionine in various foods by use of the *S. faecalis* strain noted. The assay is made following papain digestion of proteins, addition of hydrolysate to a medium containing all growth factors except amino acids, and relative growth measured by either turbidimetry or acidimetry. Although not given provisional status, a method for lysine and methionine employing *Tetrahymena pyriformis* W (ATCC 10542) was evaluated by the same collaborators. The methods employed for this organism were essentially those of Stott *et al.* (1963) where intact proteins were employed and protozoa enumerated by direct microscopic count (DMC). In a later study, Buchanan (1969) employed the *S. faecalis* strain above along with another *S. faecalis* strain to measure the nutritive value of leaf-protein preparations by determining available methionine, tryptophan, leucine, arginine, and isoleucine; and *T. pyriformis* W to determine available lysine. Unlike the previous authors, Buchanan found that the microbiological results correlated poorly with those from animal assays.

For additional applications of amino acid assays to the determination of protein quality, see the section below on Protein Quality.

B VITAMINS

While microbiological assays for amino acids are being rapidly supplanted by nonmicrobiological assay methods, conventional determination of vitamins by microbiological assays continues to be the method of choice for some vitamins in particular. A summary of some of the microorganisms employed in the assay of vitamins is presented in Ta-

TABLE 10.3. SOME MICROORGANISMS USED IN THE ASSAY OF B VITAMINS

Organisms	ATCC no.	Vitamins
Bacillus coagulans	12245	Folic acid
Brettanomyces bruxellensis	9775	Pyridoxine
Candida pseudotropicalis	2512	Nicotinic acid
Escherichia coli	12651	Pyridoxine
E. coli	10799, 14169	B_{12}
Euglena gracilis	12716	B_{12}
Gluconobacter oxydans subsp. *suboxydans*	621	*p*-Aminobenzoic acid, nicotinic acid, pantothenic acid
G. oxydans subsp. *suboxydans*	621H	Panthenol
Kloeckera apiculata	9774	Inositol
K. apiculata	18212	Inositol, thiamin
Lactobacillus casei	7469	Biotin, folic acid, folinic acid, prefolic A, nicotinic acid, pantothenic acid, pyridoxal, riboflavin
L. casei	7469a	Folic acid
L. delbruckii	9649	Pyridoxal
L. fermentum	9338	Thiamin, pyrithiamine
L. fructosus	13162	Niacinamide
L. helveticus	12046	Pantethine
L. lactis	8000, 10697	B_{12}
L. plantarum	8014	Biotin, nicotinic acid (and analogs), *p*-aminobenzoic acid, pantothenic acid, niacin
L. leichmannii	7830, 4797	B_{12}
L. viridescens	12706	Thiamin
Micrococcus luteus	11880	Biotin
Neurospora crassa	9683	Inositol
N. crassa	9278	*p*-Aminobenzoic acid
N. sitophila	9276	Pyridoxine, pyridoxal, pyridoxamine, biotin
Ochromonas danica	30004	Biotin, thiamin
O. malhamensis	11532	B_{12}
Pediococcus acidilactici	8081	Folinic acid
P. acidilactici	8042	Panthenol (dexpanthenol)
Rhodotorula lactosa	9536	*p*-Aminobenzoic acid, thiamin
Saccharomyces uvarum	9080	Inositol, pantothenic acid, pyridoxine, pyridoxal, pyridoxamine
Streptococcus faecium	8043	Folic acid, folinic acid, pyridoxal, pyridoxamine
Tetrahymena pyriformis	30008	B_6 group

ble 10.3, and a more extensive list of strains for all B vitamins as well as other aspects pertaining to vitamin assays was presented by Snell (1948). It is very important that standard or established procedures be carefully followed when assaying vitamins by microbiological methods.

The same general rules and precautions noted above for amino acids apply to the assay of B vitamins. The smallest quantity of a vitamin can affect growth response so that the cleaning and handling of glassware are of critical importance. The choice of assay organism is critical especially for those vitamins that exist naturally in several forms. In these instances, it is imperative that the standard curve be constructed by use of the form of vitamin of interest and not necessarily the synthetic or parent vitamin. The appropriate standard reference listed above should be consulted for folate assay procedures but one of the most popular assay methods is the "aseptic addition" method of Herbert (1966).

Some vitamins exist primarily in free form along with their coenzymes in natural systems (e.g., biotin, inositol, *p*-aminobenzoic acid, riboflavin) while others exist in a variety of different biologically active forms (Table 10.4). Folic acid (folacin) is known to exist in many active forms (Stokstad and Thenen 1972; Stokstad *et al.* 1977) and only three of these are noted in Table 10.4. These three forms can be assayed separately by use of three organisms as noted in Table 10.5. *Lactobacillus casei* (ATCC 7469) responds to each form noted and to mono-, di-, and triglutamates in general. When this organism is used to assay for folic acid, the determined values tend to be higher than for other strains that may be used. *Pediococcus acidilactici* (ATCC 8081) responds only to leucovorin or nonmethylated, reduced forms with three or less glutamate moieties (Stokstad and Koch 1967), while *Streptococcus faecalis* (ATCC 8043) responds to both folic acid and leucovorin. Folate assays are discussed further below.

The vitamin B_6 group consists of pyridoxine, pyridoxal, and pyridoxamine. Each can be converted to its respective 5′-phosphate (phosphorylated) and these three forms along with the three parent compounds are interconvertible in animals as well as in many microorganisms. The phosphorylated forms are generally less active than the parent compounds. Pyridoxal and pyridoxamine are found more often in animal tissue while pyridoxine exists mainly in plants (Baker and Frank 1968).

The dietary source of pantothenic acid is panthenol. Dexpanthenol is D-panthenol, the biologically active form. Pantotheine is an intermediate in the pathway of coenzyme A synthesis in mammalian liver and also in some microorganisms. Pantethine is a fragment of coenzyme A. Pantothenol, pantoyl lactone, and coenzyme A are all inactive for *T. pyriformis* and *Lactobacillus plantarum* (Baker and Frank 1968).

TABLE 10.4. SOME OF THE BIOLOGICALLY ACTIVE FORMS OF THREE B VITAMINS

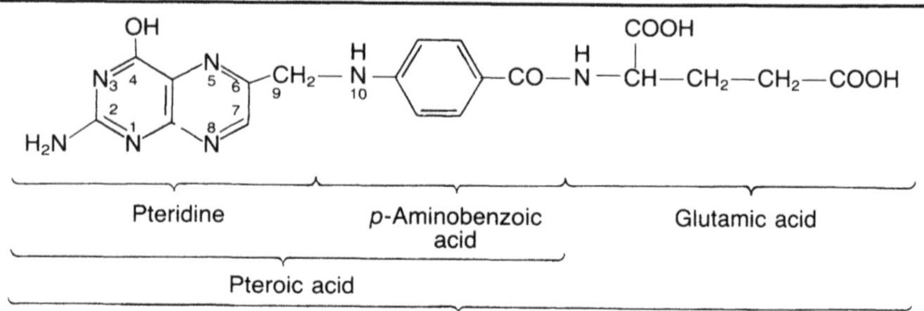

Folic acid (folacin, pteroylglutamic acid)

5-Formyltetrahydropteroylglutamic acid (leucoverin; citrovorum factor). N-5 = -HCO; N-10 = -H

5-Methyltetrahydropteroylglutamic acid (prefolic A). N-5 = -CH_3; N-10 = -H

N, CH_3, $HOCH_2$, OH, CHO

Pyridoxal

HO, O, $POCH_2$, HO, N, CH_3, OH, CHO

Pyridoxal 5-phosphate

N, CH_3, $HOCH_2$, OH, CH_2NH_2, 2HCl

Pyridoxamine · 2HCl

H_3C, N, HCl, HO, CH_2OH, CH_2OH

Pyridoxine

COOH—CH_2—CH_2—NH—C=O—HCOH—H_3CCCH_3—CH_2OH

Pantothenic acid

CM_2OH—CH_2—CH_2—NH—C=O—HCOH—H_3CCCH_3—CH_2OH

Panthenol

$HOCH_2C(CH_3)_2$—$C(OH)(H)CONHCH_2CH_2CONHCH_2CH_2SH$

Pantethine

TABLE 10.5. RESPONSE OF THREE ASSAY ORGANISMS TO THREE FORMS OF FOLIC ACID[a]

Vitamer form	*S. faecalis* ATCC 8043	*L. casei* ATCC 7469	*P. acidilactici* ATCC 8081
Folic acid	+	+	−
5-Formyltetrahydrofolic acid	+	+	+
5-Methyltetrahydrofolic acid	−	+	−

[a] Stokstad and Thenen (1972).
+ = growth; − = no growth.

Among the organisms in Table 10.3 that may be used in vitamin assays are bacteria, fungi, and protozoa. Because they are animals, the protozoa are often the group of choice for certain vitamins, notably B_{12}. Either of the following four organisms may be used for B_{12} assay: *E. coli* (ATCC 10799), *Lactobacillus leichmannii, Euglena gracilis,* and *Ochromonas malhamensis.* All four respond to biologically active B_{12} as well as to Factor I; for the clinically inactive pseudovitamin B_{12}, all but *O. malhamensis* respond. In addition, while *L. leichmannii* responds to deoxyribosides and intact DNA, the protozoan does not (Baker and Frank 1968). According to Baker and Frank, *O. malhamensis* has a B_{12} sensitivity of 1 pg/mL and its requirement for this vitamin is similar to that of man. In a similar manner, the requirement of *O. danica* for thiamin is similar to that for man and this protozoan does not respond to thiazole or pyrimidine moieties as do yeasts (Baker and Frank 1968). This organism responds to 100 pg thiamin/mL.

The protozoan *T. pyriformis* has an assay range for nicotinic acid of 1–300 ng/mL while the range for *L. plantarum* is 3–30 ng/mL. Its response to nicotinic acid is much like that of animals with no response to nicotinuric acid as does *L. plantarum* (Baker and Frank 1968). The acid and amide forms of the vitamin are utilized almost equally by this protozoan. Riboflavin and its coenzymes flavin mononucleotide (FMN) and flavin adeninedinucleotide (FAD) are all biologically active and *T. pyriformis* responds to them equally (Baker and Frank 1968).

In regards to biotin, biocytin is effective for *O. danica* while desthiobiotin competitively inhibits it.

Extraction from Foods

Critical to the assay of B vitamins in foods is their extraction from macromolecular constituents, especially proteins. Two general procedures are employed for this purpose in addition to the use of appropriate solvents. First, heating of samples is effective in releasing some vitamins such as thiamin which can be extracted by autoclaving. Avidin of egg white forms a firm complex with biotin such that the vitamin is unavailable for biological activity. Heating causes avidin to lose its biotin-combining powers and renders the vitamin free. Nicotinic acid can be extracted by autoclaving food specimen for about 30 min. The B_6 complex can be extracted by autoclaving in 0.4–0.5N HCl for 1½–to 2 hr. Gregory (1980) used a procedure similar to this in extracting B_6 from cereals. The folates bind to proteins and autoclaving for 60 min may effect deproteinization of these vitamers.

The treatment of natural products with various enzyme preparations is the other procedure employed in the extraction of vitamins from foods. The enzyme preparation Clarase may be used on samples for B_6, pantothenic acid, and riboflavin. This enzyme preparation is a mixture of amylases, proteases, lipases, and others. Takadiastase, a multispecific enzymic preparation from *Aspergillus oryzae* that has amylolytic enzymes such as diastase, is used in the extraction of thiamin and pantothenic acid. When samples are treated with enzymes, an incubation period of 24–72 hr at 37°C is usually involved. Following incubation, the enzymes are destroyed, typically by autoclaving. Samples for extraction may range from 5–25 mg of freeze-dried materials to about 2 g of dried products.

The most widely used enzyme treatment for vitamin extraction is the application of deconjugases to natural products in the assay of folic acid. Although heating can effect deproteinization of the folates as noted above, the deconjugase treatment is applied specifically to render the complex folates to their simpler forms to which the assay organisms respond (see below). Folates may disassociate from proteins and other macromolecules at pH $< 4.0–5.0$.

The Folates

As noted above, folic acid exists in multiple forms. Folic acid per se may be as little as 5% of total folates in some foods (Graham *et al.* 1980). In natural foods the pteroylpolyglutamates apparently dominate and these are the forms for which *L. casei* is least responsive (Tamura *et al.* 1972A). The number of glutamate residues is known to be as high as 7/pteroic acid moiety, and the distribution of the various folate forms in foods and animal specimens has been reviewed by Stokstad *et al.* (1977).

The accurate assay of food folates is made difficult by the existence of these vitamers in so many different forms, and the use of deconjugases is employed to render the polyglutamate forms to mono- and diglutamates. When polyglutamates form, they do so by the addition of glutamic acid residues to the gamma carboxyl group of glutamate. The most widely used conjugase preparation (Bacto-chicken pancreas) converts the polyglutamates to diglutamates as the predominant forms (Leichter *et al.* 1977). Hog kidney and raw cabbage conjugases hydrolyze polyglutamates to monoglutamates (Tamura *et al.* 1972B). In a recent study, Phillips and Wright (1983) found that hog kidney deconjugase produced consistently higher folate values than chicken pancreas deconjugase. Mammalian liver and kidneys from other animals are other proven sources of deconjugases (Stokstad and Koch 1967; McMartin *et al.* 1981). Polyglutamates undergo hydrolysis in aged meats, presumably because of bacterial action or action of meat lysosomal enzymes, and deconjugases are present in the gastrointestinal tract of man and presumably many other animals.

In regards to the simpler folates, *L. casei* responds about equally to folic acid and to the 5-formyl form but its response to the 5-methyl form is reduced to about one-half (Phillips and Wright 1982). The latter form of folate is often the most important in some foods. Phillips and Wright (1982) found that the *L. casei* assay for the 5-methyl form could be improved by lowering the initial pH of assay medium from 6.8 to 6.2 which gave results for this form comparable to those for folic acid.

Also, "positive drift" of response curves was absent when the vitamers were assayed at pH 6.2 (Phillips and Wright 1983).

The polyglutamates gave varied responses in the radioassay method (see below) while with deconjugase treatment prior to assay the microbiological assay employing *L. casei* was found to be much the better method for determining polyglutamates in natural materials (Shane *et al.* 1980). The current requirement of the HPLC assay for monoglutamates necessitates the conjugase treatment of natural products prior to assay by this method (Gregory *et al.* 1982).

The deconjugase treatment of samples allows one to determine free and total folates in foods. Free folate is that determined by extraction methods that do not employ deconjugase; while following deconjugase treatment the concentrations found represent total folate. Results obtained by Hoppner (1971) on 40 strained baby foods revealed that total folate levels exceeded free folate by a factor ~2 while in 30 defrosted frozen dinners, the mean free folate was only about one-half of total folate (Hoppner *et al.* 1973).

Of further importance in the assay of folates is the use of a reducing agent to maintain the folates in a reduced state. The compound most commonly used is ascorbic acid at levels ranging from 0.1 to 1.0% with the latter being employed by a large number of investigators including Graham *et al.* (1980) and Klein and Kuo (1981). In using 0.025%–0.1% ascorbic acid, Phillips and Wright (1982) found no differences in standard curves. Ascorbate is used in the extraction solvents and the use concentration should be maintained throughout the assay procedure by adding it to the diluent in use.

The details pertaining to the microbiological assay of folates have been outlined by Herbert (1966), by Cooperman (1967), and by AOAC (1980).

The B_6 Group

The standard dosage curve for the microbiological assay of the three free forms of the B_6 group is 1–10 ng. The three forms all have equal biological activity in man. Following a study by nine collaborators, Toepfer and Polansky (1970) recommended that the three free B_6 vitamers in food extracts be separated by use of ion exchange chromatography followed by assay with *Saccharomyces uvarum* (ATCC 9080, formerly *S. carlsbergensis*). In a more recent study by Guilarte *et al.* (1980), *Kloeckera apiculata* (ATCC 9774, formerly *K. brevis*) responded equally to all three forms over the range 2–10 ng equivalent of pyridoxine while *S. uvarum* did not. On the other hand, the differences in growth response by the two organisms diminished when the vitamer concentration range was increased to 10–50 ng.

The conventional plate assay for amino acids described above is run usually with a relatively thick layer of agar—up to 0.95 cm. Itagaki and Tsukahara (1972) assayed the B_6 vitamers with thin-layered plates (0.75-mm thickness) and found the results to be comparable to those obtained by turbidimetry using *S. uvarum*. According to these authors, the thin plate allows for faster migration of smaller molecules thus increasing the sensitivity of large plate assays.

IMPROVED MICROBIOLOGICAL ASSAY TECHNIQUES

The conventional assay procedures described above have been improved relative to accuracy, reliability, and speed by the introduction of the following methods and procedures: the use of lyophilized cultures, automation, the use of computer programs to construct standard curves, and the use of radiometric and impedance procedures to assess growth responses.

The use of lyophilized cultures for direct inoculation of assay media has been used with success for *L. plantarum* (ATCC 8014) by Gorin *et al.* (1970) and for *L. casei* (ATCC 7469) by Chen *et al.* (1978). Similarly, *S. uvarum* which was frozen in liquid nitrogen was successfully used in a pyridoxine assay by Tsuji (1966A). This investigator allowed a 1-hr period of growth-phase conditioning followed by a 6-hr incubation for the assay. The same general procedure was shown to be effective for a pantothenic acid assay using *S. uvarum* (Tsuji, 1966B). This procedure obviates the necessity to grow and prepare cultures prior to each assay run and is a time-saver for a day-to-day operation. However, freeze drying is known to effect metabolic injury in microorganisms such that their response immediately after rehydration is generally lessened when compared to fresh cultures. Maximal freeze-dry damage is seen when cells are cultured on media that contain inhibitors of varying types and whether or not an unnatural response by lactobacilli in an assay medium containing minimal quantities of growth factors occurs is not clear.

In regards to the introduction of procedures into the conventional assay methodology such that time of assay is shortened, Tennant and Withey (1972) developed an auto diluter as a means of standardizing pipetting inconsistencies that would normally exist between different technicians using pipettes manually. The device increased work output with an increased accuracy of results. A semiautomated method was developed by Berg and Behagel (1972) and was shown to be usable for a number of vitamins. This method consisted of a separate automated system for preparation of cultures, and another for turbidity measurement. Turbidimetric readings are printed and the calculations computerized. Assays done on different days by use of the device gave variation coefficients of less than 10%.

Conventional assays often lead to nonlinear response curves by turbidimetry and a solution to this problem was offered by Tsuji *et al.*

(1967), Schatzki and Keagy (1975), and Voigt *et al.* (1979A), as well as by others. The former authors employed a Fortran computer program which incorporated a given equation to convert nonlinear curves to linear. Schatzki and Keagy (1975) presented a computational method for treating response curves which are sigmoidal in ln concentrations. When applied to a folacin assay using *L. casei* turbidimetry, a response curve over a 250-fold concentration range was produced. A more complicated computer-assisted assay procedure was introduced by Voigt *et al.* (1979A). This procedure involves a four-part computer software package which includes a portion that determines when microbial growth has stabilized, and one package which evaluates the effects of inhibitors on the assay. More accurate data were produced by the computer-assisted assay than by manual methods for the same vitamins.

Radiometry

The most widely evaluated modification of the conventional microbiological assay procedures involves the use of radiometry. Two separate and distinct radiometric procedures have been introduced; one of which is microbiological radiometric assay (NR), was first developed for determining serum folate. It is based on the competitive binding to certain proteins of ^{3}H-folate and nontritiated folate. The corresponding amount of sample folate bound by the experimental procedure is determined by measuring the radioactivity of the tritiated compound with a scintillation counter. This method has been compared to the conventional microbiological assay for folates as well as to the radiometric-microbiological (RM) method and some of the many studies reported are summarized in Table 10.6. By this method, milk binding proteins were found by Rothenberg *et al.* (1972) to be sensitive to ca. 10–15 pg of the 5-methyl form of folic acid, about two times more sensitive than the *L. casei* assay method. Commercial binding kits are available for this procedure and when two such kits were compared to the conventional *L. casei* assay for serum folate by Waddell *et al.* (1976), the results were statistically comparable.

The radiometric–microbiological (RM) assay is a conventional assay procedure in which a ^{14}C-labeled metabolite is added to the assay medium so that when the assay organism utilizes the labeled metabolite, $^{14}CO_2$ is released and measured by use of a radioactivity counter. The radio-label is required to be on a carbon atom that is released as CO_2 by respiring and growing culture. The growth response by this method

TABLE 10.6. RELATIVE EFFECTIVENESS OF CONVENTIONAL MICROBIOLOGICAL, RADIOMETRIC-MICROBIOLOGICAL, NONMICROBIOLOGICAL RADIOMETRIC, AND HPLC ASSAYS IN DETERMINING FOLATE AND B_6 IN FOODS AND ANIMAL SPECIMEN

Vitamin assayed	Source	MA-7469[a]	MA-9080[b]	RM[c]	NR[d]	HPLC[e]	Synopsis of findings	References
Folic acid	3 foods	×			×	×	General agreement with cereal; wide variation with three other foods. NR not good for food use.	Gregory *et al.* (1982)
Folic acid	Spinach	×			×		NR gave higher values but MA more suitable for foods.	Klein and Kuo (1981)
Folic acid	Frozen foods	×			×		Both methods gave comparable results.	Graham *et al.* (1980)
Folic acid, 5-methyl form	Pure	×			×		NR faster, simpler, and more accurate for 5-methyl form than MA. Both methods suitable when each compound is used to construct standard curve.	Ruddick *et al.* (1978)
Folic acid		×			×		NR not suitable for mixtures of folates normally seen in biological extracts. *L. casei* better than 4 NR kits.	Shane *et al.* 1980)
Folic acid	Rat specimen	×			×		Comparable results by the two methods. NR was good for polyglutamate forms.	Tigner and Roe (1979)
B_6	7 cereals			×		×	Similar results for 5 of 7 foods. RM was excellent overall	Guilarte *et al.* (1981)
B_6	5 cereals		×		×		The 2 methods correlated well but HPLC was more satisfactory. Some extracts inhibited inoculum in RM.	Gregory (1980)

[a] Conventional microbiological assay using *L. casei* ATCC 7469.
[b] Conventional microbiological assay using *S. uvarum* ATCC 9080.
[c] Radiometric-microbiological assay (see text).
[d] Nonmicrobiological radiometric assay (see text).
[e] High performance liquid chromatography.

is assayed by CO_2 release rather than cell mass or acid production. Among radio-labeled substrates employed are D(1-^{14}C)gluconate (Chen *et al.* 1978), ^{14}C-glucose (Voigt and Eitenmiller 1979), and ^{14}C-valine (Guilarte *et al.* 1981). The RM procedure has been compared to other assays for folates and B_6 and some of these studies are summarized in Table 10.6. Of the two radiometric methods, the nonmicrobiological (NR) is faster with results for human sera obtainable in about 4 hr (Waxman *et al.* 1971). Also, when the appropriate folate-binding proteins are employed, the NR method apparently responds to a wider range of folate forms than the MR procedure (Tigner and Roe 1979). Most work to date with NR has been on clinical specimen. Results obtained by this method for foods vary among the investigators (see Table 10.6). A comparison of the relative sensitivity of three assay methods for folates is presented in Table 10.7. A further comparison by Voigt and Eitenmiller (1979) of conventional microbiological assays and RM for thiamin, pyridoxine, and pantothenic acid led them to conclude that RM was not significantly faster than turbidimetry when bacteria and yeasts were used for thiamin, but RM reduced the assay time from 4–5 days to 1 day when a protozoan (*Ochromonas danica,* ATCC 30004) was used as assay organism.

TABLE 10.7. COMPARISON OF ASSAY METHODS FOR THE DETERMINATION OF FOLACIN IN SELECTED FOODS AND OTHER BIOLOGICAL MATERIALS[a]

	Folacin concentration (μg/g)		
Sample	Radioassay	Microbiological assay	HPLC
Rat liver[b]			
0-hr autolysis	20.4	6.7	—
3-hr autolysis	20.7	19.9	—
5-hr autolysis	18.5	19.5	—
Spinach[c]			
Frozen, uncooked	4.0	2.5	—
Fresh, raw	4.2	1.6	—
Cooked	5.0	1.8	—
Brussels sprouts[d]	1.05	1.05	—
Collard greens[d]	1.08	0.93	—
Meatloaf[d]	20.1	17.9	—
Cabbage, raw[e]	1.34	0.59	2.24
Oat cereal, fortified[e]	7.83	9.30	6.19
Infant formula, fortified[e]	0.06	0.19	0.18

[a] From Gregory (1983).
[b] From Tigner and Roe (1979).
[c] From Klein and Kuo (1981).
[d] From Graham *et al.* (1980.
[e] From Gregory *et al.* (1982).

The use of impedance measurement with a Bactometer Microbial Monitoring System in the assay of folic acid with *S. faecalis* has been studied by at least one investigator (Crain 1980).

Special Assay Strains

Other improvements in conventional microbiological assays involve the use of antibiotic-resistant assay strains of bacteria which permits the detection of a vitamin in foods or specimen that may contain antibiotics. One of the most commonly used of these is a chloramphenicol-resistant strain of *L. casei* for assaying folates. In a similar vein, the choice of an assay strain may be dictated by its resistance/sensitivity to food preservatives and salts. Voigt *et al.* (1979B) found that protozoa *(Tetrahymena pyriformis, Ochromonas danica,* and *O. malhamensis),* and the yeast *S. uvarum* were adversely affected by the food preservatives sodium sorbate, sodium propionate, and sodium benzoate while in general the bacterial strains employed were not. In addition, *T. pyriformis* was affected by citrate and all protozoa were adversely affected by neutralization salts in the assay of seven different vitamins.

EVALUATION OF PROTEIN QUALITY

The official U.S. and Canadian method for assaying protein quality is the protein efficiency ratio (PER) employing rats (AOAC 1980). The method is defined by the equation,

$$\text{PER} = \frac{\text{Weight gain of rats (g)}}{\text{Protein consumed (g)}}.$$

In effect it is the slope of the line relating weight gain to protein intake (Pellett 1978), or the weight gain of a rat/g of protein consumed under specified conditions. Relative nutritive value (RNV) for microorganisms is defined by the equation,

$$\text{RNV} = \frac{\text{Sample organism count}}{\text{ANRC casein count}} \times 100.$$

Thus, it is a dose-response assay showing response vs. protein intake.

Although the rat PER is the official method, it is time consuming, relatively expensive, and requires the care and maintenance of ani-

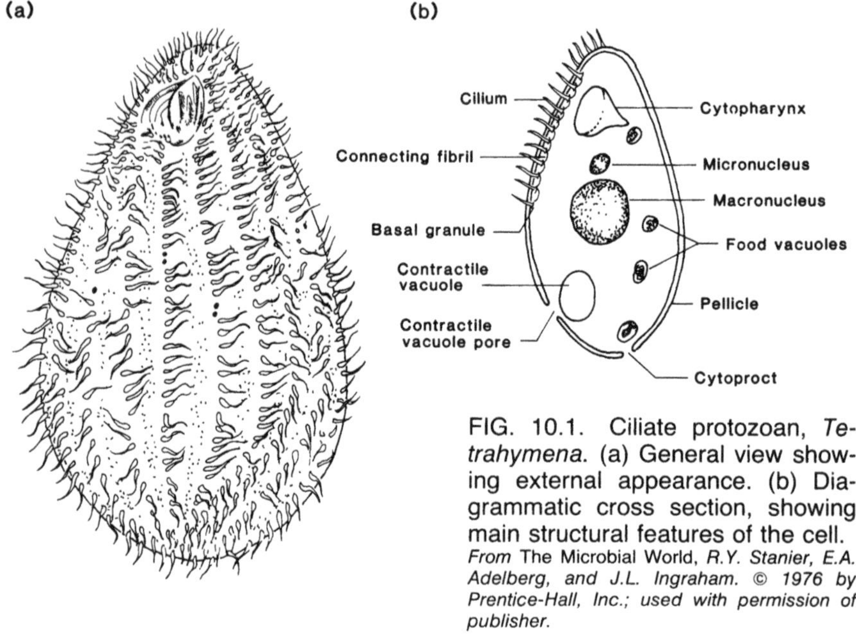

FIG. 10.1. Ciliate protozoan, *Tetrahymena*. (a) General view showing external appearance. (b) Diagrammatic cross section, showing main structural features of the cell. *From* The Microbial World, *R.Y. Stanier, E.A. Adelberg, and J.L. Ingraham. © 1976 by Prentice-Hall, Inc.; used with permission of publisher.*

mals. A more rapid and less expensive method is RNV by microorganisms. Relative nutritive value data can be converted to PER values for a more direct comparison between the two methods. These and other protein evaluation methods have been discussed by Hegsted (1977), Pellett (1978), and Harper (1981).

RNV by *Tetrahymena*

Tetrahymena pyriformis W (ATCC 10542, also *T. furgasoni;* formerly *T. geleii)* is a ciliate protozoan that is widely distributed in fresh waters. This particular strain was isolated from a pond in Massachusetts in 1940 (Kidder and Dewey 1945). It has a pear-shaped body about 50-μm long and is covered by hundreds of cilia arranged in longitudinal rows (see Fig. 10.1). Its metabolism is respiratory, requires most B vitamins except B_{12} and thiamin, is unusual in its requirement for thioctic acid, and has an absolute requirement for the 10 essential amino acids (Kidder and Dewey 1945; Frank *et al.* 1975). It is highly resistant to irradiation (Elliott 1959). As a phagotroph (holotroph), it possesses strong proteolytic properties and can thus utilize intact proteins. It is, of course, an eucaryote unlike the bacteria which are procaryotes, and its nutritional requirements may be expected to be closer to man than bacteria.

The use of this organism to assess the quality of proteins was first studied by Dunn and Rockland (1947), but the greatest interest in its use for this purpose dates back to the mid 1970s. The assay procedure most widely used is that of Stott *et al.* (1963) and Stott and Smith (1966), as modified by Landers (1975). In general, washed log phase cells, grown for 3 days in the dark in a support medium, are used as inoculum. Defatted test proteins may be employed intact or following hydrolysis by proteases such as bromelin, papain, pepsin, pancreatin, or combinations of these. Test samples containing 0.1–0.5 mg N/mL are added to *Tetrahymena* assay medium (Difco or others) and incubated in the dark at 25°C for 4 days. The substitution of dextrin for glucose in the assay medium results in better growth of the organism (Evans *et al.* 1979A). Cells are killed by use of formalin in a preservation fluid, and then enumerated in a haemocytometer. Averaged numbers of organisms per milliliter are plotted against concentration of nitrogen per milliliter in sample proteins. Casein is generally used as a reference with cell counts on this protein being assigned an arbitrary value of 100 (Pilcher and Williams 1954; Smith and Pena 1977). Landers (1975) determined the protozoan count at 0.3 mg N/mL from a plot, or a linear regression calculation employing the RNV equation above and expressed both counts per milliliter at 0.3 mg N/mL of incubation medium. By use of the RNV equation, a test protein of identical quality to the reference casein would produce a value of 100 while the value for a poorer quality protein would be less than 100.

In their original investigation of *Tetrahymena pyriformis* W (TpW) to assess protein quality, Dunn and Rockland (1947) compared lactalbumin, gelatin, and casein and found agreement by TpW with values obtained from rat, dog, and chick assays. They used acidimetry to assess TpW growth. Pilcher and Williams (1954) compared TpW with three animal assays for protein quality of three intact proteins and found that the values produced by the protozoan were consistent enough to differentiate between good and poor growth-promoting proteins (Table 10.8). In this study, TpW ranked the proteins in the same order as the rat PER. Fernell and Rosen (1956) and Rosen and Fernell (1956) compared TpW RNV of 19 foods with rat assay data and found agreement between the two methods. These investigators studied some of the parameters of growth of the organism and found that aeration of culture produced the highest cell yields. They used intact proteins at a concentration of 1 mg N/mL and found that 60–80% hydrolysis was achieved by TpW in 4 days (Rosen and Fernell 1956). They also

TABLE 10.8. COMPARISON OF *TETRAHYMENA* AND ANIMAL ASSAYS OF PROTEIN QUALITY[a]

	Tetrahymena	Animal PERs[b]		
Proteins	Casein = 100%	Rat	Mouse	Dog
Defatted whole egg	120	111	97	108
Peanut flour meal	52	51	64	58
Wheat gluten meal	11	16	36	39

[a] Pilcher and Williams (1954).
[b] Cited by Pilcher and Williams from a Rutgers Univ. Rep. 1946–1950.

noted that egg albumin was inhibitory to the assay organism and that the inhibition could be relieved by autoclaving egg white. Cell yields of 2×10^6/mL were achieved with 0.3 mg N/mL on intact proteins and 10^6/mL on amino acid media. Up to 4×10^6 organisms/mL could be achieved with 6.0 mg N/mL (Fernell and Rosen 1956). A series of protein concentrates was tested by Teunisson (1961) with TpW who found a linear response for "vitamin-free" casein up to 1.0 mg N/mL which corresponded to 10^6 organisms/mL.

The correlation between averaged TpW values and PER values corrected to the PER of casein at 2.5 are presented in Table 10.9 from data by Landers (1975). The correlation coefficient *(r)* between RNV and PER was 0.9494. According to this author, the errors in the PER and TpW assay values can be ±0.2 PER units and ±6 RNV units. One of the most extensive studies is that of Evancho *et al.* (1977) who compared PER by TpW and rat assays on 34 food samples consisting of 15 heat processed canned foods and 19 frozen foods. The calculated PER values (from RNV) are presented in Table 10.10 where r between RNV and PER was 0.90 ($p < 0.01$). According to these authors, the calculated TpW PER of 27 of the 34 samples was within 0.2 PER of rat assay while six of the seven that disagreed by more than 0.2

TABLE 10.9. AVERAGE RNV AND PER VALUES FROM EIGHT FOOD SAMPLES[a]

Sample no.	Avg RNV	PER[a]
1	20	−0.55
2	51	0.65
3	64	1.3
4	92	2.0
5	80	2.5
6	116	3.0
7	102	3.5
8	113	4.3

[a] Modified from Landers (1975).

TABLE 10.10. COMPARISON OF TpW RNV, CALCULATED TpW PER, AND RAT PER FOR 34 FOODS[a]

Sample no.	Corrected rat PER	TpW RNV	Calculated TpW PER	Sample no.	Corrected rat PER	TpW RNV	Calculated TpW PER
1	2.9	113	2.8	18	2.4	95	2.4
2	2.9	115	2.8	19	2.3	106	2.6
3	2.9	106	2.6	20	2.3	76	2.0
4	2.9	101	2.5	21	2.3	98	2.4
5	2.9	110	2.7	22	2.2	97	2.4
6	2.9	121	2.9	23	2.2	77	2.0
7	2.8	119	2.9	24	2.1	94	2.3
8	2.8	111	2.7	25	2.1	77	2.0
9	2.8	129	3.1	26	2.0	84	2.1
10	2.8	111	2.7	27	2.0	85	2.2
11	2.8	107	2.6	28	1.9	67	1.8
12	2.8	105	2.6	29	1.8	73	1.9
13	2.8	110	2.7	30	1.7	77	2.0
14	2.6	120	2.9	31	1.7	63	1.7
15	2.6	107	2.6	32	1.6	84	2.1
16	2.6	105	2.6	33	1.3	57	1.5
17	2.6	102	2.5	34	1.2	39	1.1

[a] Excerpted from Evancho *et al.* (1977).

PER were within 0.4 rat PER. These investigators concluded that the TpW assay has an accuracy comparable to the rat assay. Organisms were enumerated by both direct microscopic count (DMC) and by an electronic Coulter counter. In their study of 38 samples of meats and meat products by TpW, Evans *et al.* (1977) found a closer association with total essential amino acids and essential amino acids than with rat PER.

A large number of other investigators have employed TpW to assess protein quality and the general findings have been favorable. Evans *et al.* (1979B) used TpW to assess cooking methods on the RNV of meats; Srinivas *et al.* (1975) successfully assessed protein quality of irradiated foods by TpW; and Smith and Pena (1977) assessed protein quality of leaf protein concentrates. The latter authors noted that some plant constituents had adverse effects on the protozoan. The TpW assay was compared with a chemical method (fluorodinitrobenzene, FDNB) on the loss of protein quality by nonenzymatic browning by Warren and Labuza (1977) who found considerable variation in TpW results. The TpW and the rat assay were compared for determining protein quality in ground beef containing soy and/or single cell protein by Butrum and Kramer (1977), and their findings showed that between rat PER and TpW RNV, $r = 0.98$. These investigators concluded that TpW was essentially the same as the rat assay for the foods studied. McCurdy *et al.* (1978) successfully employed TpW to evaluate the quality of lentil proteins, and more recently, Li-Chan and Nakai (1981) used the protozoan assay to measure RNV of lysine-enriched wheat gluten. Like several investigators before them, they found that the size of protozoan cells on test proteins was one-half to one-third those grown on reference casein. Evancho *et al.* (1977) noted that cell size increased with an increase in protein/cell.

While most workers have employed *T. pyriformis* W, Baker *et al.* (1978) used *T. thermophila* (ATCC 30008) to assess the protein quality of 46 proteins and foods. The assay by this organism was compared with 28-day rat assays and excellent agreement was found. Protein samples were predigested with bromelain, and according to these authors, *T. thermophila* usefully classified the foods/proteins as poor, acceptable, or good.

In essentially all of the studies cited above on the use of *Tetrahymena* species to evaluate protein quality, the assay was found to be reliable on a wide variety of foods and proteins. The most significant study that raises questions about the use of TpW is that of Sternberg and Kim (1979) who found that TpW responded to lysinoalanine essentially as it would to lysine while mice did not. Lysinoalanine has been shown to be formed in some alkali-treated proteins through an addition reaction of dehydroalanine and lysine (Nashef *et al.* 1977). It causes nephrotoxicity in mice and rats but not in other experimental animals. The TpW and mouse PERs for wheat gluten + lysine were 1.62 and 0.85, while for wheat gluten + lysinoalanine they were 1.19 and 0.31, respectively, in the study by Sternberg and Kim. These investigators found that the L-lysine dependent mutants *E. coli* J5, *B. subtilis* 168, and *A. niger* 1794.38 grew on media where L-lysine was replaced with lysinoalanine. These findings suggest that when protein quality is determined by TpW, the possibility exists that a protein will be rated acceptable relative to lysine if lysinoalanine is present. Even if the above findings are confirmed by others, the use of TpW as a rapid and inexpensive assay for protein quality should not be ruled out. It may mean, however, that TpW could not be used on proteins processed in such a way that products such as lysinoalanine may be present.

The RNV of Proteins by Other Microorganisma

Noting that *S. faecalis* requires all 10 essential amino acids for growth, Halevy and Grossowicz (1953) examined this organism for its ability to assess protein quality. The intact proteins were hydrolyzed by pancreatin and the hydrolysates added to a medium devoid of the

TABLE 10.11. COMPARISON OF GROWTH RESPONSE OF *S. FAECALIS* AND ITS RELATIVE ACTIVITY ON FIVE PROTEIN HYDROLYSATES TO RAT PER VALUES[a]

Test hydrolysates	Relative activity, casein = 100%	Rat PER values, casein = 100%
Egg albumin	105	119
Casein	100	100
Gelatin	29	(25)[b]
Gluten	15	16
Zein	2	—[c]

[a] Modified from Halevy and Grossowicz (1953).
[b] Literature value for rat PER (Hegsted and Chang 1965).
[c] No data.

10 essential amino acids. Following inoculation with *S. faecalis,* the mixtures were incubated at 37°C for 20–24 hr, and growth assessed by turbidimetry. The growth response was related to that for casein as 100%. The relative *S. faecalis* activity compared to rat growth data (PER) is presented in Table 10.11 where it can be seen that the *S. faecalis* values compared well with the PER values for the hydrolysates evaluated by rat PER. Although the PER values for gelatin and gluten in the literature are 25 and 40, respectively, Hegsted and Chang (1965) have noted that the relative growth index for gluten is only 20, thus putting the *S. faecalis* values of Halevy and Grossowicz for these two in close agreement. The findings for wheat gluten in this study with *S. faecalis* are similar to those of Pilcher and Williams (1954) for TpW (see Table 10.8). By supplementing the assay medium with only nine of the essential amino acids at a time, the missing or limiting amino acid could be identified. Why more studies have not been carried out with *S. faecalis* in the assay of protein quality is not clear.

Bunyan and Woodham (1964) employed *S. faecalis* subsp. *zymogenes* to determine RNV of three fish meal preparations and compared the results to those obtained by rat and TpW assay. The TpW and the bacterial assays ranked the three products in the same order. The poorest quality preparation was so ranked by all three assays but the rat assay differed with the microbial assays on the two better products.

Other investigators have employed bacteria to assess protein quality including Ford (1962), Boyne *et al.* (1967), and Buchanan (1969), and summaries of their findings are presented above under Application of Amino Acid Assays.

Some investigators have employed strains of *A. flavus* to determine RNV of proteins. Mohyuddin *et al.* (1977) compared *A. flavus,* TpW, and the rat assay using four rye and wheat breads and found that *A. flavus* ranked the four products in the same order as the other two assays even though the fungus was not as sensitive as TpW. The response of the fungus was related to lysine in the products. Fungal growth was assessed by weighing dried mycelia.

Wang *et al.* (1979A) compared *A. flavus,* TpW, and the rat assay on 11 foods and measured growth response of the fungus by biomass production and by measuring diameter of fungal colony. Among replicate analyses, variations were less with *A. flavus* than with TpW. No significant correlation was found between the microbiological and rat assay values. Since *A. flavus* does not have a specific amino acid requirement, its growth response is apparently the result of the overall amino acid balance of a product.

Measuring Growth Response of Microorganisms in Protein Assays

One of the most commonly cited drawbacks to the use of TpW is the relatively high error associated with direct microscopic count (DMC) by manual means, and the time required to perform counts on a large number of samples. Acidimetry was first employed to assess cell yield (Dunn and Rockland 1947) but a 41-day incubation was required for maximum acid production (Rockland and Dunn 1949). As early as 1951, Anderson and Williams evaluated a colorimetric method to assess cell mass and found the method to be simpler and more rapid than DMC. This method is based on the reduction by respiring organisms of the colorless 2,3,5-triphenyltetrazolium chloride to its red water–insoluble triphenylformazan. The latter is read in a colorimeter. Fernell and Rosen (1956) measured ammonia production to assess TpW growth but this method appears not to have been used by others.

In a more recent study, dye-reduction, O_2-uptake, ATP assay, and DMC were compared for their ability to assess protozoal mass in TNV analyses (Wang *et al.* 1979B). The O_2-uptake was measured by a respirometer, and ATP by the luciferin–luciferase assay. Wang and coworkers found that dye reduction was the simplest of the four methods although reaction with cell populations $<10^5$/mL was poor. The 0_2-uptake requires expertise in operating a respirometer. The ATP measurement was found to be simple, very sensitive, and accurate. Automated methods for the measurement of ATP by the luciferin–luciferase assay already exist.

As a means of improving DMC, Shorrock (1976) photographed the organisms using photomicroscopy and counted them with the aid of a digital counter. This procedure, of course, allows for permanent records of counts. This investigator used papain digestion and suggested that hydrolysis was necessary to get results by TpW to agree with those by rat assays and by a chemical method for lysine employing FDNB. With this procedure, TpW and *S. faecalis* subsp. *zymogenes* assays produced the same values for available methionine but the bacterium was favored because its use gave quicker results, was more suited for use of high carbohydrate-containing samples, and because growth could be assessed by acidimetry. The bacterial assay was found to be more sensitive than TpW but the latter could be used to assay both methionine and lysine in the same extract. Shorrock (1976) also noted that TpW growth could be assessed by turbidimetry when clear hydrolysates were used.

Noting that *Tetrahymena* produces tetrahymanol (a pentacyclic terpene), Shepherd *et al.* (1977) proposed assessing TpW growth by measuring this compound by use of gas liquid chromatography (GLC). In studies with eight different proteins, TpW values assessed by GLC showed good agreement with data by chick bioassay for available lysine. Good results were obtained also for tryptophan and methionine. These investigators confirmed the work of Shorrock (1976) that turbidimetry could be used to assess TpW growth if clear hydrolysates are used. The GLC analysis of tetrahymanol was found to be more precise than DMC.

Along the lines of the above, 2-aminoethylphosphonic acid (ciliatine, AEP) is found in *Tetrahymena* and not in foods, and an assay for its presence to determine growth of TpW was proposed by Maciejewicz-Rys and Antoniewicz (1978). By this procedure, AEP is extracted from the organisms at the end of assay, and the extract passed over a Dowex column. Following elution of AEP, its quantity is determined by an analysis of its phosphorus content. Using this method, these investigators found a statistically significant correlation between RNV with growth determined by AEP analysis, RNV with growth determined by DMC, and rat assay on a variety of foods. While no improvements in correlation coefficient were observed by AEP analysis, these investigators preferred this procedure because it avoided tedious counting. Also, AEP analysis was preferred because cell size is eliminated as a factor.

ANTIBIOTICS

The microbiological assay of the activity and concentration of antibiotics and other antimicrobials is carried out essentially in the same way as for amino acids or B vitamins. The two methods that are employed are turbidimetry and plate assay either manually or automated. Although most antibiotics can be determined effectively by nonmicrobiological methods, microbiological assays give assurance that the compounds being assayed possess antimicrobial activity.

The turbidimetric method is carried out by establishing a standard curve by serially diluting the antibiotic in question in an appropriate diluent. A standardized inoculum is added followed by incubation at 37°C typically for 3–4 hr. Formaldehyde may be added to stop growth and turbidity is measured spectrophotometrically usually at 530 nm. While with an assay for a metabolite the degree of turbidity increases with increasing metabolite, with antimicrobials the opposite occurs. Automated assay procedures are common for antibiotic assays especially where large numbers of specimen are assayed. The Autoturb System (Elanco Products Co., Indianapolis, IN) is one such device used for automated assays. The history and rationale of automated microbiological assays for antibiotics have been addressed by Ferrari and Marten (1972). The turbidimetric assay for antibiotics has been modified by the use of radiometry with the $^{14}CO_2$ being read by an automated system (DeBlanc *et al.* 1972).

The plate assay for antibiotics is carried out much in the same way as described above for the large plate assay for amino acids. When serially diluted antibiotics are placed in the cylinders, zones of growth inhibition diminish as antibiotic concentrations increase. The assay strains for 42 antibiotics along with the AOAC and USP-recommended assay procedures are presented in Table 10.12. Other assay strains for these and other antimicrobials may be found in the appendix section of the ATCC Catalogue of Strains (1978).

Newer Methods for Penicillin

The determination of penicillin G in milk by a competitive assay in which added ^{14}C-penicillin competes with penicillin in the sample for enzyme sites on the cell wall of *Bacillus stearothermophilus* (ATCC 10149) has been developed and details can be found in AOAC (1982). The bacterial enzyme sites bind β-lactam antibiotics specifically and irreversibly, and the amount of bound ^{14}C is counted and compared with control samples to determine the presence of a β-lactam antibiotic. The method employs vegetative cells and is sensitive to ca. 0.01 IU of penicillin G or β-lactam equivalent.

Penicillin in milk can be quantitated by a disc method employing *B. stearothermophilus* in the same general way noted for large plate assays above. What makes the penicillin G assay unique among anti-

TABLE 10.12. SOME BACTERIA USED FOR THE MICROBIOLOGIC ASSAY OF ANTIBIOTICS IN EITHER THE AOAC OR USP PROCEDURES

Antibiotic	Test organism	ATCC no.	Assay method
Amphotericin B	*Saccharomyces cerevisiæ*	9763	Cylinder-plate
Ampicillin	*Micrococcus luteus*	9341	Cylinder-plate
Bacitracin	*Micrococcus luteus*	10240	Cylinder-plate
	Micrococcus luteus	7468	Cylinder-plate
Capreomycin	*Klebsiella pneumoniæ*	10031	Turbidimetric
Carbenicillin	*Pseudomonas aeruginosa*	25619	Cylinder-plate
Cephalexin	*Staphylococcus aureus*	6538-P	Cylinder-plate
Cephalothin	*Staphylococcus aureus*	6538-P	Cylinder-plate
Chloramphenicol	*Escherichia coli*	10536	Turbidimetric
Chlortetracycline–HCl	*Bacillus cereus*	17778	Turbidimetric
Clindamycin	*Micrococcus luteus*	9341	Cylinder-plate
Cloxacillin	*Staphylococcus aureus*	6538-P	Cylinder-plate
Colistimethate sodium	*Bordetella bronchiseptica*	4617	Cylinder-plate
Cycloserine	*Staphylococcus aureus*	6538-P	Cylinder-plate
Dactinomycin	*Bacillus subtilis*	6633	Cylinder-plate
Dicloxacillin	*Staphylococcus aureus*	6538-P	Cylinder-plate
Doxycycline	*Staphylococcus aureus*	6538-P	Turbidimetric
Erythromycin	*Micrococcus luteus*	9341	Cylinder-plate
Gentamicin	*Staphylococcus epidermidis*	12228	Cylinder-plate
Griseofulvin	*Microsporum gypseum*	14683	Cylinder-plate
Hygromycin	*Bacillus subtilis*	6633	Cylinder-plate
Kanamycin	*Staphylococcus aureus*	6538-P	Cylinder-plate
Lincomycin	*Micrococcus luteus*	9341	Cylinder-plate
Methicillin	*Staphylococcus aureus*	6538-P	Cylinder-plate
Minocycline	*Staphylococcus aureus*	6538-P	Turbidimetric
Mithramycin	*Staphylococcus aureus*	6538-P	Cylinder-plate
Monensin	*Streptococcus faecalis*	8043	Turbidimetric
Nafcillin	*Staphylococcus aureus*	6538-P	Cylinder-plate
Neomycin	*Staphylococcus epidermidis*	12228	Cylinder-plate
	Klebsiella pneumoniæ	10031	Turbidimetric
Novobiocin	*Staphylococcus epidermidis*	12228	Cylinder-plate
Oleandomycin	*Micrococcus luteus*	9341	Cylinder-plate
Nystatin	*Saccharomyces cerevisiæ*	2601	Cylinder-plate
Oxacillin	*Staphylococcus aureus*	6538-P	Cylinder-plate
Oxytetracycline	*Staphylococcus aureus*	6538-P	Turbidimetric
Penicillin G	*Staphylococcus aureus*	6538-P	Cylinder-plate
Penicillin V	*Staphylococcus aureus*	6538-P	Cylinder-plate
Polymyxin B	*Bordetella bronchiseptica*	4617	Cylinder-plate
Rifampin	*Bacillus subtilis*	6633	Cylinder-plate
Spectinomycin	*Escherichia coli*	10536	Turbidimetric
Streptomycin	*Bacillus subtilis*	6633	Cylinder-plate
	Klebsiella pneumoniæ	10031	Turbidimetric
Tetracycline	*Staphylococcus aureus*	6538-P	Turbidimetric
Tylosin	*Micrococcus luteus*	9341	Cylinder-plate
Vancomycin	*Bacillus cereus* var. *mycoides*	11778	Cylinder-plate
Viomycin	*Klebsiella pneumoniæ*	10031	Turbidimetric

biotic assays is the fact that one can determine this antibiotic specifically since specific enzymes (β-lactamase or penicillinase) exist for its destruction. When set up properly, discs containing unknown samples that produce zones of inhibition of 17–20 mm indicate the presence of penicillin if at the same time similar discs treated with penicillinase do not show zones of inhibition. The quantitative method using *B. stearothermophilus* is sensitive to ca. 0.016 IU of penicillin G/mL while the qualitative method is sensitive to ca. 0.0080 IU.

A rapid qualitative method of assaying penicillin in milk consists of acid production in the absence of antibiotic and no acid in its presence. The assay medium contains bromcresol purple and employs *B. stearothermophilus* (ATCC 10149). Penicillinase is used to confirm that the lack of acid production is due to penicillin and not another inhibitor (AOAC 1982).

The detection of penicillin in raw milk by impedance measurement of cell growth has been studied by Crain (1980) who was able to achieve results in <3 hr with *Bacillus subtilis* and <2 hr with *B. stearothermophilus* as test strains.

For antibiotic assay media and methods for the extraction of antibiotics from foods and feeds, AOAC methods (1980) should be consulted.

MUTAGENS

The testing of food additives and other chemicals for their potential mutagenic activity currently is done more by use of microorganisms than by other methods. The development of the *Salmonella*/microsome test by Ames in 1972 revolutionized mutagenesis assays and this procedure has led to the development of other microbial assays for mutagenesis testing.

The Ames Test

The Ames test is clearly the most widely used screening method for potential mutagens (Ames *et al.* 1975). The method uses specifically selected mutants of *Salmonella typhimurium* (histidine auxotrophs) which can be reverted from histidine auxotrophy back to prototrophy following exposure to a mutagen. In order to make the assay relevant to humans, homogenates of rat or human liver (S-9 mix) may be added to the assay plate to make the *in vitro* test simulate *in vivo* conditions relative to the fate of potential mutagens in the body. The test is run generally by use of petri dishes containing appropriate synthetic medium and seeded with the selected strain of *S. typhimurium*. When mutagens are placed onto the plate, they induce prototrophy in the histidine auxotrophs and these cells (revertants) grow on the histidine-free medium and form colonies. The mutagenic potential of a compound is measured by counting the number of revertant colonies that develop on test plates and subtracting from this number those that develop on control plates without the mutagen. The test may be run with or without microsomal activation (use of the S-9 mix). The number of revertants per nanomole of test substance is a measure of the mutagenic potency of the compound in question (McCann *et al.* 1975). By the Ames test, most carcinogens (ca. 85%) have been shown to be mutagens (Ames *et al.* 1975). An excellent analysis of mutagenicity vs carcinogenicity relative to the Ames test has been presented by Taylor (1982).

In regards to the various *S. typhimurium* strains, TA1535 detects mutagens causing base-pair substitutions, while TA1537 and TA1538 detect various types of frameshift mutagens. The TA98 and TA100 are more sensitive detectors of certain carcinogens such as some mycotoxins (Ames *et al.* 1975). The latter two strains appear to be the most widely used for assaying food-use chemicals for mutagenicity.

Other Microbial Systems

While the *Salmonella*/microsome system is the most widely used of microbial assays for mutagens, other organisms have been developed. *Saccharomyces cerevisiae* and *N. crassa* strains measure forward mutations in each of two genetic loci that control the synthesis of adenine. In addition, *S. cerevisiae* can detect some chromosome changes such as recombination and gene conversions. *Aspergillus nidulans,* and *Escherichia coli* WP2, as well as *S. cerevisiae* and *N. crassa,* detect chromosomal point mutations while *E. coli* Pol A detects DNA damage. None of these strains, however, has been employed as widely as the *S. typhimurium* strains.

Casciano (1982) has compared microbiological and nonmicrobiological methods of mutagenesis testing, and Pariza (1983) has presented general considerations regarding the use of mutagenesis assays for potential carcinogens in foods. Some considerations in the use of mutagenic assays for foods have been presented by Commoner *et al.* (1978). The reader who wishes to conduct assays for mutagens by the Ames test should consult Ames *et al.* (1975) and DeSerres and Shelby (1979).

MISCELLANEOUS SUBSTANCES

A rather wide variety of organic compounds as well as some inorganic elements can be assayed by use of microorganisms, and some of

TABLE 10.13. SOME MICROORGANISMS USED IN THE ASSAY OF MISCELLANEOUS SUBSTANCES

Organisms	ATCC no.	Substances
Arthrobacter flavescens	25091	Sideramines
Aspergillus niger	6275	Sulfur
A. niger	6273	Magnesium
A. niger	10581	Zn, Mg, Cu, Mo
Hansenula saturnus	9847	D-xylose
Lactobacillus acidophilus	11506	DNA
L. brevis	367, 8287	Cytosine, uracil
L. casei	7469	Glucose, azathioprine, 6-mercaptopurine, strepogenic factor
L. fructosus	13162	Fructose
L. jugurt	13866	Orotic acid
Leuconostoc mesenteroides	23386	D-lactic acid
Micrococcus luteus	272	Lysozyme
Neurospora crassa	9277	Choline
N. crassa	11063	Adenine, hypoxanthine
Rhodotorula rubra	36053	T-2 toxin
Schizosaccharomyces pombe	24971, 24972	Adenine
Streptococcus faecium	8043	Monensin
Tetrahymena furgasoni	10542	Protein

these are presented in Table 10.13. By careful selection of auxotrophic mutants, the number of nutrients, growth factors, and other compounds that can be microbiologically assayed is almost endless.

BIBLIOGRAPHY

ATCC (American Type Culture Collection). 1978. Catalogue of Strains I, 13th ed. Washington, DC.

AMES, B.N., McCANN, J., and YAMASAKI, E. 1975. Methods for detecting carcinogens and mutagens with the *Salmonella*/mammalian–microsome mutagenicity test. Mutation Res. *31,* 347–364.

ANALYTICAL MICROBIOLOGY, II. 1972. F. Kavanagh (Editor). Academic Press, New York.

ANDERSON, M.E., and WILLIAMS, H.H. 1951. Microbiological evaluation of protein quality. I. A colorimetric method for the determination of the growth of *Tetrahymena geleii* W in protein suspensions. J. Nutr. *44,* 335–343.

AOAC (Association of Official Analytical Chemists). 1980. Official Methods of Analysis, 13th ed. Washington, DC.

AOAC (Association of Official Analytical Chemists). 1982. Changes in methods. J. Assoc. Off. Anal. Chem. *65,* 450–521.

BAKER, H., and FRANK, O. 1968. Clinical Vitaminology—Methods and Interpretation. Interscience Publishers, New York.

BAKER, H., FRANK, O., RUSSOFF, I.I., MORCK, R.A., and HUTNER, S.H. 1978. Protein quality of foodstuffs determined with *Tetrahymena thermophila* and rat. Nutr. Rep. Int. *17,* 525–536.

BERG, T.M., and BEHAGEL, H.A. 1972. Semiautomated method for microbiological vitamin assays. Appl. Microbiol. *23;* 531–542.

BOLINDER, A.E. 1972. Large plate assays for amino acids, pp. 479–591. *In* Analytical Microbiology. F. Kavanagh (Editor). Academic Press, New York.

BOYNE, A.W., PRICE, S.A., ROSEN, G.D., and STOTT, J.A. 1967. Protein quality of feeding-stuffs. 4. Progress report of collaborative studies on the microbiological assay of available amino acids. Brit. J. Nutr. *21,* 181–206.

BRAEKKAN, O.R. 1960. Effect of temperature on a lag observed in the microbiological assay of pantothenic acid. J. Bacteriol. *80,* 626–627.

BUCHANAN, R.A. 1969. *In vivo* and *in vitro* methods of measuring nutritive values of leaf-protein preparations. Brit. J. Nutr. *23,* 533–545.

BUNYAN, J., and WOODHAM, A.A. 1964. Protein quality of feeding-stuffs. 2. The comparative assessment of protein quality in three fish meals by microbiological and other laboratory tests, and by biological evaluation with chicks and rats. Brit. J. Nutr. *18,* 537–544.

BUTRUM, R.R., and KRAMER, A. 1977. Comparison of rat and *Tetrahymena pyriformis* growth assays in determining nutritional quality of protein in meat and meat analogs during prolonged storage at various temperatures. J. Food Qual. *1,* 379–391.

CASCIANO, D.A. 1982. Mutagenesis assay methods. Food Technol. *36* (3), 48–52.

CHEN, M.F., McINTYRE, P.A., and KERTCHER, J.A. 1978. Measurement of folates in human plasma and erythrocytes by a radiometric microbiologic method. J. Nucl. Med. *19,* 906–912.

COMMONER, B., VITHAYATHIL, A., and DOLARA, P. 1978. Mutagenic analysis as a means of detecting carcinogens in foods. J. Food Protect. *41,* 996–1003.

COOPERMAN, J.M. 1967. Microbiological assay of serum and whole-blood folic acid activity. Am. J. Clin. Nutr. *20,* 1015–1024.

CRAIN, P.R. 1980. Fundamental aspects of monitoring food related microbes by impedance. Diss. Abstr. Int. B *41* (1), 121.

DeBLANC, H.J. JR., CHARACHE, P., and WAGNER, H.N. JR. 1972. Automatic radiometric measurement of antibiotic effect on bacterial growth. Antimicrobiol. Agents Chemo. *2,* 360–366.

DeSERRES, F.J., and SHELBY, M.D. 1979. The *Salmonella* mutagenicity assay: Recommendations. Science *203,* 563–565.

DIFCO SUPPL. LIT. 1972. Difco Laboratories, Detroit, MI.

DUNN, M.S., and ROCKLAND, L.B. 1947. Biological value of proteins determined with *Tetrahymena geleii* H. Proc. Soc. Exp. Biol. Med. *64,* 377–379.

ELLIOTT, A.M. 1959. Biology of *Tetrahymena.* Ann. Rev. Microbiol. *13,* 79–96.

EVANCHO, G.M., HURT, H.D., DEVLIN, P.A., LANDERS, R.E., and ASHTON, D.H. 1977. Comparison of *Tetrahymena pyriformis* W and rat bioassays for the determination of protein quality. J. Food Sci. *42,* 444–448.

EVANS, E., CARRUTHERS, S.C., and WITTY, R. 1977. Comparison of *Tetrahymena pyriformis* and PER responses to meats and meat products. Nutr. Rept. Int. *16,* 445–462.

EVANS, E., KHOUW, B.T., LIKUSKI, H.J., and WITTY, R. 1979A. Effects of altering carbohydrate and buffer on the growth of *Tetrahymena pyriformis* W and on the determination of the relative nutritive value of food proteins. Can. Inst. Food Sci. Technol. J. *12,* 36–39.

EVANS, E., CARRUTHERS, S.C., and WITTY, R. 1979B. Effects of cooking methods on the protein quality of meats as determined using a *Tetrahymena pyriformis* W growth assay. J. Food Sci. *44,* 1678–1680.

FERNELL, W.R. and ROSEN, G.D. 1956. Microbiological evaluation of protein quality with *Tetrahymena pyriformis* W. 1. Characteristics of growth of the organism and determination of relative nutritional values of intact proteins. Brit. J. Nutr. *10,* 143–156.

FERRARI, A., and MARTEN, J. 1972. Automated microbiological assay. Methods in Microbiol. *6B,* 331–342.

FORD, J.E. 1962. A microbiological method for assessing the nutritional value of proteins. 2. The measurement of "available" methionine, leucine, isoleucine, arginine, histidine, tryptophan and valine. Brit. J. Nutr. *16,* 409–425.

FRANK, O., BAKER, H., and HUTNER, S.H. 1975. Evaluation of protein quality with the phagotrophic protozoan *Tetrahymena,* pp. 203–209. *In* Protein Nutritional Quality of Foods and Feeds. M. Friedman (Editor). Marcel Dekker, New York.

GORIN, N., MEULENHOFF, E.J.S., and YARROW, D. 1970. Determination of niacin in orange juice with lyophilized *Lactobacillus arabinosus* ATCC 8014. Appl. Microbiol. *20,* 641–542.

GRAHAM, D.C., ROE, D.A., and OSTERTAG, S.G. 1980. Radiometric determination and chick bioassay of folacin in fortified and unfortified frozen foods. J. Food Sci. *45,* 47–51.

GREGORY, J.F., III. 1980. Comparison of high-performance liquid chromatographic and *Saccharomyces uvarum* methods for the determination of vitamin B_6 in fortified breakfast cereals. J. Agric. Food Chem. *28,* 486–489.

GREGORY, J.F., III. 1983. Methods of vitamin assay for nutritional evaluation of food processing. Food Technol. *37* (1), 75–80.

GREGORY, J.F., III, DAY, B.P.F., and RISTOW, K.A. 1982. Comparison of high performance liquid chromatographic, radiometric, and *Lactobacillus casei* methods for the determination of folacin in selected foods. J. Food Sci. *47,* 1568–1571.

GUILARTE, T.R., McINTYRE, P.A., and TSAN, M.-F. 1980. Growth response of the yeasts *Saccharomyces uvarum* and *Kloeckera brevis* to the free biological active forms of vitamin B-6. J. Nutr. *110,* 954–958.

GUILARTE, T.R., SHANE, B., and McINTYRE, P.A. 1981. Radiometric-microbiological assay of vitamin B-6: Application to food analysis. J. Nutr. *111,* 1869–1875.

GUIRARD, B.M., and SNELL, E.E. 1981. Biochemical factors in growth, pp. 79–111. *In* Manual of Methods for General Bacteriology. P. Gerhardt *et al.* (Editors). Am. Soc. Microbiol., Washington, D.C.

HALEVY, S., and GROSSOWICZ, N. 1953. A microbiological approach to nutritional evaluation of proteins. Proc. Soc. Exp. Biol. Med. *82,* 567–571.

HANDBOOK OF MICROBIOLOGY, III. 1973. pp. 1017–1051. A.I. Laskin and H.A. Lechevalier (Editors). CRC Press, Cleveland, OH.

HARPER, A.E. 1981. McCollum and directions in the evaluation of protein quality. J. Agric. Food Chem. *29,* 429–435.

HEGSTED, D.M. 1977. Protein quality and its determination, pp. 347–362. *In* Food Proteins. J.R. Whitaker and S. Tannenbaum (Editors). AVI Publishing Co., Westport, CT.

HEGSTED, D.M., and CHANG, U.-Y. 1965. Protein utilization in growing rats. I. Relative growth index as a bioassay procedure. J. Nutr. *85,* 159–168.

HERBERT, V. 1966. Aseptic addition method for *Lactobacillus casei* assay of folate activity in human serum. J. Clin. Path. *19,* 12–16.

HOPPNER, K. 1971. Free and total folate activity in strained baby foods. Can. Inst. Food Technol. J. *4,* 51–54.

HOPPNER, K., LAMPI, B., and PERRIN, D.E. 1973. Folacin activity of frozen convenience foods. J. Am. Diet. Assoc. *63,* 536–539.

ITAGAKI, T., and TSUKAHARA, T. 1972. Microbiological assay of vitamin B_6 by thin-layer cup–plate method with *Saccharomyces carlsbergensis*. J. Vitaminol. *18,* 90–96.
KIDDER, G.W., and DEWEY, V.C. 1945. Studies on the biochemistry of *Tetrahymena*. III. Strain differences. Physiol. Zool. *18,* 136–157.
KLEIN, B.P., and KUO, C.H.Y. 1981. Comparison of microbiological and radiometric assays for determining total folacin in spinach. J. Food Sci. *46,* 552–554.
LANDERS, R.E. 1975. Relationship between protein efficiency ratio of foods and relative nutritive value measured by *Tetrahymena pyriformis* W bioassay techniques, pp. 185–202. *In* Protein Nutritional Quality of Foods and Feeds. M. Friedman (Editor). Marcel Dekker, New York.
LEICHTER, J., BUTTERWORTH, C.E., and KRUMDIECK, C.L. 1977. Partial purification and some properties of pteroylglutamate hydrolyase (conjugase) from chicken pancreas. Proc. Soc. Exp. Biol. Med. *154,* 98–101.
LI-CHAN, E., and NAKAI, S. 1981. Nutritional evaluation of covalently lysine enriched wheat gluten by *Tetrahymena* bioassay. J. Food Sci. *46,* 1840–1841, 1850.
MACIEJEWICZ-RYS, J., and ANTONIEWICZ, A.M. 1978. 2-Aminoethylphosphonic acid as an indicator of *Tetrahymena pyriformis* W growth in protein-quality evaluation assay. Brit. J. Nutr. *40,* 83–90.
MANUAL OF METHODS FOR GENERAL BACTERIOLOGY. 1981. P. Gerhardt *et al.* (Editors). Amer. Soc. Microbiol., Washington, D.C.
McCANN, J., CHOI, E., YAMASAKI, E., and AMES, B.N. 1975. Detection of carcinogens as mutagens in the *Salmonella*/microsome test: Assay of 300 chemicals. Proc. Natl. Acad. Sci. *72,* 5135–5139.
McCURDY, S.M., SCHEIER, G.E., and JACOBSON, M. 1978. Evaluation of protein quality of five varieties of lentils using *Tetrahymena pyriformis* W. J. Food Sci. *43,* 694–697.
McMARTIN, K.E., VIRAYOTHA, V., and TEPHLY, T.R. 1981. High-pressure liquid chromatography separation and determination of rat liver folates. Arch. Biochem. Biophys. *209,* 127–130.
MOHYUDDIN, M., SHARMA, T.R., and NIEMANN, E.-G. 1977. Nutritive value of rye and wheat breads assessed with *Aspergillus flavus*. J. Agric. Food Chem. *25,* 200–201.
NASHEF, A.S., OSUGA, D.T., LEE, H.S., AHMED, A.I., WHITAKER, J.R., and FEENEY, R.E. 1977. Effects of alkali on proteins. Disulfides and their products. J. Agric. Food Chem. *25,* 245–251.
PARIZA, M.W. 1983. Carcinogenicity/toxicity testing and the safety of foods. Food Technol. *37* (1), 84–86.
PELLETT, P.L. 1978. Protein quality evaluation revisited. Food Technol. *32* (5), 60–79.
PHILLIPS, D.R., and WRIGHT, A.J.A. 1982. Studies on the response of *Lactobacillus casei* to different folate monoglutamates. Brit. J. Nutr. *47,* 183–189.
PHILLIPS, D.R., and WRIGHT, A.J.A. 1983. Studies on the response of *Lactobacillus casei* to folate vitamin in foods. Brit. J. Nutr. *49,* 181–186.
PILCHER, H.L., and WILLIAMS, H.H. 1954. Microbiological evaluation of protein quality. II. Studies of the responses of *Tetrahymena pyriformis* W to intact proteins. J. Nutr. *53,* 589–599.
ROCKLAND, L.B., and DUNN, M.S. 1949. Determination of the biological value of proteins with *Tetrahymena geleii* H. Food Technol. *3,* 289–292.
ROSEN, G.D., and FERNELL, W.R. 1956. Microbiological evaluation of protein quality with *Tetrahymena pyriformis* W. 2. Relative nutritive values of protein in foodstuffs. Brit. J. Nutr. *10,* 156–169.

ROTHENBERG, S.P., DACOSTA, M., and ROSENBERG, Z. 1972. A radioassay for serum folate: Use of a two-phase sequential-incubation, ligand-binding assay. New Eng. J. Med. *286,* 1335–1339.

RUDDICK, J.E., VANDERSTOEP, J., and RICHARDS, J.F. 1978. Folate levels in food—A comparison of microbiological assay and radioassay methods for measuring folate. J. Food Sci. *43,* 1238–1241.

SCHATZKI, T.F., and KEAGY, P.M. 1975. Analysis of nonlinear response in microbiological assay for folacin. Anal. Biochem. *65,* 204–214.

SHANE, B., TAMURA, T., and STOKSTAD, E.L.R. 1980. Folate assay: A comparison of radioassay and microbiological methods. Clin. Chem. Acta *100,* 13–19.

SHEPHERD, N.D., TAYLOR, T.G., and WILTON, D.C. 1977. An improved method for the microbiological assay of available amino acids in proteins using *Tetrahymena pyriformis*. Brit. J. Nutr. *38,* 245–253.

SHORROCK, C. 1976. An improved procedure for the assay of available lysine and methionine in feedstuffs using *Tetrahymena pyriformis* W. Brit. J. Nutr. *35,* 333–341.

SMITH, E.B., and PENA, P.M. 1977. Use of *Tetrahymena pyriformis* W to evaluate protein quality of leaf protein concentrates. J. Food Sci. *42,* 674–676.

SNELL, E.E. 1948. Use of microorganisms for assay of vitamins. Physiol. Rev. *28,* 255–282.

SRINIVAS, H., VAKIL, U.K.,and SREENIVASAN, A. 1975. Evaluation of protein quality of irradiated foods using *Tetrahymena pyriformis* W and rat assay. J. Food Sci. *40,* 66–69.

STERNBERG, M., and KIM, C.Y. 1979. Growth response of mice and *Tetrahymena pyriformis* to lysinoalanine-supplemented wheat gluten. J. Agric. Food Chem. *27,* 1130–1132.

STEWART, K.K. 1980. Nutrient analysis of foods: State of the art for routine analyses, pp. 1–5. *In* Proceedings, Symposium on State of the Art for Routine Analyses. K.K. Stewart (Editor). Assoc. Official Anal. Chemists, Washington, D.C.

STOKES, J.L., GUNNESS, M., DWYER, I.M., and CASWELL, M.G. 1945. Microbiological methods for the determination of amino acids. J. Biol. Chem. *160,* 35–49.

STOKSTAD, E.L.R., and KOCH, J. 1967. Folic acid metabolism. Physiol. Rev. *47,* 83–116.

STOKSTAD, E.L.R., and THENEN, S.W. 1972. Chemical and biochemical reactions of folic acid, pp. 387–408. *In* Analytical Microbiology, II. F. Kavanaugh (Editor). Academic Press, New York.

STOKSTAD, E.L.R., SHIN, Y.S., and TAMURA, T. 1977. Distribution of folate forms in food and folate availability, pp. 56–68. *In* Folic Acid: Biochemistry and Physiology in Relation to the Human Nutrition Requirement. Natl. Acad. Sci., Washington, D.C.

STOTT, J.A., and SMITH, H. 1966. Microbiological assay of protein quality with *Tetrahymena pyriformis* W. Brit. J. Nutr. *20,* 663–673.

STOTT, J.A., SMITH, H., and ROSEN, G.D. 1963. Microbiological evaluation of protein quality with *Tetrahymena pyriformis* W. 3. A simplified assay procedure. Brit. J. Nutr. *17,* 227–233.

TAMURA, T., SHIN, Y.S., WILLIAMS, M.A., and STOKSTAD, E.L.R. 1972A. *Lactobacillus casei* response to pteroylpolyglutamates. Anal. Biochem. *49,* 517–521.

TAMURA, T., BUEHRING, K.U., and STOKSTAD, E.L.R. 1972B. Enzymatic hydrolysis of pteroylpolyglutamates in cabbage. Proc. Soc. Exp. Biol. Med. *141,* 1022–1025.

TAYLOR, S.L. 1982. Mutagenesis vs carcinogenesis. Food Technol. *36* (3), 65–68, 98, 100, 102–103, 127.

TENNANT, G.B., and WITHEY, J.L. 1972. An assessment of work simplified procedures for the microbiological assay of serum vitamin B_{12} and serum folate. Med. Lab. Technol. *29,* 171–181.

TEUNISSON, D.J. 1961. Microbiological assay of intact proteins using *Tetrahymena pyriformis* W. I. Survey of protein concentrates. Anal. Biochem. *2,* 405–420.

TIGNER, J., and ROE, D.A. 1979. Tissue folacin stores in rats measured by radioassay. Proc. Soc. Exp. Biol. Med. *160,* 445–448.

TOEPFER, E.W., and POLANSKY, M.M. 1970. Microbiological assay of vitamin B_6 and its components. J. Assoc. Off. Anal. Chem. *53,* 546–550.

TSUJI, K. 1966A. Liquid nitrogen preservation of *Saccharomyces carlsbergensis* and its use in a rapid biological assay of vitamin B_6 (pyridoxine). Appl. Microbiol. *14,* 456–461.

TSUJI, K. 1966B. Liquid nitrogen preservation of *Saccharomyces carlsbergensis* and its use in a rapid biological assay of pantothenic acid. Appl. Microbiol. *14,* 462–465.

TSUJI, K., ELFRING, G.L., CRAIN, H.H., and COLE, R.J. 1967. Dose response curve linearization and computer potency calculation of turbidimetric microbiological vitamin assays. Appl. Microbiol. *15,* 363–367.

UNITED STATES PHARMACOPEIA, XIX. 1975. U.S.P. Conv., Inc., Rockville, Md.

VOIGT, M.S., and EITENMILLER, R.R. 1979. A liquid scintillation technique for microbial vitamin analyses. J. Food Sci. *44,* 1780–1781.

VOIGT, M.N., WARE, G.O., and EITENMILLER, R.R. 1979A. Computer programs for the evaluation of vitamin B data obtained by microbiological methods. J. Agric. Food. Chem. *27,* 1305–1311.

VOIGT, M.N., EITENMILLER, R.R., and WARE, G.O. 1979B. Vitamin analysis by microbial and protozoan organisms: Response to food preservatives and neutralization salts. J. Food Sci. *44,* 723–737.

WADDELL, C.C., DOMSTAD, P.A., PIRCHER, F.J., LERNER, S.R., BROWN, J.A., and LAWHORN, B.K. 1976. Serum folate levels. Comparison of microbiologic assay and radioisotope kit methods. A.J. Clin.Path. *66,* 746–752.

WANG, Y.Y.D., MILLER, J., and BEUCHAT, L.R. 1979A. Comparison of *Tetrahymena pyriformis* W, *Aspergillus flavus,* and rat bioassays for evaluating protein quality of selected commercially prepared food products. J. Food Sci. *44,* 1390–1393.

WANG, Y.Y.D., MILLER, J., and BEUCHAT, L.R. 1979B. Comparison of four techniques for measuring growth of *Tetrahymena pyriformis* W with rat PER bioassays in assessing protein quality. J. Food Sci. *44,* 540–544.

WARREN, R.M., and LABUZA, T.P. 1977. Comparison of chemically measured available lysine with relative nutritive value measured by a *Tetrahymena* bioassay during early stages of nonenzymatic browning. J. Food Sci. *42,* 429–431.

WAXMAN, S., SCHREIBER, C., and HERBERT V. 1971. Radioisotopic assay for measurement of serum folate levels. Blood *38,* 219–228.

11

Sensory Analysis as an Analytical Laboratory Tool in Food Research

R.M. Pangborn[1]

INTRODUCTION

The field of analytical measurement of the sensory attributes of foods and other consumer products remains in a state of prolonged adolescence. Although long recognized as an essential step in ingredient testing, product formulation, quality assessment, and consumer acceptance, many users of sensory procedures remain refractory to the scientific approach of measurement (Ellis 1963; Hirsh 1974–1975; Köster 1982; Pangborn 1964, 1967, 1980; Schutz 1971). It is not uncommon for chemists, clinicians, and other scientists who use careful, orthodox analytical experimentation to quantify chemical and physical properties of a substance to resort to unscientific phenomenology when faced with sensory measurement of the same material. The author attributes this dichotomy to several interrelated factors:

1. Analytical sensory science is young and undeveloped compared to other analytical fields.
2. Few scientists maintain a sustained commitment to the field due to other responsibilities or other sources of funding.
3. Few university curricula offer rigorous undergraduate or graduate academic training in sensory science [see Tilger (1971)].

[1]Department of Food Science and Technology, University of California, Davis, CA 95616

4. Several fallacious assumptions that persist among the inexperienced are listed below:

a. *It is so simple to present samples to judges, anyone can do it.* This belief fails to consider the necessity of (i) careful control of environmental conditions, (ii) sample preparation and presentation, (iii) selection of appropriate experimental designs and of test methods, and (iv) selection and training of the analytical device—the judge.

b. *We all have tongues and noses, so anyone can participate.* This simplistic observation demonstrates lack of appreciation of the extensive physiological and behavioral diversity among people (Harper 1981; Pangborn 1981; Williams 1956, 1978), which requires careful attention to sampling, and to adequate measures of reproducibility and reliability. A perusal of articles in food science journals (foreign or domestic), wherein sensory tests are reported, will demonstrate the ubiquitous use of judges interchangeably within experiments. Because they are individual measurement devices with inherent variances, judges are no more interchangeable than are analytical instruments (Cross 1978; Pangborn 1964; Pangborn and Dunkley 1964).

c. *We use colorimeters, texturometers, and GLC because sensory responses are too variable.* While there is no disagreement about human variability, instrumental analyses supplement, but can never substitute for, human measurement. By definition, colors, textures, flavors, and tastes are human sensations. Colorimeters measure absorbed or transmitted light, texturometers measure resistance to pressure or stress, and GLC is used to separate volatile compounds. One of the most valuable applications of analytical sensory measurement is the relationship to corresponding physical and chemical measures (ASTM 1968; Little 1976; Noble 1975, Szczesniak 1963; Trant *et al.* 1981).

d. *That's interesting, but which sample did they like?* The foregoing statement is made by individuals who think that consumer acceptance can be predicted from the preferences of laboratory judges. It is unfortunate that one must continuously admonish against extrapolation from the analytical sensory laboratory to the dinner table. So prevalent is the reporting of hedonic responses from laboratory experiments in the food science literature, that it merits stressing that analytical and consumer tests differ markedly. Table 11.1 outlines the analytical methods appropriate at the laboratory level to measure sensitivity and quantitative and qualitative parameters of a product. In contrast, affective methods are used to establish acceptance, preference, or degree of liking. The latter should be conducted among large

TABLE 11.1. CATEGORIZATION OF ANALYTICAL AND AFFECTIVE SENSORY PROCEDURES

I. Analytical Laboratory Tests:		
A. Discriminative	B. Quantitative	C. Qualitative[b]
1. Threshold	1. Scaling	1. Descriptive analysis
a. Method of limits	a. Ordinal (ranking)	a. "Flavor Profile"
b. Average error[a]	b. Interval (category)	b. "Texture Profile"
c. Frequency methods	c. Ratio (magnitude estimation)	c. "Quantitative Descriptive Analysis"
		d. Dilution profile
		e. Deviation-from-reference[a]
2. Discrimination	2. Duration	
a. Paired comparison	a. Time-intensity[a]	
b. Duo-trio		
c. Dual-standard		
d. Triangle		
II. Affective (Consumer) Tests:		
A. Acceptance	B. Preference	C. Hedonic
1. Accept/reject what is available	1. Select one over another alternative	1. Degree of like/dislike

[a] Discussed in detail in text.
[b] Names in quotations refer to procedures developed by private firms, i.e., the Arthur D. Little "Flavor Profile," the General Foods "Texture Profile," and the Tragon Corporation "Quantitative Descriptive Analysis."

numbers of untrained, inexperienced people who are representative of the target population of users of the product under study. Consumer testing should be conducted under normal conditions of consumption—in the home or at central locations such as fairs or other gatherings. Analytical methods, of course, are conducted with selected, trained judges under controlled laboratory conditions. Bridging the gap between the two is an important industrial endeavor (Girardot 1969; Pearce 1980; Schaefer 1979; Skinner 1980).

The remainder of this chapter is concerned with the application of three analytical procedures selected from Table 11.1: (1) *ad libitum* mixing (methods of average error) as a discrimination tool and for measurement of clinical preferences; (2) determination of stimulus strength and temporal effects (time-intensity); and (3) anchored descriptive analysis (deviation-from-reference) with subsequent multivariate treatment of the resultant data.

AD LIBITUM MIXING

In experimental psychology, *ad libitum* mixing to match the perceived intensity of a standard stimulus is called the method of average error, the method of adjustment, or the equation method. The judge receives a standard stimulus (ST) and a comparison stimulus (CO) that can be manipulated to make it stronger or weaker in order to match its intensity to the ST (Gescheider 1976). For example, ST might be a 6% sucrose solution and CO could be a solution of 15% sucrose. The judge mixes water and CO into a sample cup until the taste of the mix equals that of ST, defined as the point of subjective equality (PSE). The concentration of the mix is ascertained by an appropriate analytical method, e.g., refractive index, electrical conductivity, atomic absorption spectroscopy. Since the judge must taste and retaste ST and the mix several times, it is not possible to calculate the concentration of the mix by sample displacement.

Some of the mixes will be lower and some will be higher than ST but they will distribute themselves closely around the concentration of ST. A frequency distribution of the concentrations mixed will be fairly symmetrical, and if enough trials are run, will approximate a normal distribution. Some data require a logarithmic transformation of the responses to obtain a normal distribution. More sensitive judges will have tighter response distributions around ST. The mean of the distribution ($\bar{x}$) is the PSE, which should correspond closely to the value of ST. The variability of the judges' matches, i.e., the standard deviation of the mean, is equivalent to the just-noticeable-difference (JND) for that judge. A large standard deviation, of course, indicates poor discrimination.

As illustrated with solutions of NaCl by Woskow (1967), the method can be used to select discriminating judges. Also, the procedure can be modified to determine whether additives shift the PSE. For example, for the 6% sucrose ST mentioned above, after establishing the values for PSE and JND, the process could be repeated with ST containing 6% sucrose plus an additive such as acid or salt. A shift in PSE indicates that the additive enhanced or depressed the discrimination.

Ad libitum procedures also have been used to measure preference. Gaweki *et al.* (1976) instructed judges to add sucrose to hot tea until the preferred sweetness was obtained. Between 0.9 and 16.4% sucrose was added, as determined by refractometry of the mix. Lauer *et al.* (1976) had judges mix sodium chloride into cold tomato juice and into heated beef broth via salt shaker to the desired level of liking. Using flame photometry to measure the sodium content of the mix, Bartoshuk *et al.* (1974) directed judges to add NaCl and monosodium glutamate (MSG), separately, to unsalted tomato juice, and noted average preferences of 0.07M NaCl and 0.21M MSG.

Ad libitum salting of unsalted, commercial chicken broth was done by students ($\bar{x}=22.4$ yr) and middle-aged adults ($\bar{x}=48.2$ yr) via crystals in opaque salt shakers, and with high-salt broth solutions (Braddock and Pangborn 1983). Partial representation of these results are presented in Fig. 11.1, which depicts the average amount of sodium added to the broth, as measured by atomic absorption spectrometry. Consistently more salt was added via solution than by shakers. Salting via shaker may have been based on simple reflex shaking, not solely on gustatory criteria. A study by Greenfield *et al.* (1983) of 1900 people observed during meals in public eating places in Australia reported that less than 25% tasted their food before adding salt. Furthermore, the larger the hole area of the salt shaker, the more salt was used. This provides strong evidence for reflexive vs. gustatory influences when using shakers.

Subjects commented that they felt more confident in their additions with liquids than with shakers. Note in Fig. 11.1 that older judges added more salt than did younger judges, e.g., for NaCl, 0.315 vs. 0.37% Na^+, respectively. In corresponding hedonic scaling of chicken broth with a wide range of added salts, the younger groups demonstrated a good correlation with *ad libitum* procedures for all salts, while the older

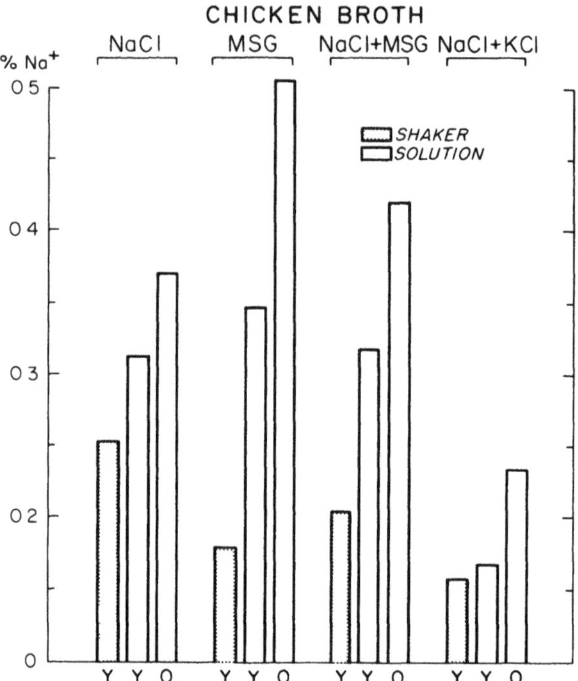

Fig. 11.1. *Ad libitum* mixing. Average amount of sodium added to unsalted chicken broth by 10 younger (Y) subjects (17–32 yr) and 20 older (0) subjects (36–66 yr), via salt shakers and by solutions, using NaCl, monosodium glutamate (MSG), and 1:1 mixtures of NaCl:MSG, and NaCl:KCl.
Source: Braddock and Pangborn 1984.

judges showed a high correlation between hedonic and *ad libitum* procedures only for NaCl. In recent experiments with 100 university students (Stone 1984), *ad libitum* addition of NaCl to unsalted beef broth ranged from 0.10 to 0.90% ($\bar{x} = 0.47\%$) Na^+, and from 5.2 to 37.4% ($\bar{x} = 13.6\%$) sucrose (measured by refractometer) in unsweetened lemonade. The foregoing were liquid-to-liquid additions. Judges who mixed to higher levels gave greater hedonic ratings to higher concentrations when presented in presalted or presweetened series. Males tended to mix to higher levels of both NaCl and sucrose.

Pangborn *et al.* (1984) instructed lean, normal-weight, and overweight female adults to mix skim milk (0% fat) and "cream" (12% fat) to preference. Responses did not differ by weight, but judges with a high-fat intake mixed significantly higher fat contents than did the low- and medium-intake groups. Furthermore, higher fat levels were mixed by judges who drank regular market milk (3.5% fat) than by those who drank skim milk. Paradoxically, none of the foregoing observations was evidenced when the same judges performed paired preference tests or scored their degree of liking for milk varying in fat content across the same range. This suggests that *ad libitum* mixing to preference may have been more representative of "true" preference than were conventional paired or scoring tests.

Table 11.2 presents a summary of the method of *ad libitum* mixing. The major advantages include its simplicity for the experimenter, as there is limited sample preparation, and for the judges, most of whom find the procedure interesting. Naturally, the procedure is limited to miscible stimuli, but the greatest limitation is the availability of a sensitive, simple chemical or physical methods for analysis of the mix. It has been noted in the author's laboratory that highly trained judges can discern smaller differences in the sweetness of sucrose solutions than the resolution of the refractometer. Mention in Table 11.2 of mischievous judges refers to the possibility that a judge might be tempted to simply pour some of the reference into the "mix" container without bothering to taste it, a possibility that could be precluded by placing a small amount of food coloring into the reference. A second

TABLE 11.2. SUMMARY OF *AD LIBITUM* MIXING PROCEDURE

Objective:	Allows the judge to determine by his own mixing, at his own speed, the concentration of a stimulus which either: a. matches a standard or b. represents his preference.
Technique:	*Discrimination:* Judge mixes a low-intensity with a high-intensity stimulus to equal intensity of a standard. *Preference:* Judge mixes a low-intensity with a high-intensity stimulus to a preferred level.
Advantages:	1. Versatile; can measure difference or preference 2. Simple to prepare and administer 3. A natural, straightforward task 4. Interesting to the judge 5. Judge has infinite response alternatives (theoretically) in contrast to responding to fixed concentrations. 6. Gives direct measure of response 7. Requires little or no training 8. Simple statistical treatment of data
Limitations:	1. Limited by availability of simple, sensitive chemical or physical analysis of the mix 2. Limited to miscible stimuli 3. Impatient judge may under- or overshoot 4. Overcautious judge takes a long time 5. "Mischievous" judge could cheat[a]

[a] See text.

discrepancy could occur when the judge is given two containers to mix duplicates, wherein one could simply divide one "mix" between the two. This can be prevented by presenting one container and removing it before presenting the second.

It is evident from the foregoing summary of the limited number of published reports on the method, that additional testing is needed in order to establish reliable recommendations on the use of *ad libitum* procedures for measurement of discrimination or of preferences. The author believes that the *ad libitum* preference procedure holds considerable promise for nutritional or clinical studies, where investigators are concerned with differences in selection, preferences, and intakes of control vs. test populations.

TIME-INTENSITY PROCEDURES

One of the most neglected parameters in analytical sensory measurement has been the temporal. The time course of perception, from onset through maximum to extinction, differs across stimuli (Kelling and Halpern 1983). For taste, compounds can vary in intensity, in quality, and in duration within a modality (sweet, sour, salty, or bitter). The temporal patterns and "aftertastes" of several bitter and sour compounds are well known. For example, individuals who are sensitive to the bitterness of the synthetic sweetener, saccharin, complain of the persistence of bitterness in sugar-free beverages. The importance of the temporal properties of several synthetic sweeteners has been emphasized in recent publications (DuBois *et al.* 1977; Swartz and Furia 1977; Swartz 1980; Larson-Powers and Pangborn 1978A).

Persistence of odors, although more difficult to measure than taste, are of considerable scientific and economic importance to the producers of flavorings and fragrances used in foods and beverages, personal care products, and in household products. Perhaps the most dramatic temporal sensations associated with food occur during mastication and swallowing, when the food structure is broken down in a regular sequence (Szczesniak 1963; Civille and Szczęsniak 1973).

Neilson (1958) was among the first to apply time-intensity (T-I) procedures to measurement of persistence, using flavorants in chewing gum, and masking agents for bitter drugs. Judges were timed with a stop watch and marked perceived intensity on a 0–3 point scale at specified time intervals during oral manipulation of the stimulus. Findings indicated, for example, that perceived bitterness of a "drug" reached lesser intensities and lasted shorter periods of time when sucrose was added to the drug, and even less so when monosodium glutamate was added. Earlier, Neilson (1957) had noted that although bitterness was a necessary and desirable characteristic of beer, if it lingered too long after swallowing, the bitterness could satiate and become undesirable. On the other hand, intense bitterness could be tolerated in beer if it were of short duration. In studies reported by McNulty and Moskowitz (1974), judges immersed their tongues in flavored oil-in-water emulsions and recorded magnitude estimates of taste intensity every 5 sec at the sound of a signal. The results demonstrated a complex effect of solute–surfactant interaction on flavor perception, which altered both the extent and the rate of flavor release.

Larson-Powers and Pangborn (1978A) utilized a procedure in which time was continuously monitored by a strip chart recorder, freeing the judge to concentrate on the stimulus without the distraction of a stop watch or of an auditory signal. The judge placed the stimulus into the mouth at "0" time while simultaneously moving the pen horizontally across the stationary paper-cutter bar, which was labeled "none" and "extremely strong" at the ends. Judges were instructed to swallow or to expectorate the sample, according to the objective of the experiment, but to continue marking perceived intensity until extinction. The moving chart, which was covered from view, resulted in a trace which represented the temporal intensity pattern of the stimulus. After digitizing the curves, several functions can be analyzed statistically, such as maximum sensory intensity, time required to reach maximum, total duration of sensation, and total area (or total perimeter) of the curve.

Originally, time-intensity techniques on a moving chart were used to measure temporal aspects of the sweetness and bitterness of sucrose, cyclamate, saccharin, and aspartame in drinks and gelatins (Larson-Powers and Pangborn 1978A). Subsequently, the procedure was used by Harrison and Bernhard (1984) to demonstrate sweetness additivity of lactose–saccharin mixtures, sweetness synergism of lactose–xylitol mixtures, and sweetness suppression of lactose–galactose mixtures. Moore and Shoemaker (1981) used the same technique to evaluate texture of ice cream. Addition of sodium carboxymethylcellulose stabilizer produced differential temporal patterns for distinct sensory parameters—iciness, coldness, viscosity, and melting time. Pangborn and Koyasako (1981) measured intensity of three parameters in canned pudding and canned crème—oral viscosity, sweetness, and chocolate flavor. As illustrated in the mean tracings in Fig. 11.2, pudding was judged to be more viscous, but less sweet than crème despite identical amounts of sweetener in the two products. The degree to which sweetness and chocolate flavor of both products persisted long

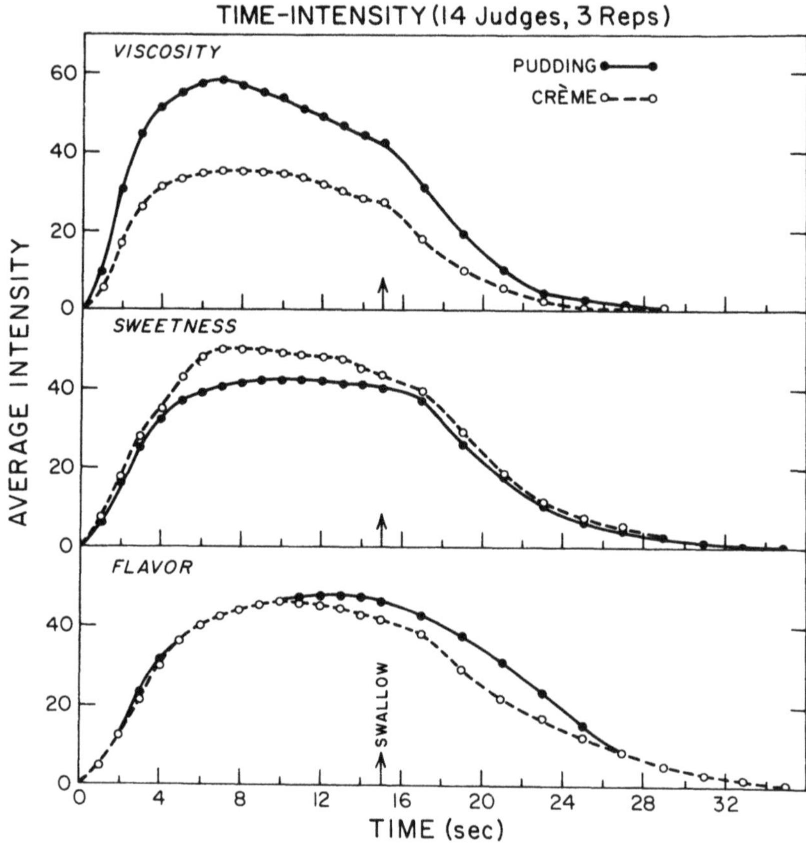

Fig. 11.2 Time-intensity curves for oral viscosity, sweetness and chocolate flavor of canned pudding and of crème. Judges placed 5-g samples into mouth at "0" time and swallowed at 15 sec and continued to record perceived intensity until extinction. Each point is the average of 42 measurements (14 judges ×3 replications).
Source: Pangborn and Koyasako 1981.

after viscosity was no longer perceived is evident from the tracing. Note also, differences in the time to reach maximum intensity of the individual attributes after placement of the 5-g sample into the mouth—viscosity reached maximum after 8.6 sec, sweetness after 11.1 sec, and flavor after 12.9 sec. The authors recommended that the TI procedure be considered for studies of the rate of breakdown of the physical structure of complex substances, such as tough, hard, chewy, sticky, cohesive, and tacky foods, to quantify release of chemical compounds responsible for tastes and flavors.

Figure 11.3 depicts a uniquely temporal taste response for bitterness which would not have been detected by simple scaling of intensity. The upper and lower tracings represent intensity of iso-α-acids (bitter principle of hops) dispersed in distilled water and in a light lager beer, respectively. Greater bitterness and a longer duration was perceived in water than in beer. Of even greater contrast are the sensory events which appear to follow swallowing (designated by the arrow) in the water solutions which are totally absent in the beer. Within 5 sec of swallowing the sample, there was a burst of bitterness in the water solutions at all concentrations tested, which was not demonstrated with beer. Although not verified, it is possible that the complexity of the beer, particularly the ethanol, CO_2, and sugars, modified bitterness intensity while "blending" the tastes (Pangborn *et al.* 1983B). The different patterns illustrated in Fig. 11.3 emphasize a useful analytical application of the time-intensity procedure, that of comparison of perceptual responses to model systems with those in food and beverage systems.

Additional studies with salts dispersed in water and in broth (Pangborn and Chung 1981) and with acid solutions varying in anionic species, pH, or total acidity (Norris *et al.* 1984) have established the presence of a lag time of 2–6 sec between perception of the taste sensation and the onset of salivation. Further experiments are underway to compare oral firmness, type of breakdown, and taste intensities of gels prepared from gelatin, carrageenan, and sodium alginate (Muñoz-Lezama 1983). Time-intensity sensory responses were compared with the temporal physical responses of these gels to compression, shearing, and puncture by an Instron Universal Texturometer.

In an innovative set of experiments, Lawless and Skinner (1979) connected a dial with a pointer to a strip chart recorder to facilitate the judge's representation of perceived taste intensity over time. They reported that the time-course of the sweetness of sucrose solutions was characterized by a rapid rise to peak intensity with an equally rapid decline. Higher peaks and longer durations were obtained with greater concentrations, but different scaling methods (category vs. ratio scaling) resulted in only small differences in the form of the T-I function.

Originally, Birch *et al.* (1980) measured time-intensity of the sweetness of sucrose and of thaumatin with stop watches to record reaction time, time to maximum intensity, time to end of maximum intensity, and total persistence time, on separate occasions. Subsequently, Birch

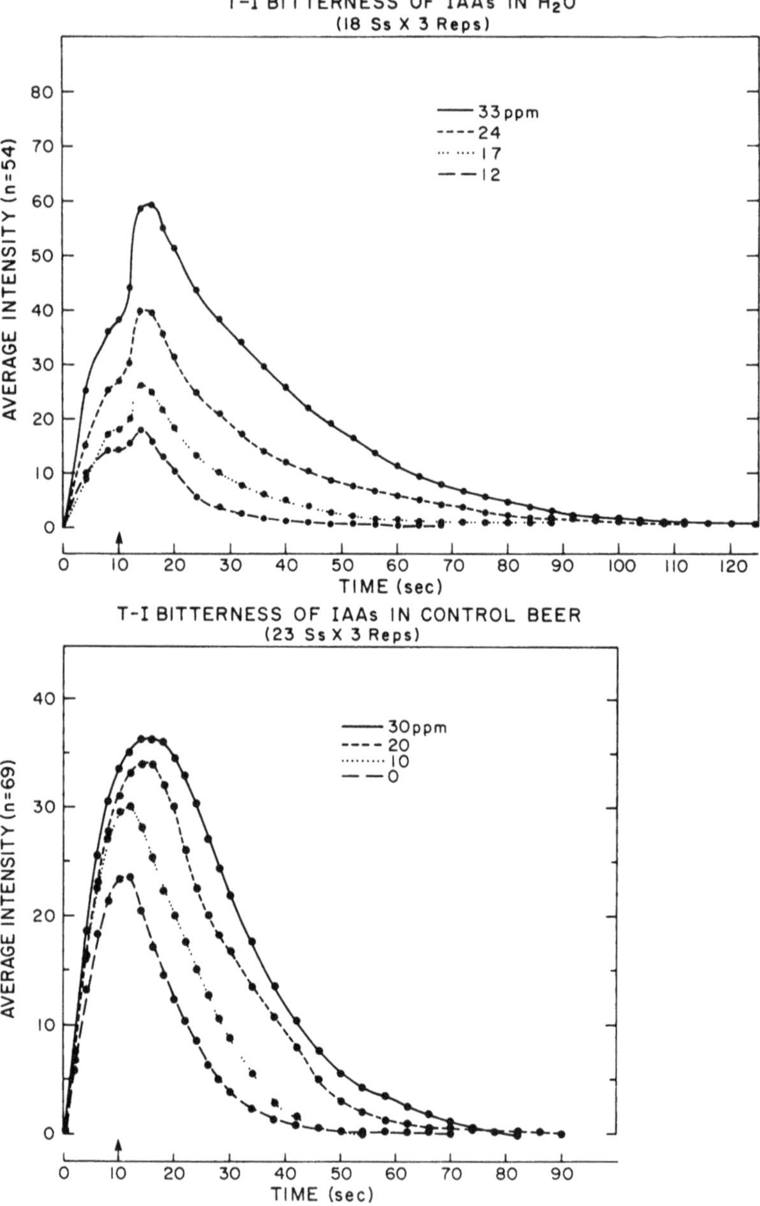

Fig. 11.3. Time-intensity curves for bitterness intensity of four concentrations of iso-α-acids dispersed in distilled water (upper graph) and in light beer (lower graph), tested at 21° and 7°C, respectively. Samples of 15 and 10 mL, respectively, were placed in the mouth at "0" time and swallowed at 10 sec (indicated by arrow), while judge continued to record bitterness until extinction.
Source: Pangborn et al. 1983.

TABLE 11.3. SUMMARY OF TIME-INTENSITY PROCEDURES

Objective:	To quantify the temporal changes in sensation that occur from the time the stimulus is placed in the mouth until extinction.
Technique:	Judge marks perceived intensity of a single parameter on a moving chart, or indirectly by manipulation of a dial or lever.
Information obtained:	
	1. Maximum intensity perceived.
	2. Time to maximum intensity.
	3. Rate and form of increase to maximum.
	4. Rate and form of decrease to extinction.
	5. Changes associated with expectoration, swallowing.
	6. Total duration of the sensation.
	7. Total area under the curve.
Advantages:	1. Provides a total temporal "picture" of sensation perceived during the critical time of oral manipulation and swallowing (or expectoration).
	a. For comparison with other oral temporal functions, e.g., salivary flow, mastication sequence, swallowing, etc.
	b. For comparison with other nonoral temporal measures, e.g., rate of physical breakdown and/or chemical reactivity of the stimulus.
Limitations:	1. Only one sensory parameter can be judged at a time.
	2. Cannot be used for slow-changing parameters such as changes in appearance or color of product.
	3. Difficult to utilize with aromas (but useful for flavors).
	4. Requires three or more practice sessions to achieve confidence in use of procedure.
	5. Requires more concentration by judge, who might be distracted by moving chart, manipulation of dial, etc.
	6. Requires some specialized equipment, e.g., recorders, etc.
	7. Unless curves can be recorded directly into a computer, the digitizing and averaging of individual curves to obtain an average curve is very labor intensive (Schmitt *et al.* 1984).

and Munton (1981) recorded temporal taste responses by having judges turn a dial proportional to perceived intensity. The dial box was positioned away from the chart recorder to prevent potential distraction. Using their Sensory Measuring Unit for Recording Flux (SMURF), the authors have characterized the time course of sweetness and flavor of chocolate solutions (Birch and Ogunmoyela 1980).

Table 11.3 provides a summary of the time-intensity procedures. At first glance it may appear that there are far more limitations than advantages to the use of this method. However, the time invested in collection of the data is more than offset by the quality and potential application of the information derived.

DEVIATION-FROM-REFERENCE DESCRIPTIVE ANALYSIS

Scoring of perceived intensities is one of the oldest of sensory methods (Amerine *et al.* 1965), and the use of adjectives to describe sensory quality is well established (Caul 1957). To improve the precision of the conventional Flavor Profile, which uses a restricted, semiquantitative scale, Quantitative Descriptive Analysis introduced an expanded, nonnumerical scale and individual replicate judgments, which lend themselves to versatile statistical analyses (Stone *et al.* 1974, 1980). With all descriptive techniques, accuracy and reproducibility of response are increased if physical references are provided for each sensory descriptor. Anchoring descriptors to chemical compounds and physical substances further enhances and standardizes communication among judges during testing, between judge and experimenter, and across experiments for comparison with literature data. A further increase in reliability can be effected when each attribute is measured in terms of degree of deviation from a standard. This reference provides the same "benchmark" for each judge and from replication to replication.

As noted some time ago, scoring in terms of deviation from an experimental control (or a commercial sample of known market quality) resulted in greater reproducibility of individual response and better agreement among judges, compared to unanchored scoring (Pangborn 1967). A good example of industrial utilization of the technique was provided by Daget (1974) for evaluating 13 commercial milk chocolates from different countries. Four texture terms and eight flavor terms were scored against a standard sample on a category scale from -1 to -4 when less than the reference and between $+1$ and $+4$ when greater. Results were presented in the form of a histogram, with bars projecting above and below the "0" line, and included deviations from the reference contrasted with itself, a measure of reliability. Similar descriptive procedures were reported by Larson-Powers and Pangborn (1978B) to compare drinks and gelatins containing synthetic sweeteners against the sucrose standard. In all media, the anchored method resulted in a greater number of descriptors which were significantly different among the four sweeteners, indicating that it had a greater resolution than did the unanchored method. Furthermore, analysis of variance of the data showed less variation due to the judges with the former compared to the latter descriptive method. Advantages and limitations of the procedure which were pointed out by Larson-Powers and Pangborn (1978B) are included in the summary in Table 11.4.

The large number of descriptors generated to describe the appearence, aroma, taste, and mouthfeel attributes of a complex food or beverage necessitates application of more than simple univariate statistical analysis. Among the multivariate statistical techniques, which have been applied to matrices of descriptive data, is principle compo-

TABLE 11.4. SUMMARY OF DEVIATION FROM REFERENCE DESCRIPTIVE ANALYSIS

Objective: To measure the degree to which specific sensory attributes of experimental samples differ from a reference (a control or a commercial sample of known quality).

Technique: Judges receive experimental samples (including a blind reference) to compare against the labeled reference (*R*) for individual attributes (*e.g.*, green color, grassy aroma, oral sourness, bitter aftertaste, etc.), using a bipolar scale:

1. *Numerical:*

Less than *R:*	Same as *R:*	More than *R:*
−5, −4, −3, −2, −1;	0;	+1, +2, +3, +4, +5

2. *Nonnumerical:*

Less than *R*	Same as *R*	More than *R*

3. *Ratio:* Specify on open-ended scale, how many times great or lesser the sample is than *R*.

Advantages:
1. Provides a fixed criterion of comparison, the same for all judges, which minimizes drifting of responses with time, or from comparison against faulty memory standards.
2. Provides internal measure of judge reliability by comparison of reference against itself as a blind sample.
3. When a large number of experimental samples necessitates the presentation of samples in an incomplete block design, comparison of each sample against the same reference increases reliability.
4. Adds reliability in storage stability studies if the same reference can be provided at each time period.
5. In product matching or product formulation, provides a relatively rapid measure of individual attributes, and hence ingredients which need to be increased or decreased to better approximate the reference.
6. Data lend themselves to several univariate and multivariate statistical methods of analysis.

Limitations:
1. Gives an indirect, rather than a direct measure of the degree of difference among experimental products.
2. Not applicable within experimental designs where a reference cannot be designated.
3. As with all descriptive methods, considerable time and effort is required to develop the descriptive terminology and the corresponding physical references, as well as to select and train judges.

nent analysis (PCA). Based upon Euclidean geometry and matrix algebra, PCA consists of calculation of orthogonal least-squares to determine the first principal axis (or factor) with subsequent axes following from successive orthogonal least-squares solutions (Vuataz 1976–1977; Vuataz *et al.* 1974).

The BMDP4M computer program for factor analysis (Dixon and Brown 1977) can be applied to a variance–covariance matrix of the original data where columns represent the experimental samples and rows are the sensory descriptors. The imput matrix is column centered by a translation (subtracting the column mean from each individual column entry), which places the origin at the centroid of the point configuration. The *X* axis is pivoted by a Varimax or Kaiser rigid rotation until it bisects the plane of elongation of the data points, to constitute the first factor. Subsequent factor loadings are orthogonal, and hence independent of the previous ones. The computer program provides rotated factor loadings for the experimental samples and for the sensory measurements. These two sets of factor loadings can be plotted on the same set of Cartesian coordinates in order to represent visually the spatial relationship of each sample with descriptor (Figs. 11.5–11.9). This multivariate analysis does not lead to a decision on an hypothesis or to an estimation of probability, but is useful (a) to study the correlations of a large number of variables by clustering them into factors, such that variables within each factor are highly correlated; (b) to interpret each factor according to the variables belonging to it; (c) to summarize many variables by a few factors (Dixon and Brown 1977).

Space limitations of the present chapter and statistical limitations of the present author preclude an extensive review of factor and principle component analysis. However, the serious reader is advised to first consult some of the early applications of the methods to sensory data, e.g., Baker (1954), Baker *et al.* (1961), Harper (1956), Harper and Baron (1951), Schutz (1964), and Woskow (1968). The assessibility of computer programs has resulted in further informative applications to sensory data, as exemplified in the articles on texture of meat (Horsfield and Taylor 1976), poultry (Frijters 1976), fish (Cardello *et al.* 1982), and other foods (Toda *et al.* 1971; Yoshikawa *et al.* 1970). Factor and principle component analysis have been utilized extensively in flavor research, for example, in studies on tea (Palmer 1974), beer (Clapperton and Piggott 1979; Hoff *et al.* 1978; Moll *et al.* 1978), wines (Noble 1978; Noble *et al.* 1983; Williams *et al.* 1980), distilled beverages (Jounela-Ericksson 1980; Piggot and Canaway 1981), and even the aroma of raw carrots (McLellan *et al.* 1983).

Several multivariate analyses facilitate visual representation of data. Figures 11.5–11.9 are illustrations of various forms of presentation of results obtained from experiments conducted in the author's laboratory, using deviation-from-reference descriptive analysis. Figures 11.4 and 11.5 derive from data obtained from 15 trained judges on the sensory attributes of experimental corn tortillas, wherein potato was substituted for part of the corn prior to preparation of the dough (Feria-Morales and Pangborn 1983). The histogram in Fig. 11.4 presents the means and standard deviations for 15 descriptors, derived from an original list of 30 descriptors, initially developed to characterize the sensory attributes of commercial corn tortillas. The variability represented by the all-corn reference sample compared to itself is a mea-

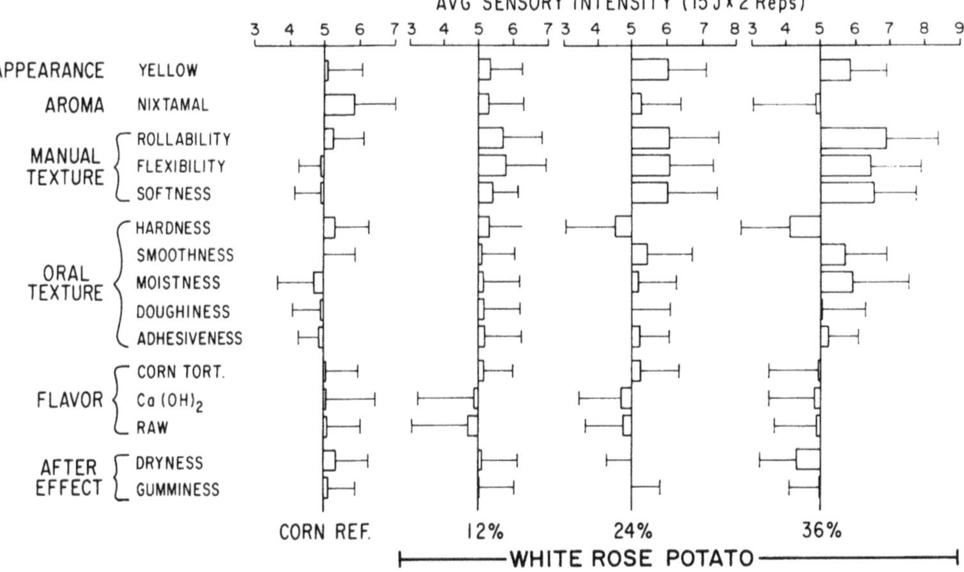

Fig. 11.4. Deviation-from-reference descriptive analysis results presented as a conventional histogram. Depicted are means and standard deviations of sensory attributes of experimental tortillas made from corn and from 12, 24, and 36% substitution with White Rose potatoes. Values are degree of deviation from the all-corn reference ($R = 5$) on a 0–10 point scale. The left column corresponds to responses to the blind reference compared against itself.
Source: Feria-Morales and Pangborn 1983.

sure of the reliability of the panel. With increasing substitution with potato, there was a corresponding increase in manual rollability, flexibility and softness, and a decrease in oral hardness. Oral smoothness and moistness, as well as yellow color, also increased slightly with increasing potato content. Representing the same data, derived from PCA, on two-dimentional Cartesian coordinates gives even better comparative visual information (Fig. 11.5). Although only 50% of the total variance is accounted for by the two dimensions, the importance of manual and oral texture as descriminators among the four tortilla samples is readily seen. With increasing concentrations of potato, there are increases in manual texture (rollability, flexibility, softness) in oral texture (moistness, smoothness, adhesiveness) and to a lesser degree in yellow color. Hardness and dryness had a high, inverse correlation with the aforementioned texture attributes. Factor loadings showed that factor 1 was highly correlated (>0.70) with manual texture, fac-

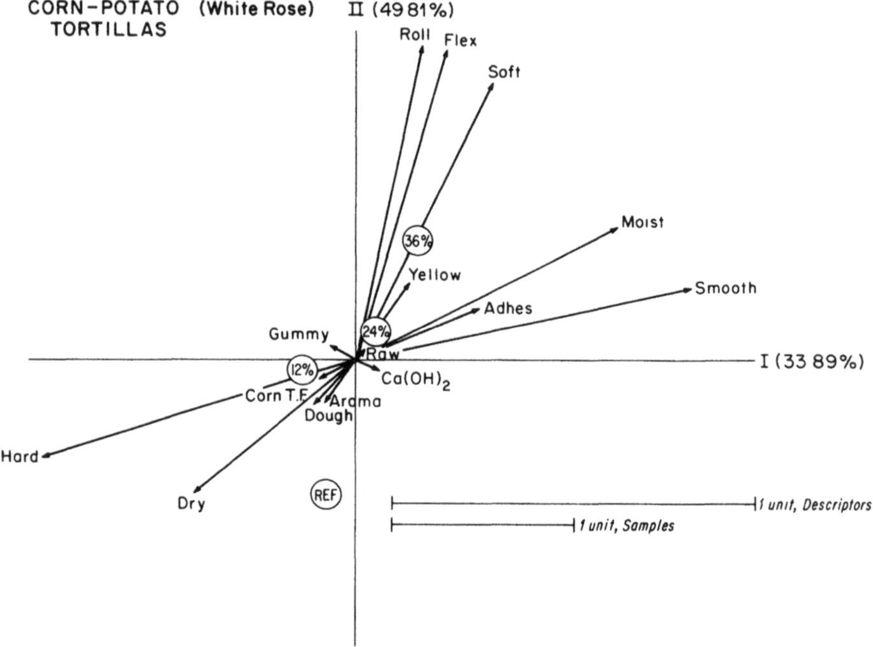

Fig. 11.5. Deviation-from-reference descriptive analysis results presented on two-dimensional Cartesian coordinates from principal component analysis of data shown in Fig. 11.4. Sensory descriptors are designated by vectors and the four experimental tortillas are indicated by circles. Parenthetical values represent the cumulative variance for the two axes.
Source: Feria-Morales and Pangborn 1983.

tor 2 with oral texture, factor 3 with flavor attributes, factor 4 with aroma, etc., up to eight factors to account for 91.5% of the total variance in the data matrix (Feria-Morales and Pangborn 1983). In the original article, of course, these visual representations are supplemented by statistical analyses, including correlation coefficients among all sensory attributes and between sensory and instrumental measurements of texture, as well as analyses of variance and least significant differences to establish statistical differences among sample means for each attribute.

Originally, Spencer *et al.* (1978) applied stepwise multiple regression analysis to compare gas liquid chromatographic (GLC) measurement of volatiles from canned cling peaches with descriptive sensory analysis of aroma and flavor. Concentrates were prepared from canned

Halford cling peaches, the conventional canning variety, and from nine experimental varieties. They were evaluated on glass capillary columns, incorporating two internal standards, tridecane and octadecane. Chromatographic peaks were numbered in order of elution time for peach variety 7-7-52, which had the most peaks. Relative concentrations of compounds (identified by mass spectroscopy and relative retention times) were normalized to the Halford variety and the data matrix subsequently was subjected to PCA. Figure 11.6 is a two-dimensional representation of the GLC data. The first axis accounts for 67.2% of the variance and represents compounds identified as linalool, geraniol, α-terpineol, isoamyl acetate, and γ-heptalactone, which are of equivalent importance. To a lesser extent, the unidentified compounds of peaks 26 and 20 and carvomenthenal also are associated with the first axis. The second orthogonal axis represents 18.3% of the data variation and is associated with compounds identified as *cis*-3-hexenyl, γ-decalactone, furfural, and unidentified peak 36. The Hal-

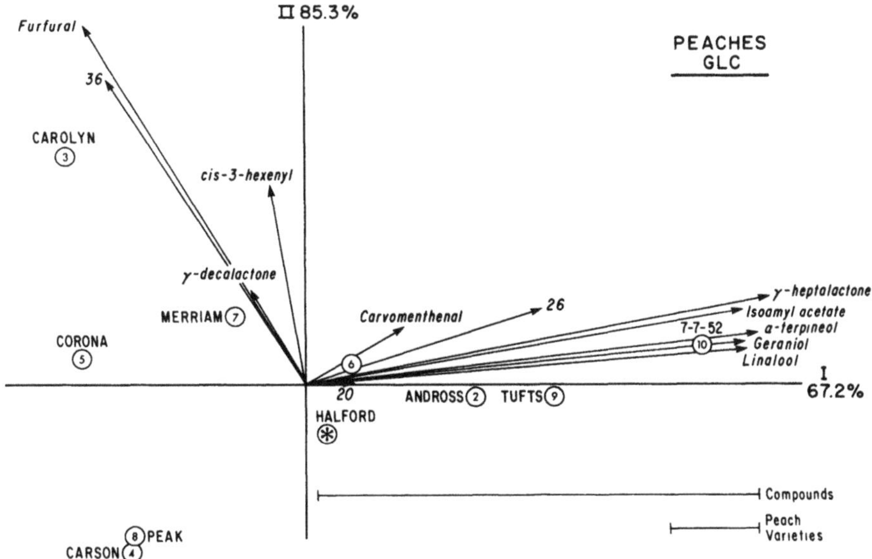

Fig. 11.6. Deviation-from-reference results presented on two-dimensional Cartesian coordinates from principal component analysis of normalized peak heights from GLC measurement of volatiles from canned cling peaches of ten varieties. The peach samples are indicated by circles for the Halford (reference) and for experimental varieties 2–10. Identified and unidentified peaks are designated by vectors.

Recalculation of original data presented previously by Spencer et al. (1978).

ford sample is located fairly close to the centroid of the configuration, indicating that it (and No. 6, Everts variety) are representative of the "average" sample among the ten evaluated. Note that peach samples 10, 9, and 2 are associated with the horizontal compound vectors, which could be interpreted as being "desirable" volatiles for peaches, while samples 7 and 3 are more closely associated with the vertical compound vectors. Sample 5 could be interpreted as lacking in the horizontal and samples 8 and 4 as lacking in the vertical compound vectors. The remaining 14% variation in the data is contained in factors three and four, which cannot be shown in two dimensions.

A projection based on principal component analysis of the descriptors for the sensory properties of the peach varieties studied by Spencer *et al.* (1978) shows that 93% of the variance was explained by the first two dimensions (Fig. 11.7). The intensities of peach odor, fruity and floral, are highly correlated, and are inversely related to attributes called dusty, woody, and overcooked. The foregoing two sets of

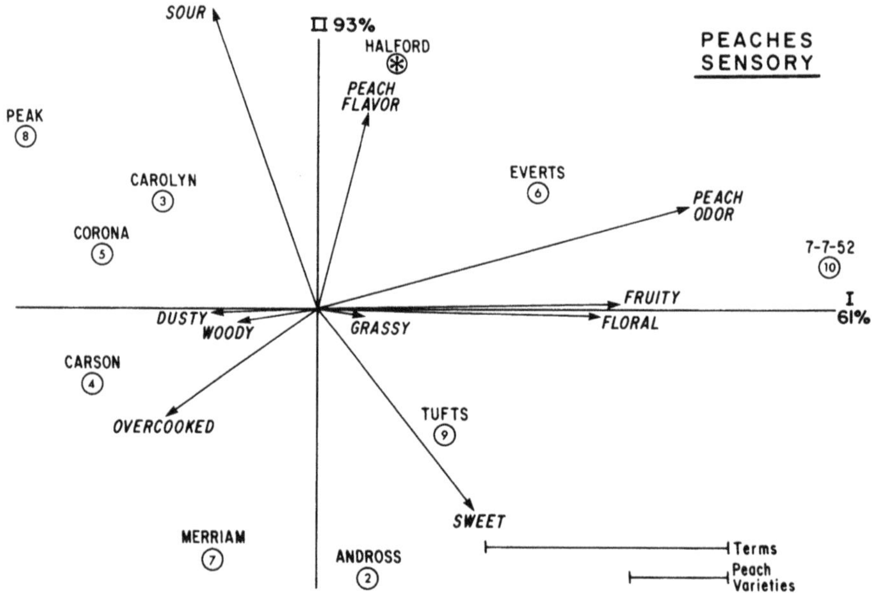

Fig. 11.7. Deviation-from-reference descriptive analysis results presented on two-dimensional Cartesian coordinates from principal component analysis of sensory evaluation of aroma and flavor of canned cling peaches of ten varieties. The peach samples are indicated by circles for the Halford (reference) and for experimental varieties 2–10. Sensory attributes are designated by vectors.
Recalculation of original data presented previously by Spencer et al. (1978).

terms are generally perpendicular to taste attributes of sour and sweet, which are inversely related to each other. Note that flavor-by-mouth is almost orthogonal, and hence independent, of peach odor. The least important attribute was grassy. Relative to the peach varieties, the Halford reference (designated by a starred circle) was not close to the centroid, i.e., it did not represent an "average" peach among the ten evaluated. However, it is closely associated with peach flavor. Samples 10 and 6, and to a lesser extent, sample 9, are associated with "desirable" peach traits of floral, fruity, and peach odor. Conversely, samples 3 and 5, and to some degree samples 4 and 8, group with "less desirable" descriptors of dusty, woody, and overcooked. Samples 7 and 2 could be interpreted as lacking in sourness and in peach flavor and, along with sample 9, are associated with varying degrees to the vector for sweetness.

Not only are the oral attributes of sourness, sweetness, and peach flavor orthogonal to those for aromas, but it is recognized that they are based, generally, on nonvolatile compounds which would not be detected by GLC. Therefore, the three oral descriptors were removed

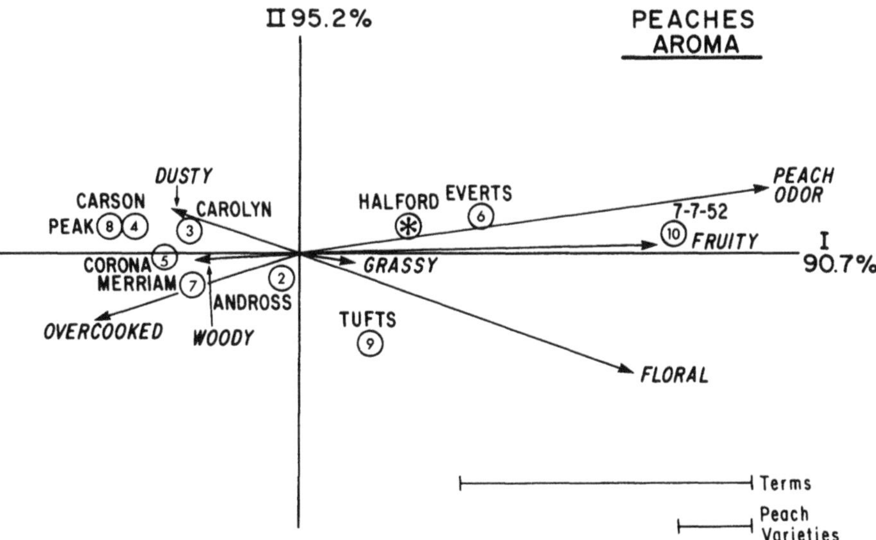

Fig. 11.8. Deviation-from-reference descriptive analysis results presented on two-dimensional Cartesian coordinates from principal component analysis of sensory evaluation of aroma, only, of canned cling peaches of ten varieties. The peach samples are indicated by circles for the Halford (reference) and for experimental varieties 2–10. Sensory attributes are designated by vectors.
Recalculation of original data presented previously by Spencer et al. (1978).

from the array and resubmitted to principal components analysis, as shown in Fig. 11.8. A striking dichotomy results, with 95.2% of the variance contained in two dimensions. The most important attributes are peach odor, fruity and floral, with high positive correlations, which in turn are negatively correlated with overcooked, dusty, and woody. The projection is unequivocable as to which peach samples correspond to the sensory descriptors.

A final integration of the peach data is represented in Fig. 11.9, where Procrustes analysis (Grower 1971, 1975; Schönemann and Carroll 1970) was applied to calculate the geometric distances between the products and between the analytical measurements of the two sets of data. Here we have confirmation of the previously intuitive extrapolation from Figs. 11.5 and 11.6. First, note that although sweetness and sourness are important negatively correlated vectors, that they are orthogonal,

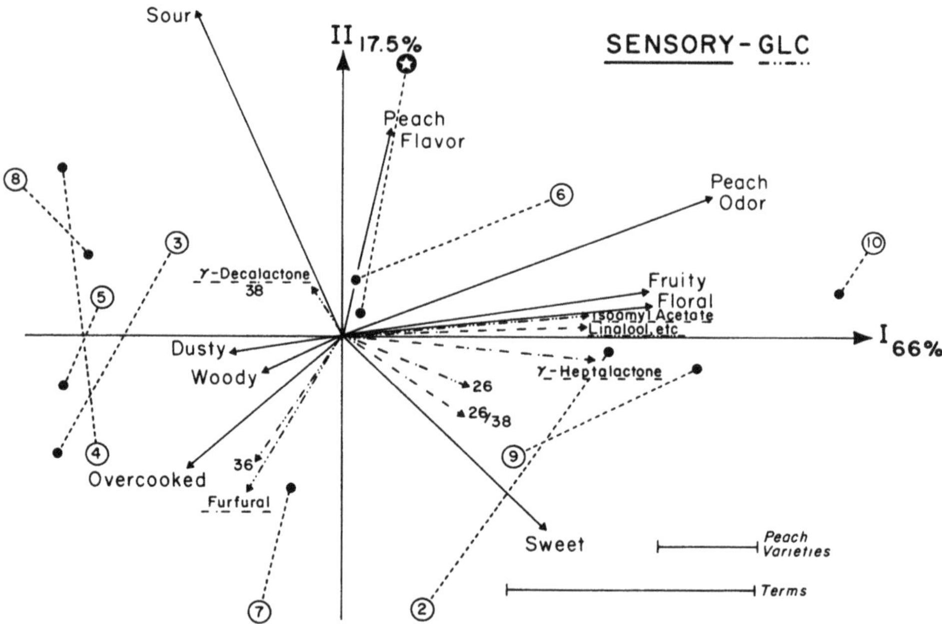

Fig. 11.9. Procrustes analysis of sensory attributes and of GLC peak heights of canned cling peaches of ten varieties plotted on the same set of Cartesian coordinates. Peach varieties are indicated by numbered circles for sensory data, connected by dots to the corresponding GLC measurement. Solid vectors represent sensory attributes and broken-line vectors represent GLC peaks. The Halford reference is indicated by a starred circle.

Recalculation of original data presented previously by Spencer et al. (1978).

and therefore independent of the main flavor attributes associated with the GLC measurements, justifying their exclusion in Fig. 11.8. It is evident that peach odors, fruity and floral, are closely associated with compounds identified as isoamyl acetate, linalool, etc. (geraniol, γ-heptalactone, and γ-terpineol, which superimposed on the same location). Furthermore, furfural and unidentified compound 36 were closely associated with the overcooked, woody, and dusty descriptors, attributed in varying degrees to peach samples 7, 4, 5, 3, and 8. Only for sample 10 was there close agreement between GLC and sensory attributes, as indicated by the relatively short distance between the numbered circle (sensory) and the connecting dot (GLC). Relatively large distances in the two types of measures are evident for samples 2 and 9, which probably were caused by sweetness of compounds which cannot be detected by GLC.

Several more interpretative paragraphs could be written about the results contained in Figs. 11.6–11.9, emphasizing the potential utility of these multivariate statistical procedures in the interpretation of multimeasurements, be they physical, chemical, or sensory. More importantly, the contribution of Procrustes analysis to the integration and cross-comparison of independent measures on the same set of products merits further application to sensory as well as to physical/chemical data. For many of us, the sophistication of the underlying statistical principles, and utilization of appropriate computer programs, necessitates extensive consultation with knowledgeable experts.

CONCLUSIONS

The foregoing introduction to three key analytical sensory methods—*ad libitum* mixing, time-intensity, and deviation-from-reference descriptive analysis—is intended to stimulate the interest of scientists working at the periphery of sensory science. Admittedly of limited use to readers with no knowledge of elementary psychophysics and statistics, the coverage in turn will seem superficial to those with extensive experience in advanced sensory and statistical analyses. The best advice to the beginner, in any analytical field, is to carefully delineate the test objective, then to judiciously select the appropriate analytical method. To the seasoned investigator, particularly those in industrial settings, a plea is made for greater dissemination of knowledge about procedures and techniques.

The author invites comment and recommendations from all who took the time to read this chapter, so that subsequent attempts to reach a broad audience will have improved breadth as well as depth. Additional examples of results obtained from application of the methods described herein would be gratefully received.

BIBLIOGRAPHY

AMERINE, M.A., PANGBORN, R.M., and ROESSLER, E.B. 1965. Principles of Sensory Evaluation of Foods. Academic Press, New York.

ASTM. 1968. Correlation of subjective–objective methods in the study of odors and taste. ASTM STP *440,* Am. Soc. Test. Mat., Philadelphia, PA.

BAKER, G.A. 1954. Organoleptic ratings and analytical data for wines analyzed into orthogonal factors. Food Research *19* (6), 575–580.

BAKER, G.A., AMERINE, M.A., and PANGBORN, R.M. 1961. Factor analysis applied to paried preferences among four grape juices. J. Food Sci. *26* (6), 644–647.

BARTOSHUK, L.M., CAIN, W.S., CLEVELAND, C.T., GROSSMAN, L.E., MARKS, L.E., STEVENS, J.C., and STOLWIJK, J.A.J. 1974. Saltiness of MSG and sodium intake. J.A.M.A. *230* (5), 670.

BIRCH, G.G., and MUNTON, S.L. 1981. Use of the "SMURF" in taste analysis. Chem. Senses *6* (1), 45–52.

BIRCH, G.G., and OGUNMOYELA, G. 1980. Effect of surfactants on the taste and flavour of drinking chocolate. J. Food Sci. *45* (4), 981–984.

BIRCH, G.G., LATYMER, S., and HOLLAWAY, M. 1980. Intensity/time relationships in sweetness: evidence for a queue hypothesis in taste chemoreception. Chem. Senses *5* (1), 63–78.

BRADDOCK, K.S., and PANGBORN, R.M. 1983. Salt and salt-substitute preferences in chicken broth as related to dietary salt intake by two age groups. Appetite (in press).

CARDELLO, A.V., SAWYER, F.M., MALLER, O., and DIGMAN, L. 1982. Sensory evaluation of the texture and appearance of 17 species of North Atlantic fish. J. Food Sci. *47* (6), 1818–1823.

CAUL, J.F. 1957. The profile method of flavor analysis. Adv. Food Res. *7,* 1–40.

CIVILLE, G.V., and SZCZESNIAK, A.S. 1973. Guidelines to training a texture profile panel. J. Texture Studies *4,* 204–223.

CLAPPERTON, J.F., and PIGGOTT, J.R. 1979. Flavour characterization by trained and untrained assessors. J. Inst. Brewing *85,* 275–277.

CROSS, H.R. 1978. Guidelines for Cookery and Sensory Evaluation of Meat. Am. Meat Sci. Assoc., Chicago IL.

DAGET, N. 1974. Profile sensory evaluation of chocolates. Paper presented at Erster Internationaler Kongress uber Kakao und Schokolade forschung, Munchen, 10 pp.

DIXON, W.J., and BROWN, M.B. 1977. BMDP-77. Biomedical Computer Programs P-Series. Univ. of California Press, Berkeley.

DuBOIS, G., CROSBY, G.A., STEPHENSON, R.A., and WINGARD, R.E. JR. 1977. Dihydrochalcone sweeteners. Synthetic and sensory evaluation of sulfonate derivatives. J. Ag. Food Chem. *25* (4), 763–772.

ELLIS, B.H. 1963. Chicanery in the flavor room. Am. Soc. Brewing Chem. 37–43.

FERIA-MORALES, A.M., and PANGBORN, R.M. 1983. Sensory attributes of corn tortillas with substitutions of potato, rice, and pinto beans. J. Food. Sci. *48,* 1124–1130, 1134.

FRIJTERS, J.E.R. 1976. Evaluation of a texture profile for cooked chicken breast meat by principal component analysis. Poultry Sci. *55,* 229–234.

GAWECKI, J., URBANOWICZ, M., JESZKA, J., and MAZUR, B. 1976. Feeding habits and their physiological determinants relationships between consumer's preferences for sweetness of drinks and certain physiological parameters. Acta Physiol. Pol. *27,* 455–460.

GESCHEIDER, G.A. 1976. Psychophysics. Method and theory. Lawrence Erlbaum Assoc., Hillsdale, NJ.

GIRARDOT, N.R. 1969. Some requirements for consumer preference testing of foods. Activities Reports, QMFCI, Chicago, *21* (2), 113–116.

GOWER, J.C. 1971. Statistical methods of comparing different multivariate analyses of the same data. *In* Mathematics in the Archeological and Historical Sciences. F.R. Hodson, D.G. Kendall, and P. Tautu (Editors), pp. 138–149. Edinburgh University Press.

GOWER, J.C. 1975. Generalized Procrustes analysis. Psychometrika *40* (1), 33–51.

GREENFIELD, H., MAPLES, J., and WILLS, R.B.H. 1983. Salting of foods—a function of hole size and location of shakers. Nature *301,* 331–332.

HARPER, R. 1956. Factor analysis as a technique for examining complex data on foodstuffs. Applied Statistics *5,* 32–48.

HARPER, R. 1981. The nature and importance of individual differences. *In* Criteria of Food Acceptance. How Man Chooses What He Eats. J. Solms and R.L. Hall (Editors). Forster Verlag AG, Zurich.

HARPER, R. and BARON, M. 1951. The application of factor analysis to tests on cheese. Brit. J. Applied Physics *2,* 35–41.

HARRISON, S.K., and BERNHARD, R.A. 1984. Time-intensity sensory characteristics of saccharin, xylitol, and galactose, and their effect on the sweetness of lactose. J. Food Sci. *49,* 780–786, 793.

HIRSH, N.L. 1974–75. Getting fullest value from sensory testing. I. Use and misuse of testing methods. II. Considering the test objectives. III. Use and misuse of test panels. Food Prod. Dev. *8* (10), 33, 35; *9* (1), 10, 13; *9* (2), 78, 80, 83.

HOFF, J.T., CHICOYE, E., HERWIG, W.C., and HELBERT, J.R. 1978. Flavor profiling of beer using statistical treatments of G.L.C. headspace data. *In* Analysis of Foods and Beverages. Headspace Techniques. G. Charalambous (Editor), pp. 187–201. Academic Press, New York.

HORSFIELD, S. and TAYLOR, L.J. 1976. Exploring the relationship between sensory data and acceptability of meat. J. Sci. Food Agric. *27,* 1044–1056.

JOUNELA-ERIKSSON, P. 1981. Predictive value of sensory and analytical data for distilled beverages. *In* Flavour '81. P. Schreier (Editor), pp. 145–164. Walter de Gruyter, Berlin.

KELLING, S.T., and HALPERN, B.P. 1983. Taste flashes: Reaction times, intensity, and quality. Science *219,* 412–414.

KOSTER, E.P. 1982. Uses and misuses of sensory analysis: psychologist point of view. Paper presented at Food Flavors Symposium, Paris, Dec. 8–10.

LARSON-POWERS, N., and PANGBORN, R.M. 1978A. Paired comparison and time-intensity measurements of the sensory properties of beverages and gelatins containing sucrose or synthetic sweeteners. J. Food Sci. *43* (1), 41–46.

LARSON-POWERS, N., and PANGBORN, R.M. 1978B. Descriptive analysis of the sensory properties of beverages and gelatins containing sucrose or synthetic sweeteners. J. Food Sci. *43* (11), 47–51.

LAUER, R.M., FILER, L.J., REITER, M.A., and CLARKE, W.R. 1976. Blood pressure, salt preference, salt threshold, and relative weight. Arch. Am. J. Dis. Child. *130,* 493–497.

LAWLESS, H.T., and SKINNER, E.Z. 1979. The duration and perceived intensity of sucrose taste. Perception & Psychophysics *25* (3), 180–184.

LEWIS, M.J., PANGBORN, R.M., and FUJII-YAMASHITA, J. 1980. Bitterness of beer: A comparison of traditional scaling and time-intensity methods. Proc., 16th Convention, Institute of Brewing (Australia and New Zealand Sect.) *17,* 165–171.

LITTLE, A.C. 1976. Physical measurements as predictors of visual appearance. Food Technol. *30* (10), 74, 76, 77, 80, 82.

McLELLAN, M.R., CASH, J.N., and GRAY, J.I. 1983. Characterization of the aroma of raw carrots (*Daucus carota* L.) with the use of factor analysis. J. Food Sci. *48* (1), 71–72, 74.

McNULTY, P.B., and MOSKOWITZ, H.R. 1974. Time-intensity curves for flavored oil-in-water emulsions. J. Food Sci. *39* (1), 55–57.

MOLL, M., VINH, T., and FLAYEUX R. 1978. Relationship between physical and chemical analysis and taste testing results with beers. *In* Flavor of Foods and Beverages. Chemistry and Technology. G. Charalambous and G.E. Inglett (Editors), pp. 329–337. Academic Press, New York.

MOORE, L.J., and SHOEMAKER, C.F. 1981. Sensory textural properties of stabilized ice cream. J. Food Sci. *46,* 389–409.

MUÑOZ-LEZAMA, A. 1983. Sensory-instrumental measurements of the physical structure of gels. M.S. thesis, Univ. of California, Davis.

NEILSON, A.J. 1957. Time-intensity studies. Drug & Cosmetic Ind. *80,* 452–453.

NEILSON, A.J. 1958. Time-intensity studies. *In* Flavor Research and Food Acceptance, pp. 88–93. Reinhold Publ. Corp., New York.

NOBLE, A.C. 1975. Instrumental analysis of the sensory properties of food. Food Technol. *29* (12) 56–60.

NOBLE, A.C. 1978. Sensory and instrumental evaluation of wine aroma. *In* Analysis of Foods and Beverages. Headspace Techniques. G. Charalambous (Editor), pp. 203–228. Academic Press, New York.

NOBLE, A.C., WILLIAMS, A.A., and LANGRON, S.P. 1983. Descriptive analysis and the quality of Bordeaux wines. *In* Sensory Quality in Foods and Beverages, Definition, Measurement, and Control. A.A. Williams and R.K. Atkin (Editors), pp. 324–334. Ellis Horwood Ltd., Chichester.

NORRIS, M.B., NOBLE, A.C., and PANGBORN R.M. 1984. Human saliva and taste responses to acids varying in anions, titratable acidity and pH. Physiology and Behavior *32* (2), 237–244.

PALMER, D.H. 1974. Multivariate analysis of flavour terms used by experts and non-experts for describing teas. J. Sci. Food Agric. *25,* 153–164.

PANGBORN, R.M. 1964. Sensory evaluation of foods: A look backward and forward. Food Technol. *18* (9) 63–67.

PANGBORN, R.M. 1967. Use and misuse of sensory methodology. Food Quality Control *15,* 7–12.

PANGBORN, R.M. 1980. Sensory science today. Cereal Foods World *25* (10), 637–640.

PANGBORN, R.M. 1981. Individuality in response to sensory stimuli. *In* Criteria of

Food Acceptance. How Man Chooses What He Eats. J. Solms and R.L. Hall (Editors), pp. 177–219. Forster Verlag AG, Zurich.

PANGBORN, R.M., and CHUNG, C.M. 1981. Parotid salivation in response to socium chloride and monosodium glutamate in water and in broths. Appetite: J. Intake Research *2,* 380–385.

PANGBORN, R.M., and DUNKLEY, W.L. 1964. Laboratory procedures for evaluating the sensory properties of milk. Dairy Science Abstracts *26* (2), 55–62.

PANGBORN, R.M., and KOYASAKO, A. 1981. Time-course of viscosity, sweetness, and flavor in chocolate desserts. J. Texture Studies *12,* 141–150.

PANGBORN, R.M., LEWIS, M.J., and YAMASHITA, J.F. 1983. Comparison of time-intensity and category scaling of bitterness of iso-α-acids in model systems and in beer. J. Inst. Brewing. *89,* 349–355.

PANGBORN, R.M., BOS, K.E.O., and STERN, J.S. 1984. Relation of body size of adult females to dietary intake and taste responses to fat in milk. Appetite (in press).

PEARCE, J. 1980. Sensory evaluation in marketing. Food Technol. *34* (11), 60–62.

PIGGOTT, J.R. and CANAWAY, P.R. 1981. Finding the word for it—methods and uses of descriptive sensory analysis. *In* Flavour '81, P. Schreier (Editor), pp. 33–46. Walter de Gruyter, Berlin.

POWERS, J.J. and MOSKOWITZ, H.R. 1976. Correlating sensory objective measurements /New methods for answering old problems. ASTM STP *594,* Am. Soc. Test. Mat., Philadelphia, PA.

SCHAEFER, E.G. 1979. Manual on Consumer Sensory Evaluation, ASTM STP *682,* Am. Soc. Test. Mat., Philadelphia, PA.

SCHMITT, D.J., THOMPSON, L.J., MALEK, D.M. and MUNROE, J.H. 1984. An improved method for evaluating time-intensity data. J. Food Sci. *49,* 539–542, 580.

SCHÖNEMANN, P.H. and CARROLL, R.M. 1970. Fitting one matrix to another under choice of a central dilation and a rigid motion. Psychometrika *35* (2), 245–255.

SCHUTZ, H.G. 1964. A matching-standard method for characterizing odor qualities. Ann. N.Y. Acad. Sci. *116* (2), 517–526.

SCHUTZ, H.G. 1971. Sources of invalidity in the sensory evaluation of foods. Food Technol. *25* (3) 249, 252–253.

SKINNER, E. 1980. Sensory evaluation in distribution. Food Technol. *34* (11), 65–66.

SPENCER, M.D., PANGBORN, R.M., and JENNINGS, W.G. 1978. Gas chromatographic and sensory analysis of volatiles from cling peaches. J. Agric. Food Chem. *26* (3) 725–732.

STONE, H., SIDEL, J.L., and BLOOMQUIST, J. 1980. Quantitative descriptive analysis. Cereal Foods World *25,* 642–644.

STONE, H., SIDEL, J., OLIVER, S., WOOLSEY, A., and SINGLETON, R.C. 1974. Sensory evaluation by quantitative descriptive analysis. Food Technol. *28,* 24–33.

STONE, L.J. 1984. Influence of selected personality traits on dietary intake and on preference for sweet and salty stimuli. M.S. thesis, Univ. of California, Davis.

SWARTZ, M. 1980. Sensory screening of synthetic sweeteners using time-intensity evaluation. J. Food Sci. *45,* 577–581.

SWARTZ, M.L. and FURIA, T.A. 1977. Special sensory panels for screening new synthetic sweeteners. Food Technol. *31* (11) 51–55.

SZCZESNIAK, A.S. 1963. Classification of textual characteristic. J. Food Sci. *28,* 385–389.

TILGNER, D.J. 1981. A retrospective view of sensory analysis and some considerations for the future. Adv. Food Res. *19,* 215–277.

TODA, J., WADA, T., YASUMATSU, K., and ISHII, K. 1971. Application of principal component analysis of food texture measurements. J. Texture Studies *2,* 207–219.

TRANT, A.S., PANGBORN, R.M., and LITTLE, A.C. 1981. Potential fallacy of correlating hedonic responses with physical and chemical measurements. J. Food Sci. *46* (1), 583–588.

VUATAZ, L. 1976/77. Some points of methodology in multidimensional data analysis as applied to sensory evaluation. Nestlé Research News (Lausanne, Switzerland), pp. 57–71.

VUATAZ, L., SOTEK, J., and RAHIM, H.M. 1974. Profile analysis and classification. Proc., IV Int. Congr., Food Sci. and Technol., Madrid, 1, 68–71.

WILLIAMS, A.A., BAINES, C.R., and ARNOLD, G.M. 1980. Towards the objective assessment of sensory quality in inexpensive red wines. Centennial Symposium, Dept. of Viticulture and Enology, Univ. of California, Davis.

WILLIAMS, R.J. 1956. Biochemical Individuality. The Basis for the Genetotrophic Concept. John Wiley & Sons, New York.

WILLIAMS, R.J. 1978. Nutritional individuality. Human Nature *1* (6), 46–53.

WOSKOW, M.H. 1967. Some new methods of flavor evaluation. MBAA Tech. Quart. *4,* 68–72.

WOSKOW, M.H. 1968. Multidimensional scaling of odors. *In* Theories of Odors and Odor Measurement. N. Tanyolac (Editor), pp. 147–188. Robert College, Istanbul.

YOSHIKAWA, S., NISHIMARU, S., TASHIRO, T., and YOSHIDA, M. 1970. Collection and classification of words for description of food texture. III. Classification by multivariate analysis. J. Texture Studies *1,* 452–463.

12

Determination of Flavor Components

Gary A. Reineccius[1]

INTRODUCTION

Flavor is a very complex sensation consisting primarily of aroma and taste but also complemented by responses to texture, the warmth of spices such as pepper and cooling of menthol. The sensation of taste is relatively limited. Normally taste is considered to be limited to salt, sweet, sour, bitter, and perhaps, metallic responses. The major contributor to flavor perception is aroma. It is estimated that one can discriminate between thousands of different odors. Due to the importance of aroma to flavor, it is not unexpected that a large proportion of analytical flavor studies has focused on the volatile constituents of foods. This emphasis on the volatile constituents can be detrimental. Researchers who study only the volatile components of products such as Cheddar cheese will never duplicate the flavor of this product. The mouthfeel (e.g. graininess) and taste (e.g. bitterness, astringency, and acidity) are essential components of the flavor of a good aged Cheddar cheese. It is doubtful that any nice clear bottle of aroma constituents will ever come very close to conjuring up the flavor of Cheddar. Nevertheless, this chapter focuses on what has occupied the efforts of a majority of flavor chemists, the aroma portion of food flavor.

To appreciate the purpose of the following analytical methods, it is worthwhile commenting briefly on the unique problems faced by the flavor chemist. An initial concern is for the complexity of food flavor. In excess of 700 volatile components have been identified in meat flavor. It is estimated that in excess of 10,000 different volatiles will one

[1]Department of Food Science and Nutrition, University of Minnesota, St. Paul, MN 55108

day be identified in food products (Rijkens and Boelens 1975). These volatiles represent a large variety of functional groups and chemical classes of compounds. A second concern is that the nose is extremely sensitive to some odors. Some nitrogen-containing heterocyclics and sulfur-containing compounds have an odor threshold in the parts per billion or trillion concentration range. This requires that the flavor must be concentrated prior to analysis. Concentration of the flavor then necessitates that the flavor be isolated from the food so that concentration is possible. Now we have the task of extracting an extremely large number of volatiles representing numerous functional groups from a food and concentrating this extract for instrumental analysis. The separation, quantification, and interpretation of the analytical data generated from a flavor isolate are again a formidible task. Gas chromatography (GC) is typically the method of choice for analysis of flavor isolates. However, despite the tremendous resolving powers and sensitivity of GC, separation and detection of individual flavor chemicals is difficult. Significance of GC data is complicated by the variation in sensitivity and aroma character perceived by the nose. For example, ethanol has little or no aroma while an equivalent concentration of ethanethiol would have a very intense and characteristic aroma. A noncritical analyst using GC might determine that these two chemicals are of equal importance to the flavor of a food.

This chapter considers some of the more commonly used methods for flavor isolation and analysis. Emphasis is placed on those techniques that are relatively new or innovative. Quite complete reviews on methods for the isolation of flavors from foods are available (Jennings and Rapp 1983; Reineccius and Anandaraman 1983; Maarse and Belz 1981; Reineccius and Anandaraman 1982; Teranishi *et al.* 1981; Bemelmans 1979).

SAMPLE PREPARATION

We simply cannot put an apple or an orange into a GC and obtain a flavor profile. Often we must grind, homogenize, or otherwise treat the sample prior to even starting the process of flavor isolation. There is substantial documentation in the literature supporting the fact that foods are very dynamic systems. They contain very active enzyme systems which will alter the flavor profile once cellular organization is disrupted (Drawert *et al.* 1965; Fleming *et al.* 1968; Kazeniak and Hall 1970). Inactivation of enzyme systems via high temperature-short time thermal processing and homogenization with alcohol may be accomplished.

Nonenzymatic reactions may also cause chemical changes or microbial growth during sample preparation and/or the flavor isolation procedure. Oxidation, nonenzymatic browning, or acid-catalyzed alterations in the flavor profile are possible. Often the sample must be neutralized; isolation temperatures must be limited to <60°C and sample processing must be done under a vacuum in order to minimize chemical changes in the flavor profiles. Microorganisms are often inhibited by chemical means (e.g., sodium fluoride) or thermal processing (Ribereau-Gayon *et al.* 1975).

ISOLATION OF FOOD FLAVORS

The flavor isolation method selected will be determined by a number of factors, including the purpose of the flavor study, the availability of equipment, the amount of sample, and the time available. An on-line quality control situation would demand a rapid method of flavor isolation (e.g., headspace sampling, solvent extraction, or direct injection). These techniques would not give a complete flavor profile but only the major components. If the major components (perhaps not even those responsible for the flavor) would provide the necessary information, then one of the rapid methods of analysis would be acceptable. However, if a complete flavor profile is necessary, then more time consuming methods, such as distillation or an efficient solvent extraction, would be required.

The proper sample selection, preparation, and flavor isolation method must be carried out. Excellent analytical techniques cannot correct errors made earlier in preparation of the flavor isolate. A general rule is that the final flavor isolate should be examined by sensory evaluation to insure that the components of interest (e.g., an off-flavor or characteristic desirable flavor) are contained in the flavor isolate.

With these considerations in mind, let us briefly review some of the more interesting and useful methods for the isolation of flavors from foods.

FLAVOR ISOLATION

Headspace Methods

One of the simplest methods of isolating flavor from foods is by direct injection of the headspace vapors above a food product. Unfortunately this method has not provided the sensitivity needed for trace analysis and is not well suited to modern capillary column gas chromatography. Packed column GC is typically limited in headspace injection volume to 10 ml or less. This would limit the method to volatiles present in the headspace at concentrations greater than 10^{-7} g/L headspace (Schaefer 1981). Jennings (1983) has suggested an approach to make headspace sampling more compatible with capillary column GC. This eliminates the need for inlet splitting and greatly enhances the detection limits at only minimal loss in chromatographic resolution.

Direct headspace sampling has been used extensively where rapid analysis is necessary and major component analysis is satisfactory. Examples of method applications include measurement of hexanal as an indicator of oxidation (Seo and Joel 1980; Sapers *et al.* 1973) and 2-methyl propanal, 2 methyl butanal, and 3-methyl butanal as indicators of nonenzymatic browning (Buttery and Teranishi 1963). Peterson *et al.* (1973), Kazeniak and Hall (1970), and Yamashita *et al.* (1977) have used headspace analysis to study the major volatile constituents of a variety of food products.

Headspace concentration techniques have found wide usage in recent years (Charalambous 1978). This concentration method may involve simply passing headspace vapors through a cryogenic trap or alternatively, a more complicated extraction and/or adsorption trap.

The simple cryogenic trap (Fig. 12.1) offers some advantages and disadvantages. The cryo trap quite efficiently (if properly designed and operated) will collect headspace vapors irrespective of compound polarity and boiling point. However, water is typically the most abundant volatile in a food product and therefore one collects an aqueous distillate of the product aroma. This distillate must be extracted with an organic solvent, dried, and then concentrated for analysis. These additional steps add analysis time and provide opportunity for sample contamination.

Recently the use of adsorbent traps has become the most common means to concentrate headspace vapors. Adsorbent traps offer the advantages of providing a water-free flavor isolate (trap material typically has little affinity for water) and are readily automated in in-line isolation analysis GC systems (Fig. 12.2). The adsorbent initially used for headspace trapping was charcoal (Paillard 1965; Jennings and Nursten 1967; Tang and Jennings 1967; Grob 1973). The charcoal was either solvent extracted (CS_2) or thermally desorbed with backflushing (inert gas) to recover the adsorbed volatiles. Initial work suggested potential artifact formation and incomplete desorption (Palamand *et al.* 1968; Bailleul *et al.* 1962). While Grob (1973) has shown that these problems can be overcome, the recent use of charcoal has been limited.

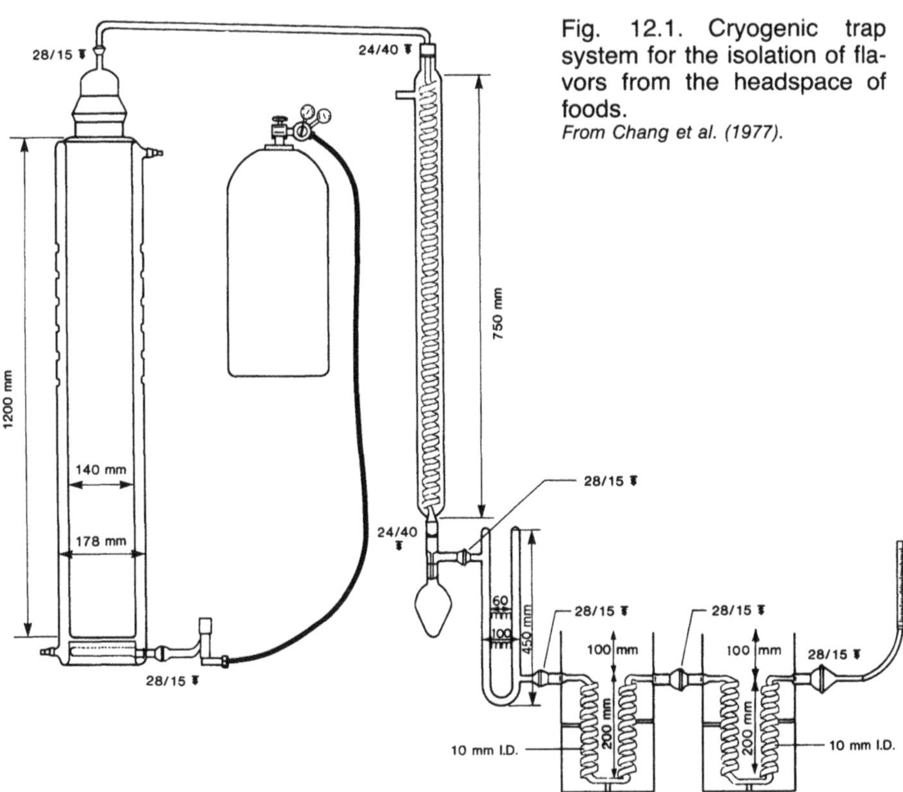

Fig. 12.1. Cryogenic trap system for the isolation of flavors from the headspace of foods.
From Chang et al. (1977).

The use of synthetic porous polymers as headspace trap material now dominates. Initially Tenax (porous polymer) was most commonly used; however, Tenax combinations are now seeing greater application. These polymers exhibit good thermal stability and reasonable capacity (Table 12.1). Adsorbent traps are generally placed in a closed system and loaded, desorbed, etc., via the use of automated multiport valving systems. The automated closed system approach provides reproducible GC retention times and quantitative precision necessary for some studies.

The primary disadvantage of adsorbent traps is their differential adsorption affinity and limited capacity. Buckholz *et al.* (1980) have shown that the most volatile peanut aroma constituents will "break through" two Tenax traps in series after purging at 40 mL/min for only 15 min. Therefore, the GC profile one obtains will be very dependent upon purge time, flow rate, and volatile type in the headspace.

Jennings (1979A) has provided an interesting alternative to head-

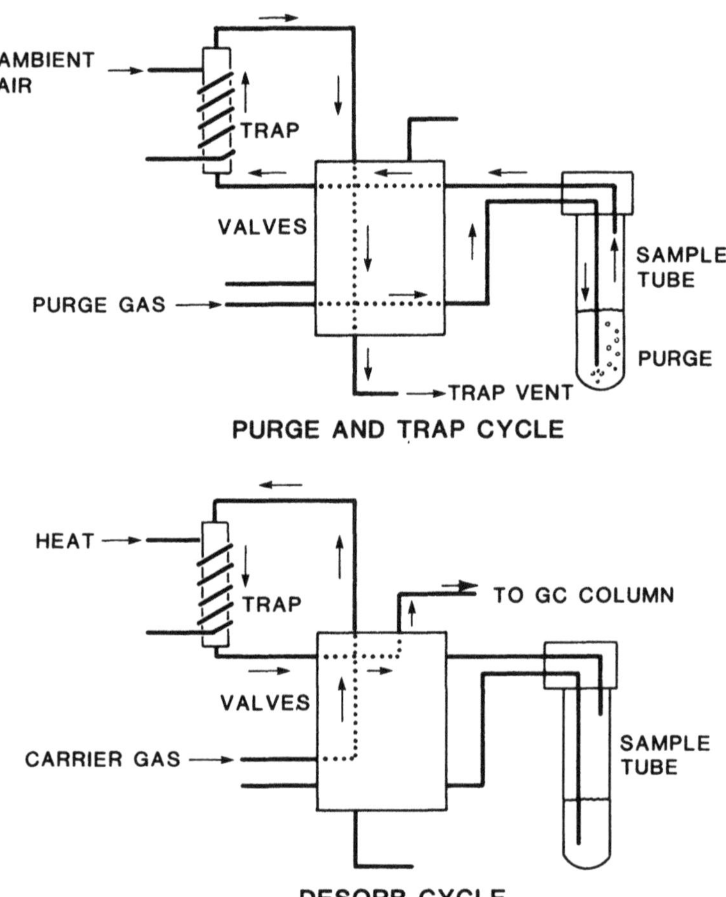

Fig. 12.2. Diagram of an automated purge and trap system during the purge and desorption cycles.
From Ruen (1980).

space trapping (Fig. 12.3). This system actually extracts headspace vapors with refluxing Freon as it passes through the reflux column. While this method will also trap water vapor, a modified toluene distillation head could be placed above the Freon reflux flask (but below the reflux column) and the condensed water could be taken off this side arm. The sample flask would then remain free of water. This headspace technique would offer the advantage of efficient trapping without interference from water vapor.

Jennings and Filsoof (1977) have provided a comparison of direct

TABLE 12.1. ADSORBENTS USED FOR HEADSPACE CONCENTRATION TECHNIQUES[a]

Adsorbent	Structure	Surface area, m^2/g	Adsorbent capacity[b]	
			Ethanol	Benzene
Charcoal (PCB)	Coconut carbon	1150–1250	7.9	24.7
Porapak P	Styrene–ethyl vinyl–benzene–divinyl benzene	50–100	NA[c]	0.28
Porapak Q	Ethyl vinyl benzene–divinyl benzene	550–650	0.18	NA
Tenax GC	Diphenyl–phenylene oxide	18.6	NR[d]	0.53
Slicia Gel	Si–O	—	13.1	NR
XAD-1	Styrene–divinyl benzene	100	NA	0.36
XAD-2	Styrene–divinyl benzene	300	0.023	1.8
XAD-4	Styrene–divinyl benzene	849	0.40	2.9
XAD-7	Acrylic ester	445	0.90	1.8
XAD-8	Acrylic ester	212	0.50	1.0
XAD-9	Sulfoxide	70	0.70	0.82
XAD-12	Polar N, O group	20	NR	0.28

[a] Sydor and Pietrzyk (1978).
[b] Percentage of weight of polymer.
[c] NA = not available.
[d] NR = not retained.

headspace and headspace concentration techniques for the isolation of flavors from model systems (Fig. 12.4). The lack of sensitivity and poor quantitative results exhibited by all headspace techniques illustrates the limitations of headspace approaches.

Distillation Methods

Virtually the only parameter that aroma compounds have in common is volatility. A compound must have sufficient volatility to reach the olfactory system and be detected. Therefore, it is reasonable that distillation methods may be used as an initial step in flavor isolation. The distillation process typically uses either product moisture or outside steam to heat and codistill the volatiles from a food product. This means that a very dilute aqueous solution of volatiles results, and a solvent extraction must be performed on the distillate in order to permit flavor concentration for analysis.

The distillation method most commonly used today is some modification of the original Nickerson-Likens distillation head (Fig. 12.5). This simultaneous distillation/extraction apparatus offers the advantages that it can be operated under vacuum, is extremely efficient for

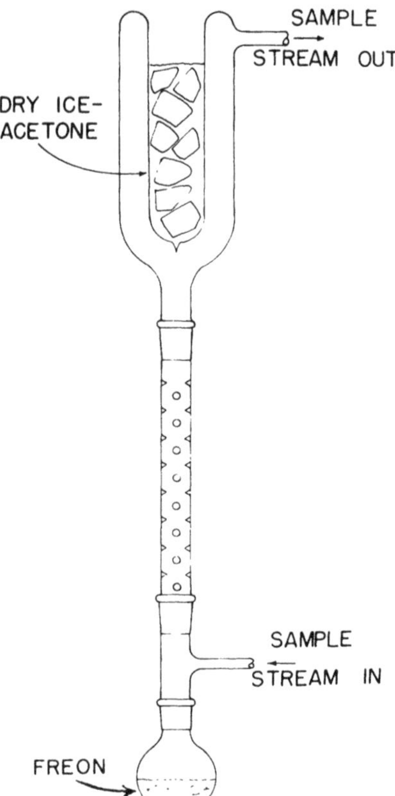

Fig. 12.3. Apparatus for the continuous isolation of volatile organics from headspace vapors by Freon reflux.
From Jennings (1979A).

extraction of the distillate, and uses minimal solvent volume. Reduced pressure operation minimizes thermally induced artifact formation. The very intimate contact between solvent and distillate vapors affords high extraction efficiencies. Minimal solvent volume reduces the problem with artifacts entering the system from solvent impurities.

As shown in Fig. 12.4, the Nickerson-Likens method offers quite efficient flavor recoveries. It provides better qualitative and quantitative data than either direct headspace or headspace concentration methods. Shortcomings of the method include isolation time (generally 2½ hr) and potential for artifacts. Artifacts may enter the flavor isolate from solvent impurities, rubber tubings, vacuum grease, anti-

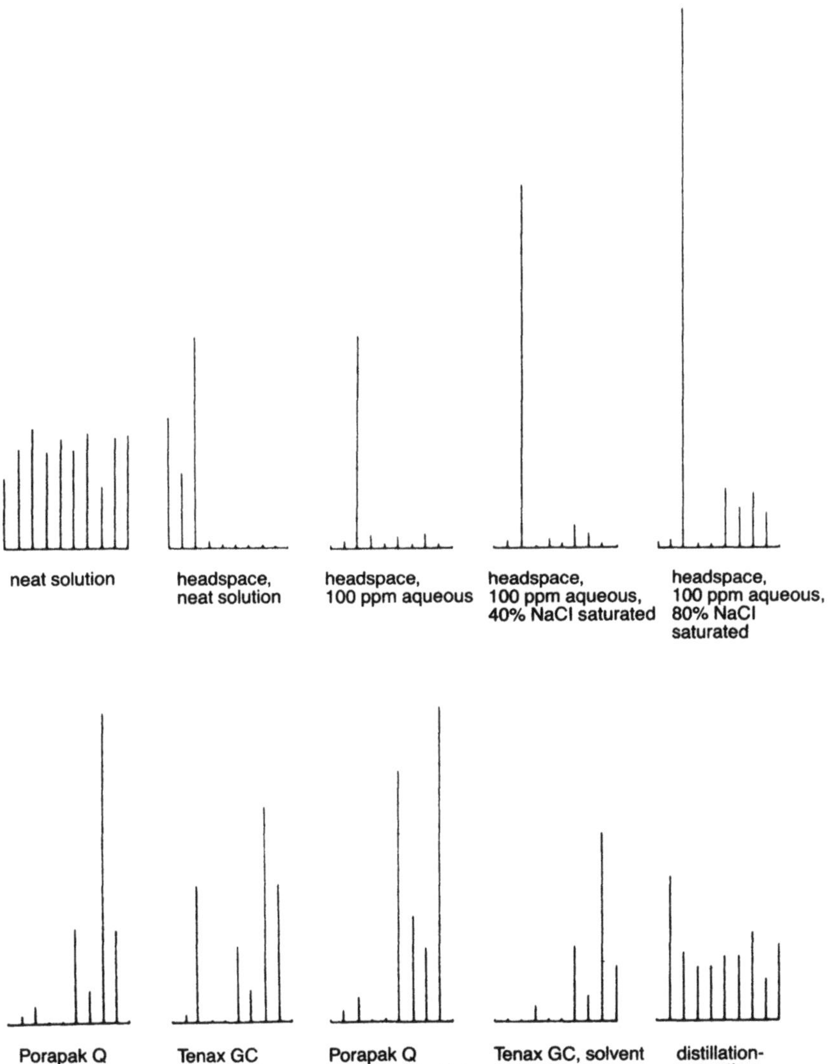

Fig. 12.4. Comparison of methods for the isolation of volatiles from a model system. Relative integrator response (left to right) ethanol, 2-pentanone, heptane, pentanol, hexanol, hexyl formate, 2-octanone, limonene, heptyl acetate and γ-heptalactone.
From Jennings and Filsoof (1977).

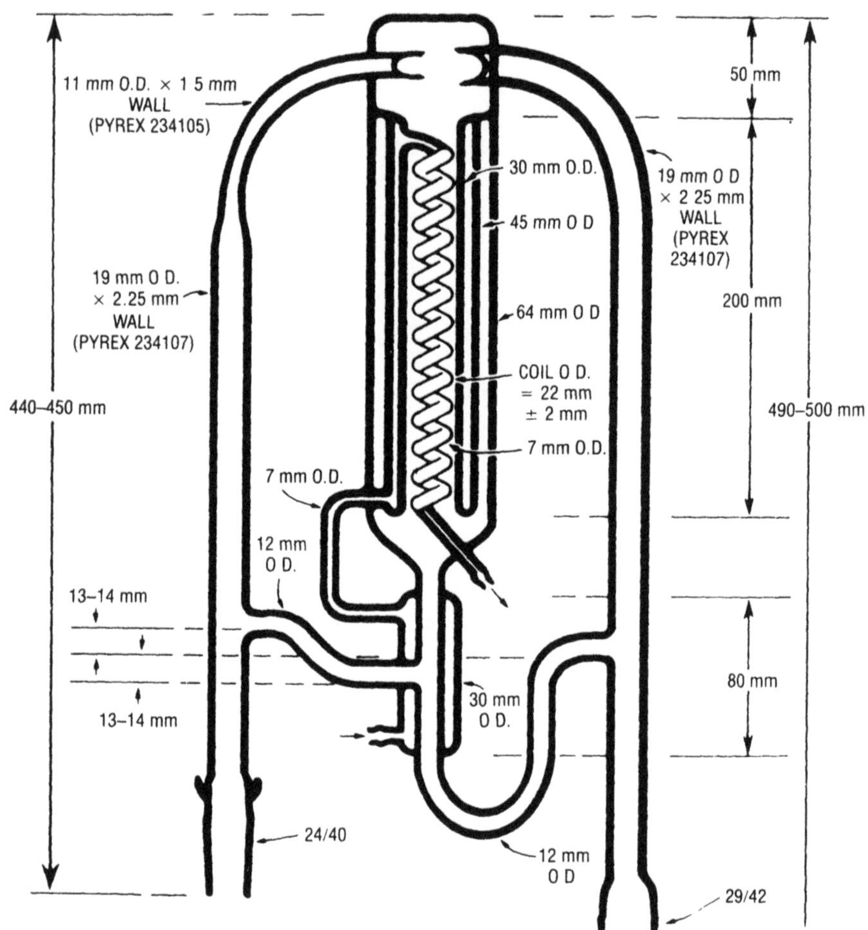

Fig. 12.5. Modified Nickerson-Likens distillation head.
From Schultz et al. *(1977).*

foam agents, steam supply (contaminated water), thermally induced chemical changes, and leakage of contaminated laboratory air into the system. It is common to find trace amounts of laboratory solvents in the flavor isolates. We often can determine what solvents were used in the laboratory on any given day of flavor isolation.

Solvent Extraction

Solvent extraction has the potential to yield the most accurate qualitative and quantitative data on flavor profiles of foods. Since most flavors are preferentially soluble in organic solvents, an efficient extraction can yield nearly complete recovery of flavors from a food system. However, the method is typically limited to the isolation of flavors from fat-free foods (e.g. wines, some breads, fruit and berries, some vegetables, and alcoholic beverages). If the food contains a significant amount of fat, the fat will be coextracted with the flavor and will interfere with subsequent concentration and analytical procedures. Therefore, an additional isolation step would be required to separate the fat from flavor constituents.

The most interesting solvent for flavor isolation is liquid CO_2. Liquid CO_2 extraction requires specialized equipment since high pressures (320–800 psi) are required to effect a phase change and produce suitable solvent properties. While liquid CO_2 extraction is not a new technique (Schultz *et al.* 1974; Vitzthum *et al.* 1975; Jennings 1979A) has provided a small-scale extractor for obtaining flavor isolates from solid materials (Fig. 12.6). Shortcomings of this extractor are the limited sample size and requirement that the sample be a solid. The greatest advantages lie in the low extraction temperatures, inert extraction atmosphere, excellent recovery of very volatile components, and solvent-free flavor isolate. A solvent-free flavor isolate makes sensory evaluation possible without solvent interference. Liquid samples can be accommodated in this system by initially loading a flavor on charcoal or one of the porous polymers (headspace trapping) and then extracting the loaded trap in the CO_2 extractor.

Solvent extractions may be carried out in quite elaborate equipment such as the CO_2 extractor just described or can be as simple as a batch process in a separatory funnel. Batch extractions can be quite efficient if multiple extractions and extensive shaking is used (Reineccius *et al.* 1972). The continuous extractors (liquid/liquid) are more efficient but require more costly and elaborate equipment. Two rather innovative approaches to solvent extraction are the Jennings bottle (Jennings 1981) and the extraction bomb of Blakesley and Loots (1977).

The Jennings bottle (Fig. 12.7) is a quick method of sampling an aqueous system for volatiles (e.g. apple essence quality control). The aqueous material is introduced through the wide bore neck followed by a small volume (100 μL) of organic solvent. The bottle is shaken well and then water is added to bring the solvent into the capillary neck (centrifugation may be necessary to bring the solvent into the bottle neck). The solvent is sampled via syringe and injected directly into a GC for analysis. While efficient recovery is not achieved, the method can be quite reproducible if properly applied and may be very suitable for quality control situations.

The extraction bomb (Fig. 12.8) uses Freon 12 as the extracting sol-

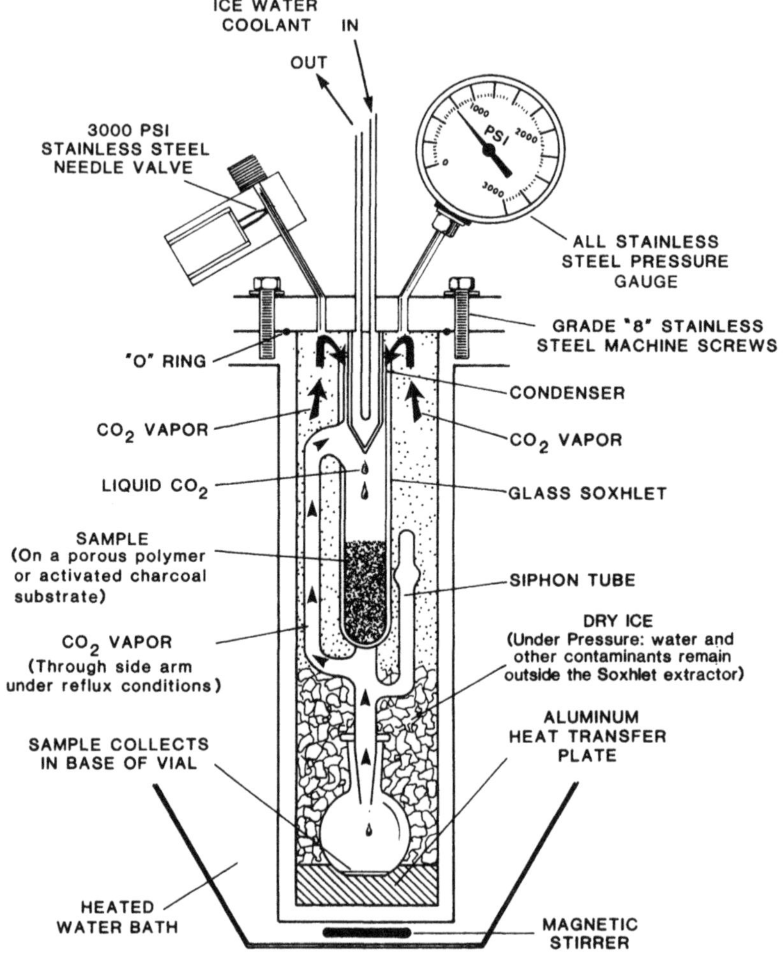

Fig. 12.6. Research scale liquid CO_2 extractor.
Courtesy of J & W Scientific, Orangevale, CA.

vent. The sample and Freon are added to a chilled apparatus; the apparatus is stoppered and then put on a rotary shaker at room temperature. After 3 hr, the sample is chilled to freeze the food, the Freon decanted into the upper chamber, and then the bottom of the apparatus is chilled to distill the Freon back to the bottom of the chamber. The extraction is then continued using the distilled Freon. This method is quite attractive since the sample contacts only glass and Teflon, receives minimal heat treatment (room temperature), has limited air

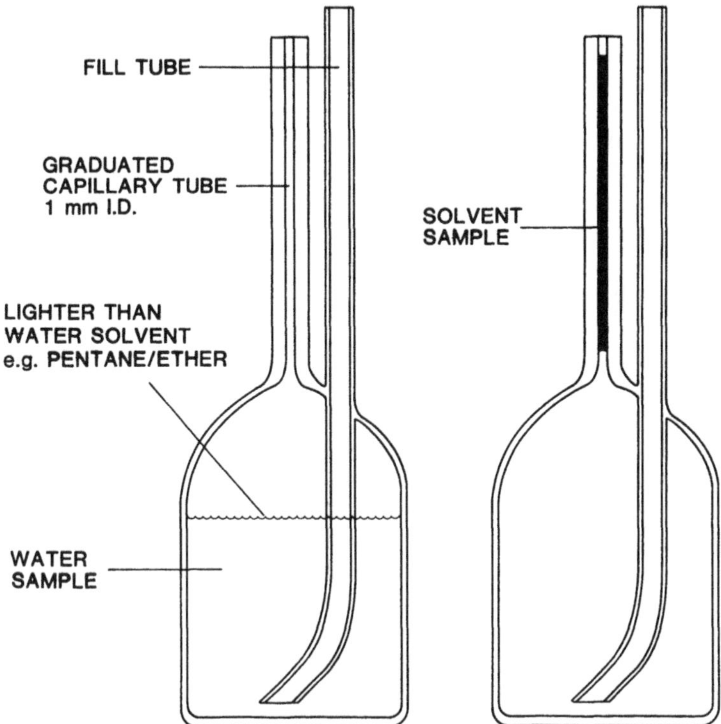

Fig. 12.7. Modified Babcock bottle for the solvent extraction of aqueous samples.
Courtesy of J & W Scientific, Orangevale, CA.

contact and requires little operator time. The primary drawback is limited sample capacity.

The solvent extraction techniques discussed to this point all assume that the food sample contains little or no fat. If fat is present, steam distillation, high vacuum stripping, or dialysis is required to separate the flavor from the fat. Steam distillation of the fat/flavor extract has been reported in the literature; however, this seems inappropriate since one could have simply done a Nickerson-Likens extraction on the food initially. High vacuum stripping is occasionally used for volatile recovery. The fat/flavor isolate would be passed through either a high-vacuum thin-film swept surface stripper or a short path length cold finger apparatus. The thin-film swept surface apparatus offers the advantage of being a continuous process so large fat/flavor volumes can be processed.

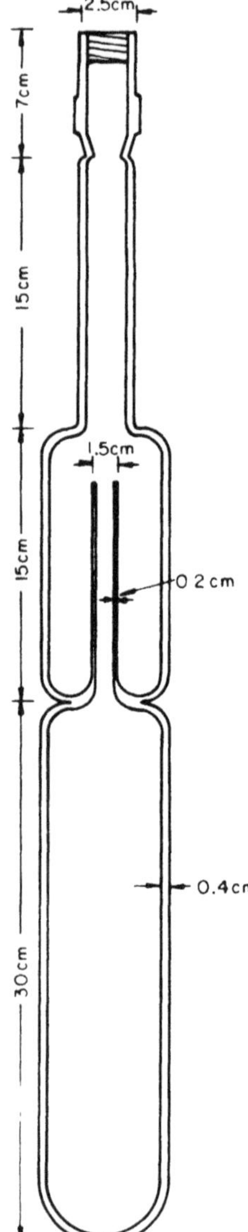

Fig. 12.8. Glass bomb for the extraction of flavors from foods using Freon 12.
From Blakesley and Loots (1977).

An alternative method recently developed in our laboratory involves dialysis of the solvent/fat/flavor isolate against organic solvent [1% water in diethyl ether (Chang 1982)]. This procedure, when initially developed by Benkler and Reineccius (1979, 1980), had prob-

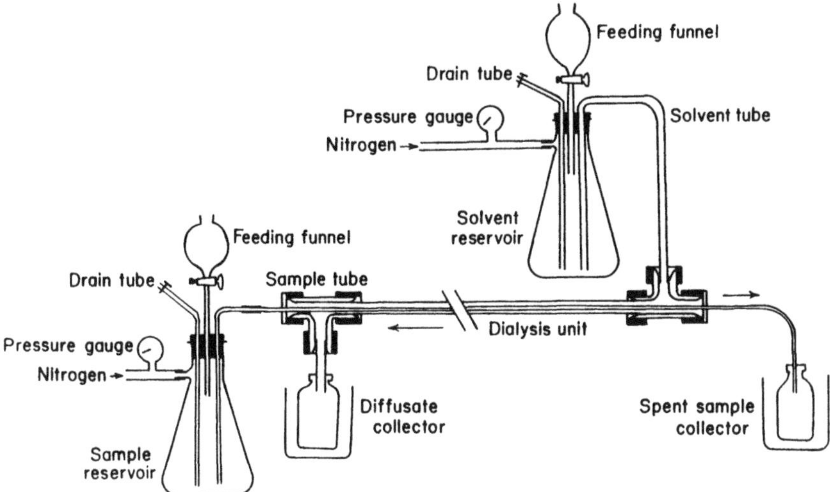

Fig. 12.9. Countercurrent continuous dialysis system for the separation of flavors from solvent extracts of fatty foods.
From Chang (1982).

lems with the adsorption of basic compounds onto the dialysis membrane (perfluorosulfonic acid polymer), poor recoveries (batch process), and artifacts were formed via acid catalyzed changes in the flavor. Recent modifications by Chang (1982) have corrected these problems by deactivating the membrane via water addition to the solvent and using a countercurrent continuous dialysis system (Fig. 12.9). This procedure offers the advantage of effectively separating low molecular weight volatiles from fat without involving a distillation process (vacuum stripping or Nickerson-Likens). The primary disadvantages are the difficulty of setting up the apparatus and poor recoveries of higher molecular weight volatiles (20% recovery for C_{10}–C_{14} hydrocarbons).

Direct Injection

It is theoretically possible to do flavor analysis by direct injection of a food into a gas chromatograph. Assuming one can inject 2–3 μL samples into a GC and the GC has a detection limit of 0.1 ng (0.1 ng/2 μL), one could detect volatiles in the sample at concentrations greater than 50 ppb. Problems with direct injection arise due to thermal degradation of the nonvolatile food constituents, damage to the GC column, or decreased separation efficiency due to water in the food sample, contamination of the column and injection port by nonvolatile materials, and reduced column efficiency due to slow vaporization of volatiles from the food. (Injection port temperatures are reduced to minimize thermal degradation of the nonvolatile food constituents.)

Direct injection is commonly used to determine the flavor quality of vegetable oils (Dupuy *et al.* 1971; Legendre *et al.* 1979). A relatively large volume of oil (50–100 μL) can be directly injected into an injection port of a GC which has been stuffed with glass wool. Since vegetable oils are reasonably thermally stable and free of water, this method is particularly well suited to oil analysis. Legendre *et al.* (1979) designed an injection port system which will isolate the food constit-

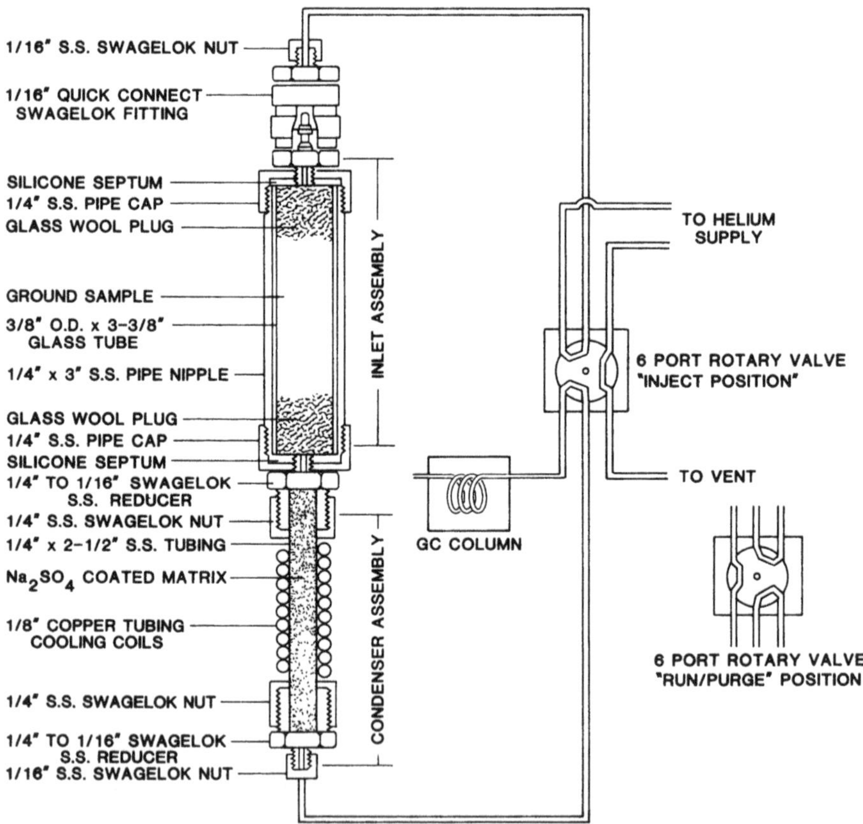

Fig. 12.10. Gas chromatography inlet system used for the analysis of food flavors by direct injection.
From Legendre et al. (1979).

uents from the GC column and trap out the water (Fig. 12.10). They used this inlet system for the analysis of salad oils, vinegar, and corn–soy food blends.

Summary

This section would not be complete without a reminder that there are numerous methods available for the isolation of volatile flavor constituents from foods. Each method has its own unique assets and limitations. There is no isolation method that is without compromise. One must always be certain that the flavor character desired for study has been isolated from the food prior to starting the GC analysis. This involves sensory analysis of the flavor isolate.

ANALYSIS OF FLAVOR ISOLATES

Once a suitable flavor isolate has been obtained, it must then be analyzed for composition. While HPLC has found some application (Reineccius and Anandaraman 1983), GC has most commonly been used for flavor work. Gas chromatography is particularly well suited for flavor analysis because of its tremendous separating abilities, its sensitive universal or specific detectors, and the volatility of components being analyzed. Recent developments in GC continue to enhance the suitability of this technique for flavor work.

Mass spectrometry (MS) is the method of choice when identification of unknown flavor components is desired. While other spectral methods (e.g., UV, IR, and NMR) may contribute substantial structural information, these techniques typically lack the sensitivity to be useful in flavor studies.

Gas Chromatography

The most revolutionary development in GC in recent years has been the invention of fused silica capillary columns. Capillary columns typically produce 3000–5000 theoretical plates/meter which provide the tremendous separating powers which are needed for the analysis of very complex food flavors. Capillary column GC was originally a method available to only a few researchers who had the talents to coat their own metal columns or straighten the ends and install a fragile glass capillary. Fused silica has made capillary column GC available to nearly everyone. The conversion of packed column GC to capillary columns is facilitated by the durability and flexibility of fused silica columns. It is often possible to thread the column *inside* the old GC column-to-detector plumbing which virtually eliminates dead volume and reactive metal surfaces. The availability of all glass inlet systems (J & W Scientific) readily converts the inlet system to a capillary inlet. The conversion of a packed column instrument to a capillary GC is readily accomplished.

In addition to being easy to work with, fused silica columns are extremely inert. Fused silica contains only a few ppm inpurities (reactive materials) as compared to the percentage levels found in typical glass columns. Unlike the previous glass capillaries, the fused silica columns are flexible so it is unnecessary to heat the column to straighten the ends. Heating the column creates reactive glass sites. The inertness of fused silica is essential for the analysis of sulfur and phenolic flavor compounds.

Recently, bonded fused silica columns have been developed. These phases are bonded to the fused silica columns and, additionally, are internally cross-linked. While this produces a column that can be solvent flushed to remove contaminants, the most interesting advantages to flavor work are phase stability to on-column and splitless injection and the manufacture of thick phase columns. With nonbonded columns splitless and on-column injection can eventually result in migration of liquid phase (depending on solvent choice and liquid phase) into pools at the condensation or injection site. This nonuniformity of phase and resultant "bare" spot in the column can reduce column efficiency. The bonded phase cannot be washed into pools, so solvent choice is not limited and column efficiency is maintained.

The major shortcoming of capillary columns for flavor work is their limited capacity (typically a few nanograms per component). Bonded phases permit the manufacture of thick film columns. Commercially, columns with 1-μm phase thickness are available which will accommodate up to about 100 ng per component. Experimental columns have been made with up to 8-μ phase thickness (Grob and Grob 1983). These very thick film columns will handle approximately 1μg of material. This enhanced column capacity makes splitting with smelling or multiple detectors feasible as well as trapping for analysis by other spectral methods. Excellent discussions of GC are available by Jennings (1979A, B, 1980); Chapter 13 of this volume; and Freeman (1981).

Gas Chromatography—Computers

Newer GC equipment have built-in computer systems and data processing options. Older GC equipment often may be interfaced with a microcomputer. These interfaces or new equipment options offer a new dimension to the chromatographer. Computer analysis of GC data permits specialized report formats, totally automated operation (i.e., the computer can analyze the data and make control decisions based on the data), GC run averaging, and numerous other mathematical data manipulations.

An interesting application of GC computer systems is the development of sensory estimations based on GC flavor profiles. The original work in this area involved stepwise discriminant analysis (SDA). Gas chromatography profile data were fed into a large computer along with the results from taste panel studies. The computer program would then select the GC peaks which would best predict some flavor attribute. For example, this could be used to discriminate between Coke and Pepsi or the degree of oxidation of potato chips. The interface between a GC and computer now offers the opportunity to make similar flavor profile evaluations (sensory quality) in the research lab. One option is to take the original data to a large computer and determine (SDA) the peaks best indicating the sensory quality of interest. The lab computer may then be programmed to look at those peaks and combine them in the proper order (SDA equation) to make the sensory evaluation.

Klopp and co-workers (1983) worked on an alternative approach to determining GC–sensory correlations. This approach involves computerized subtraction of a GC run from a food product with one sensory quality from a second with a different sensory quality. The peaks which differ between the two products show up as either positive or negative on subtraction while those that do not change negate each other. This process selects differences between two samples but does not yield any significance estimation. The next step is to apply statistical methods to determine a combination of peaks selected (i.e., develop an equation) to predict flavor quality. This entire process can be done on a small personal computer. An example of the subtraction method is shown in Fig. 12.11 for flavor changes in spray-dried orange oil. We are currently working on the method of relating the GC peaks selected by subtraction to sensory quality.

Mass Spectrometry

As mentioned earlier, MS is the method of choice when a flavor component must be identified. Mass spectrometry is particularly valuable because it is readily interfaced with a GC and has excellent sensitivity (low femtogram level). The earlier time-of-flight or mag-

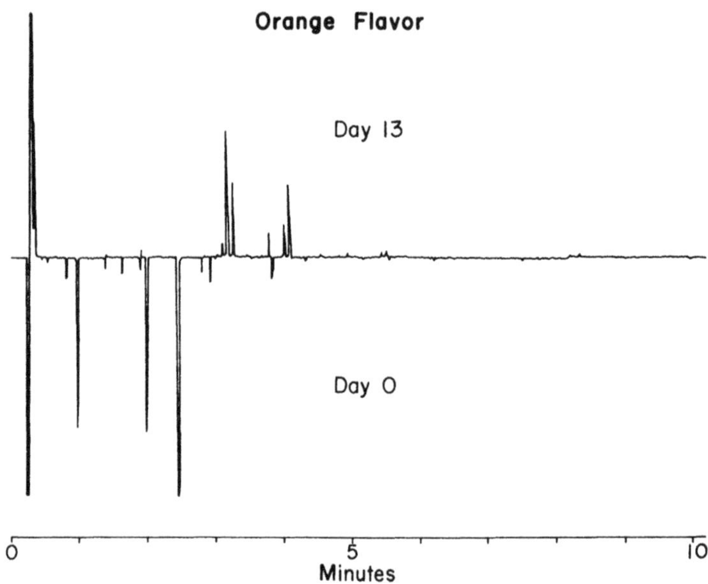

Fig. 12.11. Computerized subtraction of the chromatogram of oxidized orange oil from fresh orange oil.
Klopp et al. (1983).

netic sector mass selectors were quite suitable for packed column GC/MS. However, recent demands for rapid scanning necessitated by capillary column GC/MS have required changes in magnet design and made the quadrupole mass selectors more common. New magnetic sector instruments use a laminated magnet which develops the magnetic field necessary for mass selection but yet permits rapid scanning. The quadrupole mass selector can very rapidly scan a wide mass range and thus is particularly well suited to capillary column GC.

Some new developments in MS which are of particular interest to the flavor area include negative ion MS, atmospheric pressure ionization (API), soft ionization techniques, and dual MS (or MS/MS). Negative ion MS has only recently become popular. Traditional ionization techniques produce only about 1/1000 the quality of negative ions as positive ions. Therefore, negative ion MS has lacked the necessary sensitivity to be useful in trace analysis. However, recent developments in ionization methods have made this technique quite competitive with positive ion MS. This means that the flavor chemist will soon have more MS data (both + and − ion MS) to assist in identifications.

Atmospheric pressure ionization techniques involve ionizing the target compound at atmospheric pressure rather than the traditional high vacuum. Consequently API results in many more molecular collisions in the ion source and, subsequently, a higher level of target compound ionization. In fact API techniques yield approximately a tenfold increase in ionization efficiency and, therefore, sensitivity.

Soft ionization techniques (chemical, field desorption, and fast atom ionization) assist the flavor chemist by providing molecular weight information. The traditional electron impact ionization technique at 70 eV often yields little or no parent ion. Without a parent ion, one may not be able to determine molecular weight of the unknown. Molecular weight information is very valuable in determining compound identification. While electron impact ionization may be accomplished at low energies (15–20 eV) and this would enhance the probability of observing a parent ion, sensitivity drops off quickly at low ionization energies. The soft ionization techniques offer excellent sensitivity plus a high probability of observing the parent ion.

Probably the most exciting development in MS is the technique of Mass Spec/Mass Spec (MS/MS) which involves the use of two MS in tandem (Fig. 12.12). The first MS generally uses chemical ionization to produce a mass chromatogram of a sample. The potential of MS/MS to flavor research was demonstrated in an excellent piece of work by Davis and Cooks (1982) who presented results of a MS/MS study on the volatile constituents of nutmeg (Fig. 12.13). Note that there was no preliminary GC separation of the nutmeg volatile components. The sample (ground nutmeg) was placed directly in the solid probe inlet and heated. The first MS functioned as an instant separation tool. The second MS then can take a typical ionization spectra of each selected ion (daughter ion spectra). The potential of this technique in flavor analysis and general trace analysis is tremendous. The first separation, based on mass, can provide qualitative and quantitative data *instantly*. Separations, which may be quite difficult by GC, may be simple using this technique. The primary limitations of the method are that compounds of the same nominal mass (low resolution system) or same elemental composition (high resolution system) cannot be separated in the first MS, and there is currently little daughter ion MS

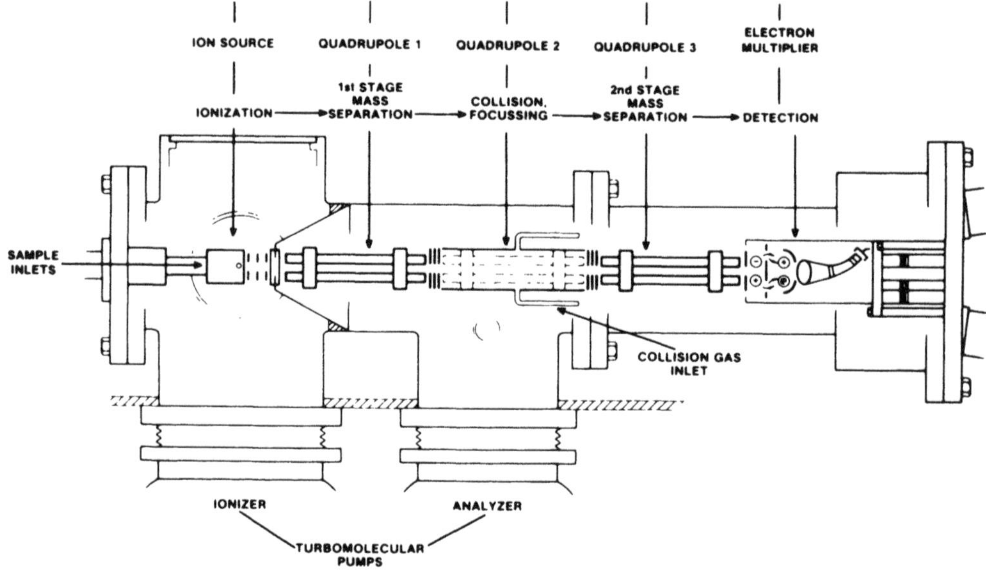

Fig. 12.12. Cutaway view of a triple stage quadrupole GC/MS/MS system.
Courtesy of Finnigan MAT, Sunnyvale, CA.

data available. While it may be difficult to circumvent the problem of resolution of compounds with the same mass, the second problem (data on daughter ion spectra) will probably be solved as the technique becomes more widely available.

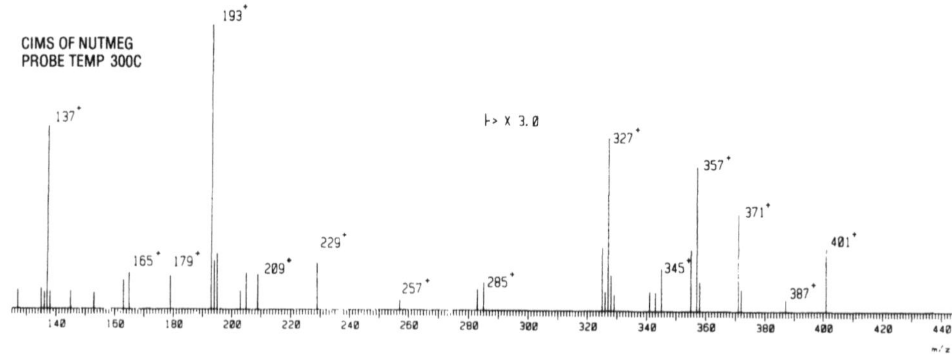

Fig. 12.13. Chemical ionization mass spectrum of nutmeg taken on a triple quadrupole mass spectrometer.
From Davis and Cooks (1982).

BIBLIOGRAPHY

BAILLEUL, G., BRATZLER, K., HERBERT, W., and VOLLMER, W. 1962. Active Kohle und ihre industrielle ver wendung. F. Enke (Publisher). Stuttgart.

BEMELMANS, J.M.H. 1979. Review of isolation and concentration techniques. *In* Progress in Flavour Research. D.G. Land and H.E. Nursten (Editors). Applied Science Publishers, London.

BENKLER, K.F., and REINECCIUS, G.A. 1979. Separation of flavor compounds from lipids in a model system by means of membrane dialysis. J. Food Sci. *44,* 1525–1528.

BENKLER, K.F., and REINECCIUS, G.A. 1980. Flavor isolation from fatty foods via solvent extraction and membrane dialysis. J. Food Sci. *45,* 1084–1085.

BLAKESLEY, G.N., and LOOTS, J. 1977. A convenient method for multiple extraction of volatile flavor components from food slurries and pulps using a two chambered glass bomb extractor and dichlorodifluoromethane. J. Agric. Food Chem. *25,* 961–963.

BUCKHOLZ, L.L., WITHYCOMBE, D.A., and DAUN, H. 1980. Application and characteristics of polymer adsorption method used to analyze flavor volatiles from peanuts. J. Agric. Food Chem. *28,* 760–765.

BUTTERY, R.G., and TERANISHI, R. 1963. Measurement of fat oxidation and browning aldehydes in food vapors by direct injection gas–liquid chromatography. J. Agric. Food Chem. *11,* 504–507.

CHANG, Y.I. 1982. Refinement of the membrane dialysis technique for the separation of flavors from foods. M.S. thesis, Univ. of Minnesota, St. Paul.

CHANG, S.S., VALLESE, F.M., HUANG, L.S., HSIEH, L.S., HSIEH, O.A.L., and MIN, D.B.S. 1977. Apparatus for the isolation of trace volatile constituents from foods. J. Agric. Food Chem. *25,* 450–455.

CHARALAMBOUS, G. 1978. Analysis of Foods and Beverages, Headspace Techniques. Academic Press, New York.

DAVIS, D.V., and COOKS, R.G. 1982. Direct characterization of nutmeg constituents by mass spectrometry–mass spectrometry. J. Agric. Food Chem. *30,* 495–504.

DRAWERT, F., HEIMANN, W., ENBERGER, R., and TRESSL, R. 1965. Enzymatische Veranderung des naturlichen Apfelaromass bei der Aufarbeitung. Naturwissenschaften *52,* 304–305.

DUPUY, H.P., FORE, S.P., and GOLDBATT, L.A. 1971. Elution and analysis of volatiles in vegetable oils by gas chromatography. J. Am. Oil Chem. Soc. *48,* 876.

FINNIGAN MAT CORP., Sunnyvale, CA. Topic number 8004.

FLEMING, H.P., COBB, W.Y., ETCHELLS, J.L., and BELL, T.A. 1968. The formation of carbonyl compounds in cucumbers. J. Food Sci. *33,* 572.

FREEMAN, R.R. 1981. High Resolution Gas Chromatography. 2nd ed. Hewlett Packard, Avondale, PA.

GROB, K. 1973. Organic substances in potable water and in its precursor. Part I. Methods for their determination by gas–liquid chromatography. J. Chromatogr. *84,* 255–273.

GROB, K., and GROB, G. 1983. Capillary columns with very thick coatings. J. High Resol. Chrom. and Chrom. Comm. *3,* 133–139.

JENNINGS, W.G. 1979A. Vapor-phase sampling. J. High Resol. Chrom. and Chrom. Comm. *2,* 221–224.

JENNINGS, W.G. 1979B. The use of glass capillary columns for food and essential oil analysis. J. Chrom. Sci. *17,* 636–639.

JENNINGS, W.G. 1980. Gas Chromatography with Glass Capillary Columns. 2nd ed. Academic Press, New York.
JENNINGS, W.G. 1981. Recent developments in high resolution gas chromatography. *In* Flavour '81. P. Schreier (Editor). de Gruyter, New York.
JENNINGS, W.G. 1983. Personal communication. Dept. of Food Sci., Univ. of California—Davis, Davis, CA 95616.
JENNINGS, W.G., and FILSOOF, M. 1977. Comparison of sample preparation techniques for gas chromatographic analysis. J. Agric. Food Chem. *25,* 440–445.
JENNINGS, W.G., and NURSTEN, H.E. 1967. Gas chromatographic analysis of dilute aqueous systems. Anal. Chem. *39,* 521–523.
JENNINGS, W.G., and RAPP, A. 1983. Sample Preparation for Gas Chromatography Analysis. Huethig Publishing, New York.
KAZENIAK, S.J., and HALL, R.M. 1970. Flavor chemistry of tomato volatiles. J. Food Sci. *35,* 519–530.
KLOPP, D.J., PHILLIPS, R.J., and REINECCIUS, G.A. 1983. Chromatographic addition and subtraction. Application Note. Hewlett Packard, Inc., Avondale, PA.
LEGENDRE, M.G., FISHER, G.S., FULLER, W.H. DUPUY, H.P., and RAYNER, E.T. 1979. Novel technique for the analysis of volatiles in aqueous and nonaqueous systems. J. Am. Oil Chem. Soc. *56,* 552–555.
MAARSE, H., and BELZ, R. 1981. Handbuch der Aroma Forschung. Akademie-Verlag, Berlin.
MacLEOD, A.J., and CAVE, S.J. 1975. Volatile flavour components of eggs. J. Sci. Food Agric. *26,* 351–360.
PAILLARD, N. 1965. Analyse des produits volatils emis par les pommes. Fruits *20,* 189–197.
PALAMAND, S.R., MARKL, K.S., and HARDWICK, W.A. 1968. Some techniques used for the concentration of volatile compounds in flavor research. Amer. Soc. Brew Chem. Proc. (Annu. Meet.) pp. 75–78.
PETERSON, E.E., LORENTZEN, J., and FLINK, J. 1973. Influence of freeze-drying parameters on the retention of flavor compounds in coffee. J. Food Sci. *38,* 119–122.
REINECCIUS, G.A., and ANANDARAMAN, S. 1983. Analysis of volatile flavors. *In* Recent Advances in the Chromatographic Analysis of Organic Compounds in Foods. J. Lawrence (Editor). Marcel Dekker, New York.
REINECCIUS, G.A., KEENEY, P.G., and WEISBERGER, W. 1972. Factors affecting the concentration of pyrazines in cocoa beans. J. Agric. Food Chem. *20,* 202–206.
RIBEREAU-GAYON, P., BOIDRON, J.N., and TERRIER, A. 1975. Aroma of muscat grape varieties. J. Agric. Food Chem. *23,* 1042–1047. Cited by Rijkens and Boelens (1975).
RIJKENS, F., and BOELENS, H. 1975. The future of aroma research. *In* Aroma Research: Proceedings of the International Symposium on Aroma Research. H. Maarse and P. J. Groenen (Editors). Wageningen, The Netherlands.
RUEN, W.W. 1980. Analysis of the fruity off-flavor in milk using headspace concentration–capillary column gas chromatography. M.S. thesis, Univ. of Minnesota, St. Paul, MN.
SAPERS, G.M., PANASIUK, O., and TALLEY, F.B. 1973. Flavor quality and stability of potato flakes: Effects of raw material and processing. J. Food Sci. *38,* 586–589.
SCHAEFER, J. 1981. Isolation and concentration from the vapour phase. *In* Handbuck der Aroma Forschung, pp. 44. H. Maarse and R. Belz (Editors). Akademia-Verlag, Berlin.
SCHULTZ, T.H., FLATH, R.A., MON, R., EGGLING, S.B., and TERANISHI, R.

1977. Isolation of Volatile Components from a Model System. J. Agric. Food Chem. *25* (3), 446–449.

SCHULTZ, W.G., SCHULTZ, T.H., CARLSON, R.A., and HUDSON, J.S. 1974. Pilot plant extraction with liquid CO_2. Food Technol. *28* (6), 32–36.

SEO, E.W., and JOEL, D.L. 1980. Pentane production as an index of rancidity in freeze-dried pork. J. Food Sci. *45,* 26–29.

SYDOR, R., and PIETRZYK, D.J. 1978. Comparison of porous copolymers and related adsorbents for the stripping of low molecular weight compounds from a flowing air stream. Anal. Chem. *50,* 1842–1847.

TANG, C.S., and JENNINGS, W.G. 1967. Volatile components of apricot. J. Agric. Food Chem. *15,* 24–28.

TERANISHI, R., FLATH, R.A., and SUGISAWA, H. 1981. Flavor Research: Recent Advances. Marcel Dekker, New York.

VITZTHUM, O.G., WERKHOFF, P., and HUBERT, P. 1975. Volatile components of roasted cocoa: Basic fraction. J. Food Sci. *40,* 911–916.

YAMASHITA, I., IRINO, K., NEMOTO, Y., and YOSHIKAWA, S. 1977. Studies on flavor development in strawberries. 4. Biosynthesis of volatile alcohol and esters from aldehyde during ripening. J. Agric. Food Chem. *25,* 1165–1168.

13

Gas Chromatography

Walter Jennings[1]

INTRODUCTION

Gas chromatography (GC) is the world's most widely used analytical technique. It is sobering to realize that perhaps only 3% of its myriad users realize the full potential of the method, another 3% approach that potential, and well over 90% waste time, effort, and money to produce inferior results that still deliver some useful information; the technique is powerful, and power corrupts.

The purpose of this chapter is to explore the state of the art in gas chromatography, and at the same time to develop some understanding of the prospects and advantages. This will require examining (1) some of the basic principles of gas chromatography; (2) some of the newer developments in gas chromatographic hardware; and (3) a few recent developments in techniques.

BASIC PRINCIPLES OF GAS CHROMATOGRAPHY

It is generally recognized that open tubular, or "capillary" columns yield superior separations, but unfortunately, this fact is usually accepted and dismissed. Those individuals requiring superior separations may resort to capillary chromatography, but those who are attaining satisfactory separations on packed columns—and who do 90+% of the chromatography in the world today—overlook the other advantages offered by the capillary system.

[1]Department of Food Science and Technology; University of California, Davis, CA 95616

The reason for the superior separation, of course, is that the range of retention times exhibited by identical molecules of each solute undergoing separation is narrower for the capillary column than for the packed column. As the ranges of retention times exhibited by identical molecules of the various solutes broaden, incomplete separation and multicomponent peaks result.

The retention time of a solute is the sum of the time spent in the liquid phase plus the time spent in the gas phase. It is readily apparent that the packed column exhibits a multiplicity of flow paths, leading to a broader range of times that identical molecules spend in the gas phase; the capillary, on the other hand, exhibits a single flow path. The packed column contains much more liquid phase, which has a tendency to agglomerate at points where particles of solid support contact each other or the column wall. This leads to a highly irregular "film" thickness; solute molecules dissolving in a thick portion are more retained than identical molecules dissolving in a thin portion, leading again to a broader range of retention times. In the capillary column, the liquid phase is present as a much thinner and more uniform film. A third point that should be made in rationalizing the superior separation of the capillary system emerges from the fact that packing materials are generally inefficient at heat transfer. Hence a range of temperatures must exist across any transverse section of the packed column; the magnitude of that range is a function of both the radius of the column and the rate of temperature change, if any. The vapor pressure of a solute is an exponential function of the absolute temperature; there must then exist a range of volatilities (and a range of partition ratios, *k)* for identical solute molecules located in any transverse section of the packed column. In the capillary column, the thinner, more uniform film of liquid phase is deposited directly on a low thermal mass column wall. Provided that the column is heated only by the oven air (i.e., does not possess a line of sight to the oven heater, and is therefore shielded from radiant heat), the liquid phase temperature is constant through any transverse section of the column.

It is because of these narrower ranges of retention times that the capillary column delivers much higher separation efficiencies. What is frequently overlooked is that in those cases where these high efficiencies are not required and the separation delivered by a packed column is adequate, very short capillaries can be substituted to great advantage. Not only does this substitution permit the same equipment to perform ten to fifteen times as many analyses in the same time, but the shorter chromatographic time per analysis means that the peaks suffer less diffusion; hence sensitivities are much higher.

On a pro-rata basis, the short section of capillary column required usually represents a lower purchase price than does the packed column. The end result is a considerable savings, both in the initial cost of the column and in the cost per analysis of labor, energy, gases, chart paper, and equipment depreciation. Hence the capillary system can be expected to deliver superior separations, shorter analysis times, and higher sensitivities, all in a more inert system and at lower cost (Jennings 1981A, 1982). Because separations are much improved, it is not unusual to find that the coefficients of correlation in quantitative analysis are increased by an order of magnitude for capillary as compared to packed column analyses.

Under any given set of circumstances, the degree to which components are separated (which, as has been noted, will affect the degree to which one can manipulate the analysis time and sensitivity) is essentially a function of

1. the degree to which the peak maxima are separated, and
2. the "sharpness" of the peaks (Jennings 1980).

The former is primarily affected by column contributions, including column length; by operational parameters such as the type of carrier gas and the conditions of its use (e.g., velocity, temperature, pressure drop); and by solute properties, including their partition ratios and relative retentions. The "sharpness" of the peaks is affected by column contributions (column length and the height equivalent to a theoretical plate), by operational parameters (e.g., carrier gas choice, conditions of use, temperature), by solute properties (e.g., partition ratios), and by the efficiency of the injection process.

In summarizing this area, we can conclude that for any given set of circumstances, resolution is maximized when

1. the band beginning the chromatographic process is as narrow as possible;
2. all the identical molecules of each solute behave in the same manner, i.e., both the range of times they spend in the gas phase and the range of times they spend in the liquid phase are as small as possible; and
3. the interaction(s) between the molecules of each of the solutes to be resolved and the liquid phase are as different as possible.

These factors, the complications of their interrelationships, and methods for their exploitation have been considered elsewhere (Jennings 1983).

PROBLEM AREAS IN FOOD ANALYSES

Bearing in mind the concepts we have just developed, let us turn our attention to some specific problem areas in food analysis, and begin by distinguishing between what has been termed "total volatile analysis," and "direct vapor analysis," or "headspace analysis" (Weurman 1969). The latter represents those volatiles overlying a food, while the former includes all of the volatiles that can be isolated from that food, including those dissolved in fat globules or adsorbed to particulate matter. The headspace volatiles make the major contribution to food aroma, and the primary interest of the flavor analyst is generally the qualitative and quantitative comparisons of headspace volatiles.

One of the major problems that must be faced in the direct injection of headspace volatiles relates to earlier considerations: the band beginning the chromatographic process must occupy the shortest possible length of column. This seriously limits the size of the injection that can be made, and in headspace injections the major volatile is usually air and the next most abundant is water; the concentrations of the volatiles of interest are often too low to activate the detector. Consequently, some type of pre-concentration step is usually required: Various procedures involving extractions and distillations have been widely used, and a variety of adsorbants, including activated carbon and the porous polmers such as the Porapaks and Tenax GC, are often employed to isolate entrained volatiles from large volumes of "headspace" (Jennings and Rapp 1983). In this latter case, both solvent elution and thermal desorption (e.g., "purge and trap") are used to recover the adsorbed volatiles. The low volatility of some high-boiling (or highly solvated) solutes on the one hand and "breakthrough" of some highly volatile solutes on the other are problems with some substances; in addition, the efficiencies of recovery from the trapping substrate may be poor.

DEVELOPMENTS IN GC HARDWARE/METHODOLOGY

Recent developments in columns, in injector hardware, and in methodology combine to offer the flavor chemist opportunities for high sensitivity headspace injections that can eliminate the tedium and errors inherent in the above procedures. The introduction of highly efficient fused silica columns in which the liquid phase is present as a cross-linked bonded nonextractable film (Jenkins and Wohleb 1980) have made stable thick-film water-tolerant columns commonly available. From these same workers came the concept of using a syringe fitted with a needle of fused silica rather than stainless steel; not only is the sampling device now much more inert, but the use of an on-column injector with a glass alignment device permits injections to be made directly inside a column as narrow as 0.25-mm i.d. (Jennings 1981B). Because of the flexibility of the fused silica column, it is a simple matter to open the oven door, bend a portion of the inlet end of the column to a U-configuration, and force it into a Dewar flask of liquid nitrogen (Fig. 13.1). Using a special fitting fabricated by the author that adapts a fused silica needle to a 1-mL gas-tight syringe, a 0.5–1-mL headspace sample is now injected into the column, via the on-column injector. The sample matrix—air—passes through, but the entrained solutes are cold trapped in the liquid phase. In addition, the solute band is narrowed, because in that trapping process, the front of the solute band encounters a colder environment and is decelerated to a greater degree than is the rear of the band; the band is subjected to a "thermal focus" (Jennings and Rapp 1983).

This permits the injection of much larger volumes of headspace gas than can normally be tolerated without overwhelming the sample-accepting volume of the system. In addition, the volatiles in that larger volume are now concentrated in the form of a narrow band on the beginning of the column. For many samples, this simple method not only eliminates the need for purge and trap, but circumvents the difficulties and uncertainties that accompany such methods, and yields superior results.

Several factors can influence the results obtained: the volatilities of the solutes in the sample, the temperature to which the trapping section of the column is cooled, the length of column subjected to cooling, the velocity of the carrier gas during sample injection, and the speed with which the injection is made—i.e., the introduction velocity—all play interrelated roles. Extremely volatile solutes still have a tendency to "breakthrough," which is evidenced by malformed peaks at the front of the chromatogram. This tendency is reduced by the use of lower coolant temperatures, as illustrated in Fig. 13.2. Even under the best of circumstances, the cold-trapped band is not stationary, and moves, albeit slowly, under the impetus of the carrier gas flow. Anything that prolongs the period over which the trapped sample must be retained (e.g., an excessively slow injection) or that lessens its retention under the trapping conditions (e.g., thinner films, higher temperatures) will lead to degradation of the starting band. Once that band traverses the trap and begins ascending the positive temperature

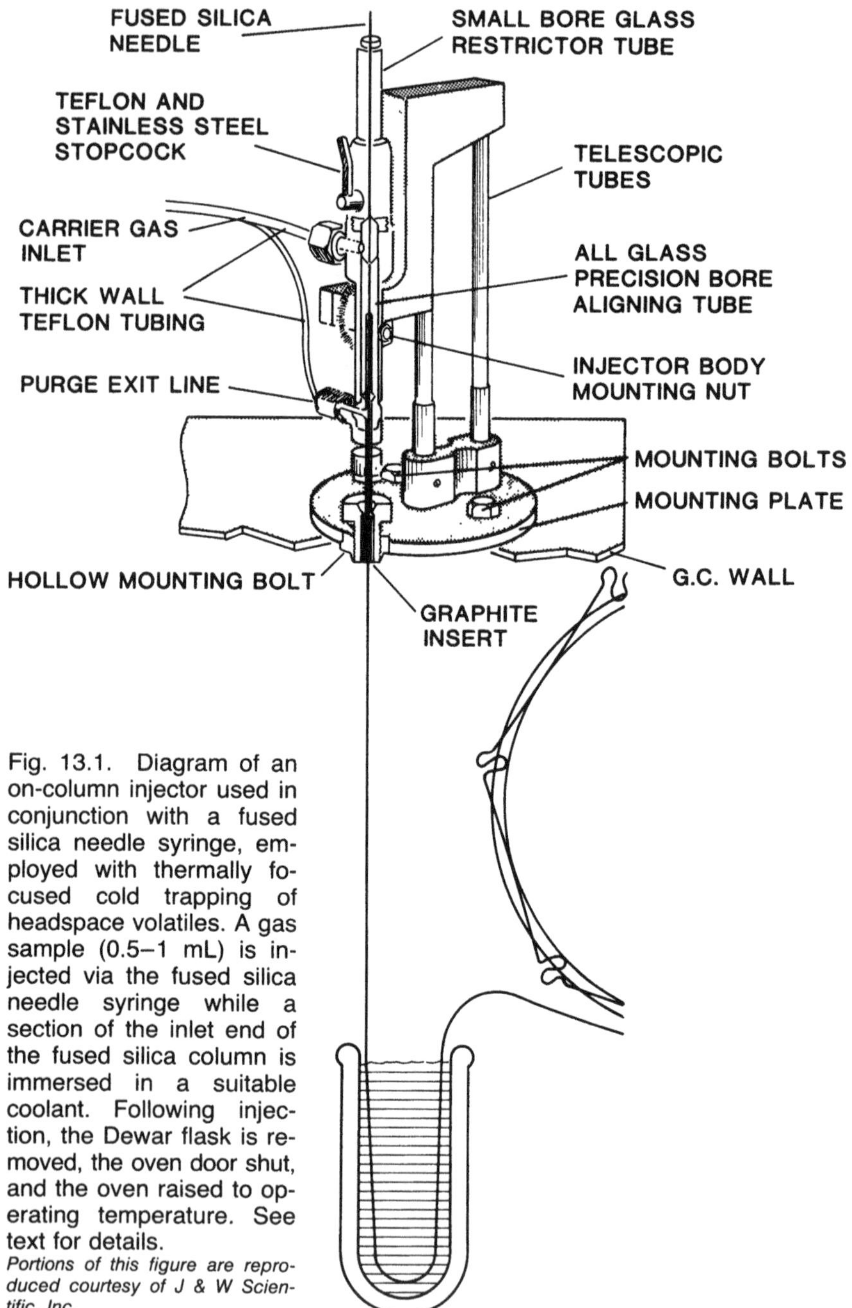

Fig. 13.1. Diagram of an on-column injector used in conjunction with a fused silica needle syringe, employed with thermally focused cold trapping of headspace volatiles. A gas sample (0.5–1 mL) is injected via the fused silica needle syringe while a section of the inlet end of the fused silica column is immersed in a suitable coolant. Following injection, the Dewar flask is removed, the oven door shut, and the oven raised to operating temperature. See text for details.

Portions of this figure are reproduced courtesy of J & W Scientific, Inc.

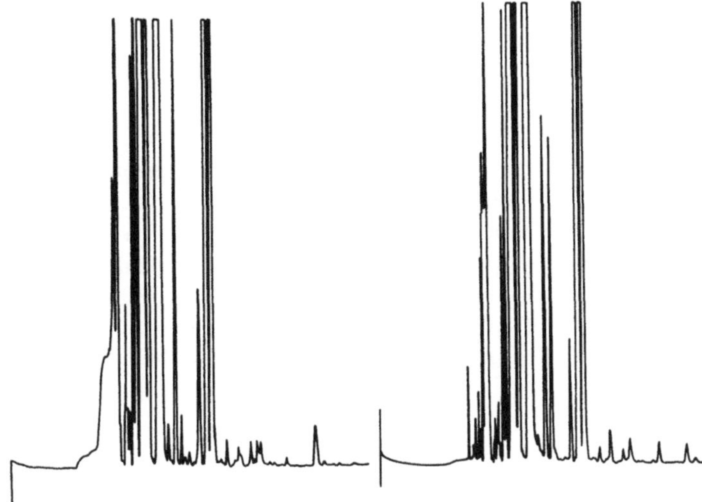

FIG. 13.2. Chromatograms produced by injection of 1 mL of headspace overlying 2 g cocoa at 65°C. Column, 30 m × 0.25 mm fused silica with a 1.0-μm film of DB-5; sample introduced at room temperature, and oven programmed from 50° to 200°C at 10°/min. Left, dry ice–ethanol coolant; right, liquid nitrogen coolant.

ramp at the other side, it undergoes a "reverse thermal focus" (Jennings and Takeoka 1982), with disastrous effects on the chromatographic results. On the other hand, if the injection is made too rapidly, the velocity of the solutes through the cooled section of column may be too great to allow for efficient trapping. More volatile solutes give better results with lower trapping temperatures and thicker film. Most of the solutes detected by headspace analysis are reasonably volatile, and as a general rule, liquid nitrogen is the most effective coolant. A length of column of approximately 20 cm should be looped and immersed in the coolant, and ca. 3–7 sec should be taken to inject the vapor sample. Immediately following injection, the syringe needle should be withdrawn, closing the entry stopcock as soon as the needle clears that point, and the coolant is removed. The low thermal mass of the column permits it to assume oven temperature almost immediately, thus avoiding the positive temperature ramp.

The introduction of a vapor sample to the column as performed above is very different from accomplishing that sample introduction by any other injection mode. Where the sample is injected into a chamber and then carried forward to the column, it must undergo dilution with the carrier gas; more time is required to carry it to the column, and the prospects for less efficient on-column trapping and band broadening are increased.

Our goal in establishing these guidelines is to permit the injection of larger amounts of the solutes to be detected in spite of the fact that these are present at very low concentrations in a gaseous matrix, while adhering to two of the conditions that we earlier established as essential to good gas chromatography: (1) the injection band must occupy the shortest possible length of column as it begins the chromatographic process; (2) the chromatographic conditions must be such that all of the molecules of each solute exhibit the narrowest possible range of retention times. Figure 13.3 shows typical results as obtained on a headspace injection from a freshly opened cola drink. Food materials whose volatiles occur in lower concentrations can be placed in a loosely stoppered flask and suspended in an oil bath at an elevated temperature so that the vapor pressures of those volatiles is increased to a point where they dominate that equilibrium atmosphere. Water is a major constituent of that atmosphere, but properly selected cross-linked bonded phase fused silica columns will even tolerate direct water injection; condensed water that may appear in the syringe prior to injection is of no concern. Figure 13.4 shows a headspace chromatogram

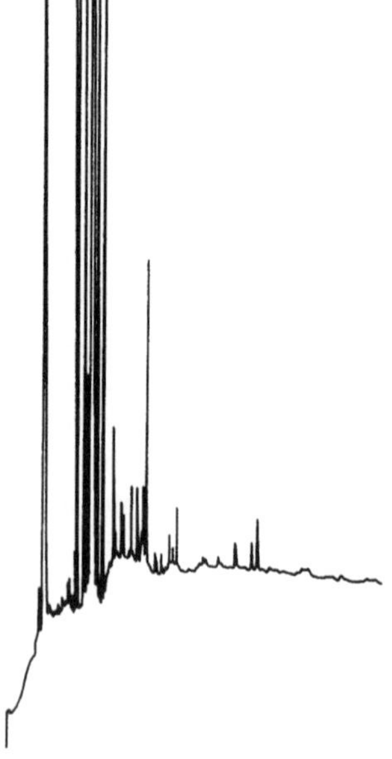

FIG. 13.3. Chromatogram produced by injection of 1 mL of headspace from a freshly opened cola drink. Same conditions as Fig. 13.2, right.

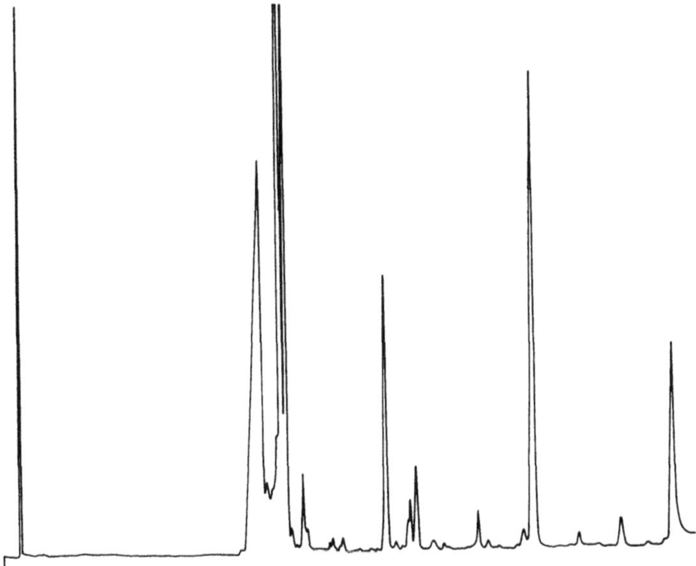

FIG. 13.4. Chromatogram produced from a 1-mL injection of headspace from a single badly off-flavored walnut, crushed and held in a loosely stoppered flask at 65°C for 15 min. Injection conditions same as Fig. 13.2, right. Note that the most volatile solutes have "broken through" and exhibit malformed peaks due to the ascent of the positive temperature ramp. Their conformation could be improved by lower trapping temperatures, a longer trapping section, a shorter injection time, or a lower carrier gas velocity; some of these variables could exert, however, negative influences on other solutes (see text).

generated from a single off-flavored walnut, crushed and "equilibrated" at 65°C.

In our earlier discussions, we also established one other criterion leading to increased resolution: the interaction(s) between the molecules of each of the solutes to be resolved and the liquid phase should be as diverse as possible. This differentiation between the retention behavior of two solutes, of course, is best described as their "relative retention" for which we use the Greek α. The resolution equation, which can be expressed as

$$R_s = \sqrt{n}/4\ [(\alpha - 1)/\alpha][k/(k+1)]$$

specifies that the resolution between any two solutes, R_s, increases with the number of theoretical plates (n), the partition ratio (of the second solute) (k), and the relative retention of those solutes (α) (Jennings 1980).

The number of theoretical plates can be bolstered both by increasing column length, and/or decreasing column diameter. However, either route leads to a higher pressure drop through the system, which adversely affects both the van Deemter curve and the overall separation (Ingraham *et al.* 1982A). One method of overcoming these limitations is by recycling many times through short columns (Jennings *et al.* 1979A). However, while more than 2,000,000 theoretical plates were developed by the system cited above, it was necessary to have specially manufactured at great expense the hardware required for its construction. At some time in the future, this approach to large plate numbers will have practicality, but that time is not yet here (Jennings *et al.* 1979B). Our work in this area is continuing.

For a given liquid phase and a given solute, the partition ratio k is most strongly affected by (1) the temperature and by (2) the phase ratio of the column. The advent of highly cross-linked bonded phase columns has led to the availability of stable thick-film columns whose utility for cold-trapping of headspace volatiles we discussed earlier. With all other factors constant, the partition ratio of a solute increases in direct proportion to the thickness of the liquid phase film; this relationship is often exploited now to yield improved separation, particularly of highly volatile (i.e., normally low-k) solutes. Figure 13.5 shows a typical application.

The "polarity" of the liquid phase influences to the greatest degree the relative retentions of solutes. Polarity, as it is used in gas chromatography, is not a well-defined term, and the author uses it with reluctance. Most chromatographers would define a nonpolar liquid phase as one in which solutes are retained only through forces of dispersion. Any other interaction, including those involving dipoles and hydrogen bonding, would increase the "polarity" of that liquid phase. Obviously, both the liquid phase and the solute are involved in any such interactions, and this is really the root of the problem in discussions of "polarity." DB-17, a bonded form of OV-17, is a 50% phenyl polymethyl siloxane. It has been described as "equivalent in polarity" to DB-1701 (or OV-1701), a 7% phenyl–7% cyanopropyl polymethyl siloxane. For some compounds, this is probably true, but solutes with aromaticity would regard the former as more polar than the latter, while solutes with a strong hydrogen-bonding tendency would regard the latter as more polar than the former.

At any rate, we can take advantage of these properties of the liquid phase, either by intelligent selection if we do know the functional groups

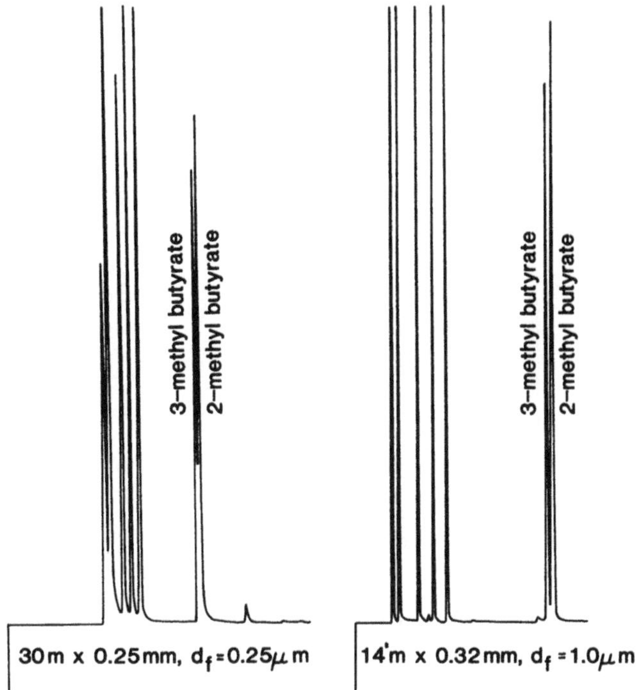

FIG. 13.5. Effect of liquid phase film thickness on component resolution. Test solutes, 2-methyl butyrate and 3-methyl butyrate, both columns DB-5. Column dimensions: left, 30 m × 0.25 mm with 0.25-μm film; right, 14 m × 0.32 mm with 1.0-μm film. Resolution is enhanced because the average partition ratio of the test solutes has been increased from ca. 1 (left) to ca. 4.7 (right).
Chromatogram generated by D.F. Ingraham.

in the solutes, or empirically (i.e., trial, error, and selection) if we do not. Figures 13.6–13.9 show examples where these interactions are exploited to achieve increased separation.

In those circumstances where the interactions between each of the solutes and two or more liquid phases can be determined individually and independently, the selectivity of the liquid phase can be exploited to an even greater degree. This approach is limited to mixtures that can be chromatographed under isothermal conditions, and is based on the fact that the distribution constant of a solute in a binary liquid phase mixture is the sum of the products of the volume fraction of each liquid phase multiplied by the distribution constant of the solute

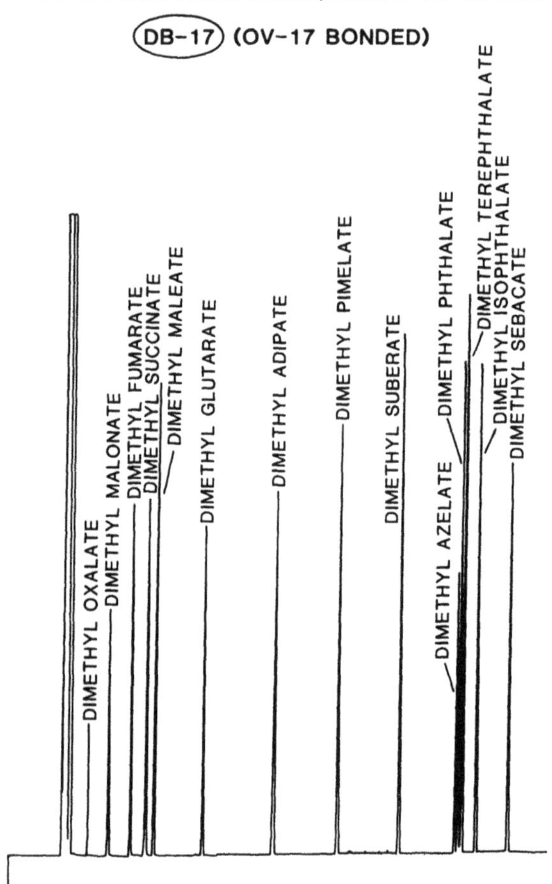

FIG. 13.6. Dimethyl esters of dicarboxyllic acids on a 30 m × 0.25 mm fused silica column coated with a 0.25-μm film of DB-17, a cross-linked, bonded 50% phenyl methyl silicone liquid phase. Split injection, programmed 120°–200°C at 10°/min. Components as in Fig. 13.10 except 10, azelate; 11, phthalate; 12, terephthalate; 13, isophthalate.

in the pure phase (Maier and Karpathy 1962). Methods for estimating the optimum binary mixture for a given set of circumstances have been proposed (e.g., Laub and Purnell 1976), as have empirical (Ingraham *et al.* 1982B) and precise (Takeoka *et al.* 1983) methods of compensating for complications engendered by the pressure drop through the system.

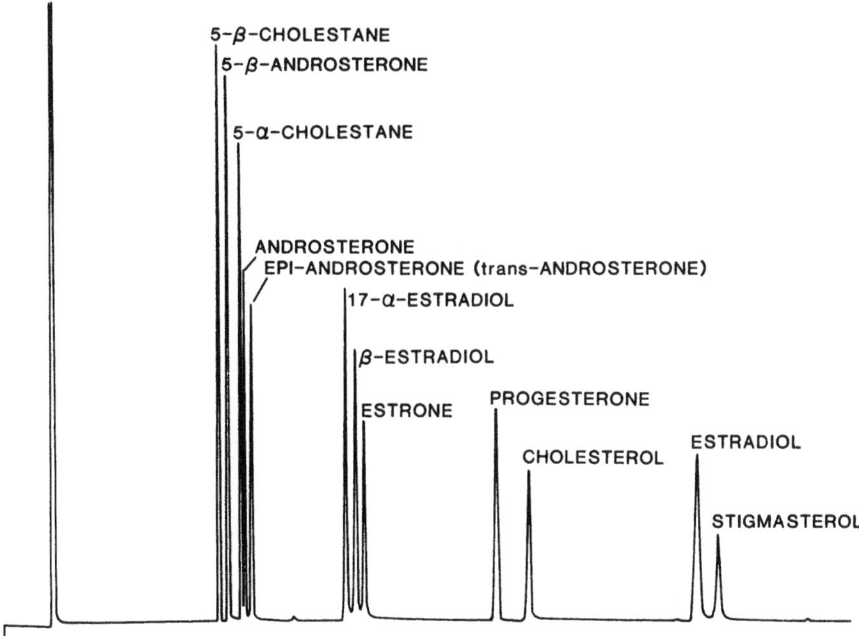

FIG. 13.7. Free underivatized sterols and steroids on a 30 m × 0.25 mm fused silica column coated with a 0.15-μm film of DB-17. Split injection at 260°C isothermal.

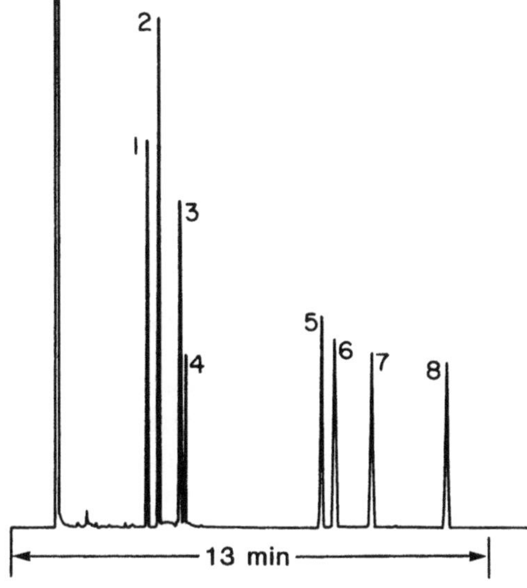

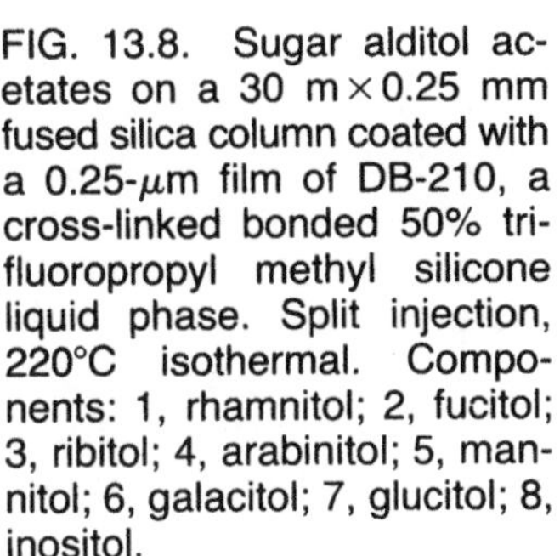
FIG. 13.8. Sugar alditol acetates on a 30 m × 0.25 mm fused silica column coated with a 0.25-μm film of DB-210, a cross-linked bonded 50% trifluoropropyl methyl silicone liquid phase. Split injection, 220°C isothermal. Components: 1, rhamnitol; 2, fucitol; 3, ribitol; 4, arabinitol; 5, mannitol; 6, galacitol; 7, glucitol; 8, inositol.

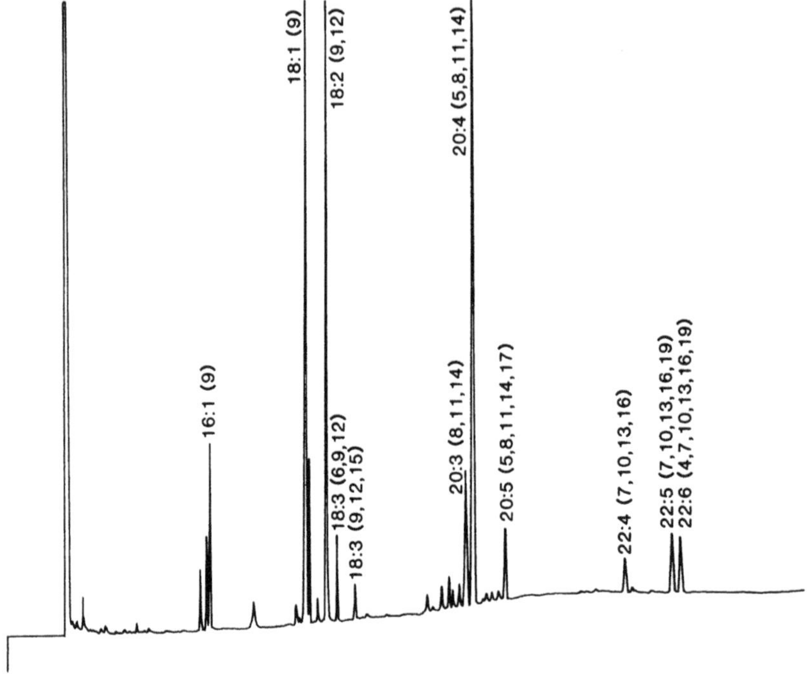

FIG. 13.9. Polyunsaturated fatty acid methyl esters on a 30 m × 0.25 mm fused silica column coated with a 0.15-μm film of DB-225 (see Fig. 13.6). Split injection, programmed from 180° to 220°C at 4°/min. Components: 1, 16:1 (9); 2, 18:1 (9); 3, 18:2 (9,12); 4, 18:3 (6,9,12); 5, 18:3 (9,12,15); 6, 20:3 (8,11,14); 7, 20:4 (5,8,11,14); 8, 20:5 (5,8,11,14,17); 9, 22:4 (7,10,13,16); 10, 22:5 (7,10,13,16,19); 11, 22:6 (4,7,10,13,16,19).

These esoteric concepts offer real utility in the solution of some of our practical problems. Films and plastics offer many advantages as food packaging materials, but the possible migration of unreacted monomers, plasticizers, and residual solvents into that food is real. In spite of considerable effort (e.g., Kolb *et al.* 1981), the separation of a mixture of the solvents commonly used in food packaging films has not been achieved on a single-gas chromatographic column. Figure 13.10 shows a chromatogram in which complete separation was achieved when the concepts discussed above were combined in a computer program to define the optimum liquid phase mixture for the separation of this solvent mixture (Ingraham *et al.* 1982B). Figures 13.11–13.14 illustrate how the separation of another mixture of interest to segments of the food industry was optimized by these methods.

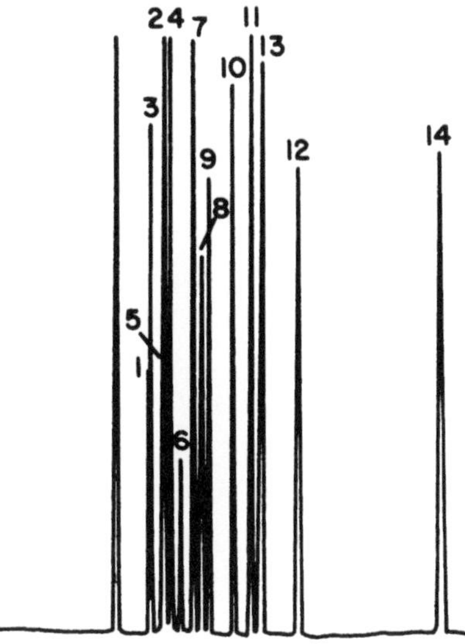

FIG. 13.10. Chromatogram of a mixture of solvents used in food packaging films, achieved on serially coupled column composed of a 13.1-m segment of fused silica coated with the polyethylene glycol Carbowax 20 M, followed by a 13.35-m segment of fused silica coated with polymethylsiloxane DB-1. Split injection, 60°C isothermal. Components: 1, methanol; 2, ethanol; 3, acetone; 4, isopropanol; 5, methyl acetate; 6, dichloromethane; 7, 2-butanone; 8, ethyl acetate; 9, tetrahydrofuran; 10, isopropyl acetate; 11, benzene; 12, *n*-propyl acetate; 13, *n*-heptane; 14, *n*-octane.

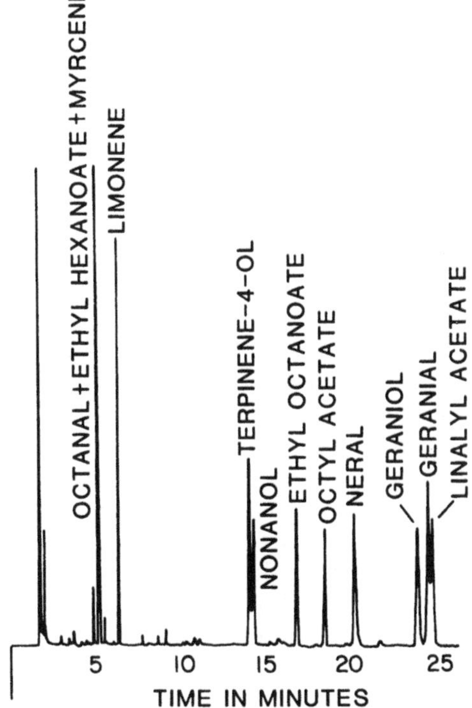

FIG. 13.11. Chromatogram of a test mixture on a 30 m × 0.33 mm fused silica column coated with a 0.25-μm film of DB-1. Note that octanal, ethyl hexanoate, and myrcene emerge as a single peak, and the separation of terpinene-4-ol and nonanol, and of geranial and linalyl acetate, are incomplete; analysis time, 25 min.
From Takeoka et al. 1983.

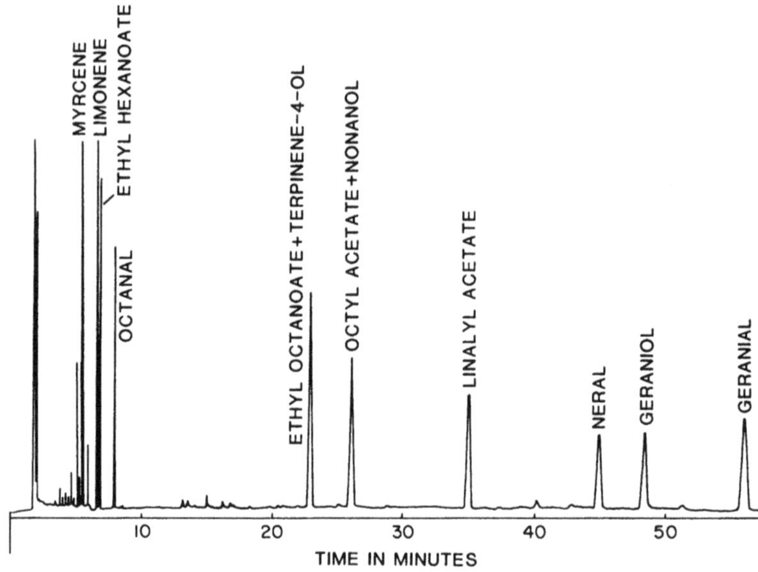

FIG. 13.12. Chromatogram of the test mixture shown in Fig. 13.11 on a 30 m × 0.25 mm column coated with DB-1701. Note that while the components co-eluting above are now separated, ethyl octanoate and terpinene-4-ol now form a singlet, as do octyl acetate and nonanol; analysis time, 57 min.
From Takeoka et al. 1983.

RECENT ADVANCEMENTS

Finally, there is another area of separation science that the author is convinced will soon assume increased importance, but which has not been discussed. Gas chromatography and earlier liquid chromatography have been examined. Certain arguments have been employed to explain why smaller diameter (and open tubular) columns delivered superior performance in gas chromatography; these same arguments are valid in liquid chromatography, but their application in the case of the latter is limited by two interrelated factors. (1) First, the methods of detection common to liquid chromatography are much less sensitive. For those solutes that have particularly strong properties of absorption, flourescence, or high indices of refraction, sensitivities are adequate, but many solutes possess these properties to such a limited extent that their detection becomes more difficult. (2) Second, because solute diffusivity in the mobile phase is much lower in

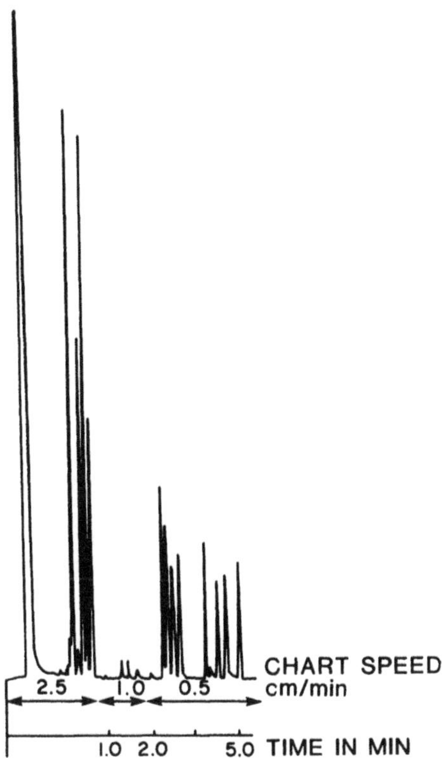

FIG. 13.13. Chromatogram of the test mixture shown in Figs. 13.11 and 13.12 on a serially coupled column composed of 6.8 m of the column used in Fig. 13.11, plus 7.3 m of the column used in Fig. 13.12. Note that the chart speed has been step programmed, producing a nonlinear time scale. Separation is complete; analysis time, 4.9 min.
From Takeoka et al. 1983.

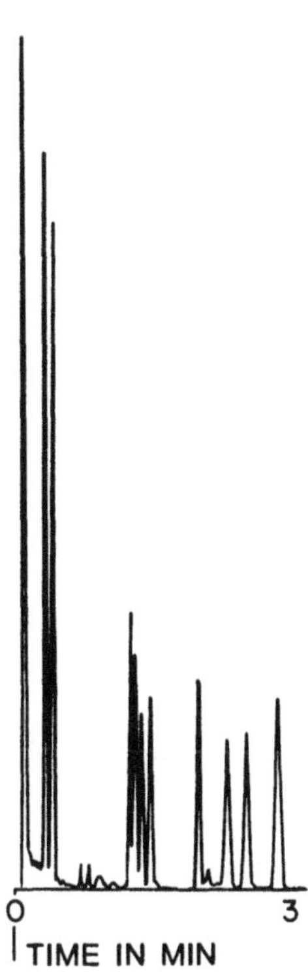

FIG. 13.14. Chromatogram of the test mixture shown above on a coupled column composed of 2.9 and 3.1 m segments, and operated at the optimum practical gas velocity. Analysis time, 2.9 min.
From Takeoka et al. 1983.

liquid chromatography, the mass transport step of the van Deemter equation requires either long analysis times, or extremely fine diameter columns; the latter pose what have been impossible demands on the volume and (bearing in mind the sensitivity limitations) configuration of the detector cell. Two different solutions to this impasse are being explored.

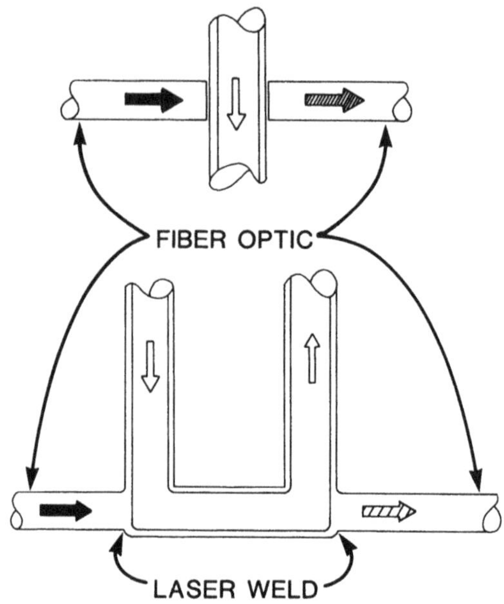

FIG. 13.15. Schematic of a zero dead volume, micro volume fused silica detector flow cell, designed to employ optical waveguides for spectrometer interfacing. By laser welding the fiber optics directly to the side wall of the "column," the latter becomes in effect the end wall of the cell; this permits greatly increased sensitivity because of the larger path length/volume ratio. Although designed for liquid chromatography, the cell may well have application for both gas and super critical fluid chromatography (patents pending).

The first solution, which is still under development (vide infra), involves a micro-volume zero-dead-volume fused silica flow cell (Fig. 13.15), designed for liquid chromatography, but which may also have applicability as a gas chromatographic detector. The cell is intended to interface to a computer-spectrometer via fiber optic waveguides. Fiber optic material currently available, however, rapidly loses its properties of transmission under these conditions; newer materials now under exploration may rectify this difficulty (De Lucca 1983). Because the cell design permits larger path length/volume ratios, than can be achieved by scanning transversely across the column, higher sensitivity will be possible.

The second solution entails a process known as supercritical fluid chromatography. When a gas is subjected to high pressure, it tends to liquify unless it is held above its critical temperature; under these conditions, it forms a supercritical fluid, and possesses both some of the properties of a gas, and some of the properties of a liquid: properties such as solvation, solution, and diffusivity are all high. Interesting results are being achieved by several workers (e.g., Novotny *et al.* 1981; Peaden and Lee 1982); the utility of adapting normal gas chromatographic equipment to this end is under exploration (Grob 1983), and prototype equipment is now available from a major GC supplier. While there are theoretical reasons to doubt that supercritical fluid chromatography will replace gas and/or liquid chromatography, it will almost surely assume increased importance as an alternative method of analysis in the coming years.

BIBLIOGRAPHY

DE LUCA. C. 1983. Personal communication. Spectran, Inc., P.O. Box 650, Sturbridge, MA 01566.

GROB, K. 1983. An attempt to extend ordinary capillary gas chromatography to supercritical fluid chromatography (SFC). HRC & CC *6,* 178–184.

INGRAHAM, D.F., SHOEMAKER, C.F. and JENNINGS, W. 1982A. Computer comparisons of variables in gas chromatography. HRC & CC *5,* 227–235.

INGRAHAM, D.F., SHOEMAKER, C.F. and JENNINGS, W. 1982B. Optimization of liquid phase mixtures. J. Chromatogr. *239,* 39–50.

JENKINS, R.G., and WOHLEB, R.H. 1980. The use of bonded phases in fused silica capillary gas chromatography. Presented at the Int. Symp. Adv. in Chromatogr., October 6–9, Houston TX.

JENNINGS, W. 1980. Gas Chromatography with Glass Capillary Columns, 2nd ed. Academic Press, New York.

JENNINGS, W. 1981A. Comparisons of Fused Silica and Other Glass Columns in Gas Chromatography. Huethig Publishing Co. Heidelberg.

JENNINGS, W. 1981B. Recent developments in high resolution gas chromatography. *In* Flavour '81. P. Schreier (Editor), pp. 233–251. de Gruyter Publishing Co., Berlin.

JENNINGS, W. 1982. New developments in capillary columns for gas chromatography. *In* Advances in Chromatography. J.C. Giddings, E. Grushka, J. Cazes, and P.R. Brown (Editors). Vol. 20, pp. 197–215. Marcel Dekker Inc., New York.

JENNINGS, W. 1983. Ultra-high resolution gas chromatography: constraints, compromises and prospects. *In* Ultrahigh Resolution Chromatography. ACS Symp. Ser. 250, pp. 27–36. S. Ahuja (Editor). American Chemical Society, Washington, DC.

JENNINGS, W., and RAPP, A. 1983. Sample Preparation for Gas Chromatographic Analysis. Huethig Publishing Co., Heidelberg.

JENNINGS, W. and TAKEOKA, G. 1982. Some theoretical aspects of capillary on-column injections. Chromatographia *15,* 575–576.

JENNINGS, W., SETTLAGE, J.A., and MILLER, R.J. 1979A. Multiple sort pass glass capillary gas chromatography. HRC & CC *2,* 441–443.

JENNINGS, W., SETTLAGE, J.A., MILLER, R.J., and RAABE, O.G. 1979B. New approach to chromatographic optimization in glass capillary glass chromatography. J. Chromatogr. *186,* 189–196.

KOLB, B., POPISIL, P., and AUER, M. 1981. Quantitative analysis of residual solvents in food packaging printed films by capillary gas chromatography with multiple headspace extraction. J. Chromatogr. *204,* 371.

LAUB, R.J. and PUNELL, J.H. 1976. Optimization of gas chromatographic analysis of complex mixtures of unknown composition. Anal. Chem. *48,* 1720–1724.

MAIER, H.J., and KARPATHY, O.C. 1962. Prediction of separation and specifications of chromatographic columns. J. Chromatogr. *8,* 308–318.

NOVOTNY, M., SPRINGSTON, S.R., PEADEN, P.A., FJELDSTED, J.C., and LEE, M.L. 1981. Capillary supercritical fluid chromatography. Anal. Chem. *53,* 407A–414A.

PEADEN, P.A., and LEE, M.L. 1982. Supercritical fluid chromatography: methods and principles. J. Liq. Chromatogr. *5* (Suppl. 2), 179–221.

TAKEOKA, G., RICHARD, H.M., MEHRAN, M., and JENNINGS, W. 1983. Optimization via liquid phase mixtures in capillary gas chromatography. HRC & CC *6,* 145–151.

WEURMAN, C. 1969. Isolation and concentration of volatiles in food odor research. J. Ag. Food Chem. *17,* 370–384.

14

Recent Developments in the Analysis of Pesticides

Gunter Zweig[1]

INTRODUCTION

The widespread use of pesticides in the production of food and fiber has resulted in the occurrence of pesticide residues in raw agricultural crops (fruit and vegetables), milk and dairy products, and processed foods. This fact should not surprise any citizen in 1983 and certainly does not imply that our food supply is not safe for human consumption. The Federal Government, mainly through two agencies, the FDA and EPA, has recognized that the presence of small amounts of pesticides might be inevitable and has established, therefore, the concept of so-called "tolerances" or "food additives" (Sections 402, 408, and 409 of the Federal Food, Drug and Cosmetics Act).

Pesticide residues in foods can arise from the purposeful application of pesticides during the growing stages of the crops, during storage, or during long distance shipment of commodities. These are "direct residues" and tolerances have been established for all pesticides which are used on crops and commodities for human consumption. Secondly, pesticide residues may originate in milk and other dairy products, for example, from the ingestion by dairy animals of feeds which have been treated with pesticides sometime during the growing season. These feeds may consist of grass or so-called by-products, such as corn husks, almond shells, or pineapple greens. When pesticide residues are found in milk, they are usually considered to be illegal

[1]Environmental Protection Agency, Office of Pesticide Programs, Washington, DC 20460. Present address: School of Public Health, University of California, Berkeley, CA 94720

or "indirect" residues, as for example the recent finding of heptachlor and heptachlor epoxide at above-tolerance levels in Hawaiian milk. Highly contaminated products are usually destroyed.

Thirdly, there is a possibility of contamination of foods by pesticides from the ubiquitous occurrence of pesticides in the environment (mostly persistent, organochlorine pesticides) or pesticide-like compounds, such as PCB's, or trace contaminants, such as dioxins.

Each of these cited cases can be monitored through analysis of pesticide residues by modern analytical techniques. The particular need for the analysis determines the method of choice. For example, the grower of fruits and vegetables knows what pesticides have been applied to his crops and may be required by the food processor to certify which pesticide residues are present. These analyses are the simplest and most direct to carry out, because the analyst is dealing with known chemicals and can direct his strategy towards recommended methods for those pesticides sprayed on the grower's plots.

If food processors, especially baby food manufacturers, want to make certain that measurable amounts of pesticide residues are absent, their quality control chemists should analyze the raw agricultural crops as well as the processed foods. The chemists may not be aware of all of the pesticides that have been used during production of the crops and must devise "multiresidue" methods for the important classes of pesticides that are most commonly used on particular crops. Their task is thus commonly more arduous than that of the grower's analysts.

The government analysts, especially the FDA analysts, are faced with an even more complicated task. They are charged with the analyses of "total diet" samples in which they may find any number of over 1000 commonly used pesticides. The FDA chemist uses multiresidue methods, which will be described later.

The methods used to analyze pesticide residues range from relatively simple methods for one or several pesticides, usually involving chromatographic techniques, to the proverbial search for the "needle in the haystack" involving GC or HPLC coupled to mass spectrometry. Selection of the method depends upon the purpose of the analysis and the client for whom the analysis is being performed.

METHODS OF PESTICIDE ANALYSIS

Review of Classical Methods

In a pragmatic review of classical methods of analysis for pesticides, one does not have to go back to the beginning of analytical chemistry since the watershed in the analysis of submicrogram residue quantities was in 1960 with the advent of various types of chromatographic techniques, especially gas–liquid chromatography. The classical methods of pesticide analysis were discussed by Zweig (1963) in Vol. I of Analytical Methods for Pesticides, Plant Growth Regulators, and Food Additives. Most analyses for pesticides were conducted by spectrophotometric techniques (colorimetric, fluorometric, and ultraviolet spectrophotometric), which are rarely used today. Infrared spectrophotometry has not achieved an important status as an analytical technique for pesticide residues. However, this may change over the next several years due to the development of Fourier-transform IR interferometry. Further discussion of these techniques are presented below.

Spectrophotometric methods depend upon the development of a visible color or a fluorescent reaction product, but the presence of interfering "false positivies" means that blanks are usually high, and rigorous cleanup has to be undertaken. Furthermore, the spectrophotometric method usually does not allow the simultaneous analysis of metabolities or degradation products of the pesticide sought. Fluorescence techniques are also limited, because few pesticides fluoresce or lend themselves to reactions which produce a suitable fluorescent derivative, without the interference of naturally fluorescent co-extractives. However, there are four commercial pesticides for which fluorometric methods have been developed: Co-Ral, Guthion, DEF, and Zinophos (now a discontinued product).

Paper chromatography achieved some popularity during those years, especially due to the ease of analysis and the low cost of equipment. Paper chromatography was the first assay system with which one could detect, during the same analysis, pesticides of similar structures and their metabolites. The method was invaluable in providing information on the contamination of milk and dairy products by chlorinated hydrocarbon pesticides. The method usually suffers from the lack of quantitative precision due to the inherent heterogeneity of the chromatographic medium.

In those "Middle Ages" of pesticide analysis, many chlorinated pesticides were analyzed by "total halogen" methods using the Schoeniger combustion flask, or "carrying it to the extreme," neutron activation of butter samples in a nuclear reactor.

Organophosphate insecticide residues were analyzed by biochemical techniques involving the enzyme acetylcholinesterase. Obviously, these techniques were nonspecific; however, they were useful as screening methods for the analysis of foods or crops with a known pesticide spray history.

Finally, bioassays using flies, mosquito larvae, or plants (for plant growth regulators) provided valuable data on the possible contamination of foods but in most cases were incapable of giving accurate information on the identity of the pesticide contaminant.

A final word about analytical techniques in the late 1950s and early 1960s. Gas chromatography, which up to that time was mostly in the domain of petroleum chemists, had suddenly become an analytical tool for the analysis of pesticides due mostly to the pioneering work of Coulson *et al.* (1959) who showed that high-melting chlorinated hydrocarbons, like DDT, dieldrin, aldrin, and heptachlor, could be gas chromatographed at elevated temperatures with the proper column packing without decomposition. Coulson and Cavanagh (1960) developed the first and truly quantitative and specific detector for halogen-containing compounds, the microcoulometric titration cell. Their work and the effort of others ushered in the modern era of pesticide residue analysis which will now be briefly reviewed.

Modern Gas Chromatographic Analysis of Pesticides

Column Packings and Optimum Experimental Conditions. Most pesticides will chromatograph on packed columns of silicone compounds coated at low concentrations on acid- or base-washed diatomaceous earths which are commercially available under trade names like Chromosorb®, Gas Chrom®, Supelcoport®, or Anakrom®. To minimize the on-column decomposition of the chlorinated insecticide endrin, the columns should be treated by silylation before use. The current usage of glass columns instead of stainless steel columns also minimizes the decomposition and/or transformation of many pesticides.

Table 14.1 is a list of liquid phases commonly employed for the gas chromatography of most classes of pesticides. Table 14.2 lists several combinations of liquid phases and column temperatures giving optimum performance for the stated classes of pesticides. Diethyleneglyl-

TABLE 14.1. STATIONARY LIQUID PHASES FOR COLUMN PACKINGS RECOMMENDED FOR PESTICIDES[a]

Name	Chemical composition	Alternative names
DC-200	Methyl silicone	OV-1; OV-101; SE-30
QF-1	Trifluoropropylmethyl silicone	OV-210; SP2401
SE-52	Methyl silicone; 10% phenyl subst.	OV-3
XE-60	Cyanoethylmethyl silicone	OV-225
OV-17	Methyl silicone; 50% phenyl subst.	SP-2250
OV-7	Methyl silicone; 20% phenyl subst.	—
DEGS	Diethyleneglycol succinate	—

[a] Thompson and Watts (1981).

TABLE 14.2. RECOMMENDED COLUMN PACKINGS FOR GLC PESTICIDE ANALYSIS[a]

Packing	Optim. column temp. (°C)	Application
1.5% OV-17/1.95% OV-210	200	Organochlorines
4% SE-30/6% OV-210	200	Organochlorines and OP
5% OV-210	180	Organochlorines
10% DC-200	200	Organochlorines, OP, N-cmpds
5% DC-200/7.5% QF-1	200	Organochlorines, OP, N-cmpds
1.6% OV-17/6.4% OV-210	200	OP, Organochlorines, N-cmpds
10% OV-210	200	OP
1.5% OV-17/1.95% OV-210	200	OP
2% DEGS	165–200	OP & Phenoxy esters (Anon. 1980A)

[a] Thompson and Watts (1981).

col succinate (DEGS) is suitable for phenoxyalkanoic esters and other polar compounds, but retention times for this type of packing have been found to be erratic and not easily reproducible (Thompson and Watts 1981).

The interested reader should consult Thompson and Watts (1981), the Pesticide Analytical Manual (Anon. 1980A), and Follweiler and Sherma (1984) for suitable columns and experimental conditions to resolve the particular class or combination of pesticides which are of concern.

While wall-coated open tubular (WCOT) capillary columns possess a much higher resolution power than packed columns, there are few references on the use of WCOT technology for pesticide analyses (Jennings 1980). McLean and Truman (1983) have recently demonstrated that excellent resolution could be achieved for a mixture of vinclozolin, endosulfan I and II, and endosulfan sulfate on a capillary column (see Fig. 14.1). Capillary columns have a low loading capacity and can reduce the sensitivity of the analyses. This can be partially overcome by using the more sensitive electron capture detectors.

Detectors

The development of element-selective or -specific detectors has made gas chromatography a very successful analytical tool for pesticide analysis (Holland and Greenhalgh 1981).

Electron Capture Detector (EC). There are two major recent modifications of the EC detector, one with tritium, the other with a ^{63}Ni source of electrons. The ^{63}Ni-detector appears to be more popular because it can be used up to 400°C and thus possesses greater versatil-

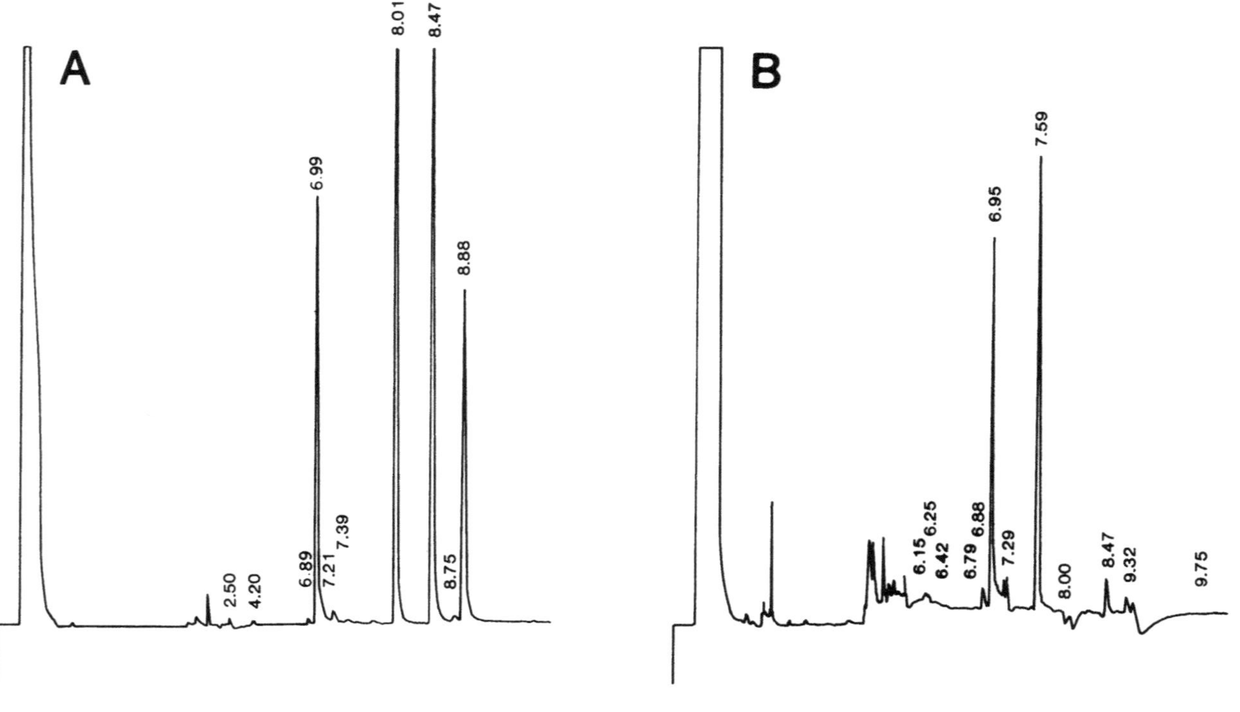

FIG. 14.1. WGOT gas chromatogram of mixture of pesticides: (A) mixture of standards, 0.1 ng/μL; vinclozolin (6.99 min); endosulfan I (8.01); endosulfan II (8.47); endosulfan sulfate (8.88); (B) extract of field dosimeter showing 0.07 ng vinclozolin; unknown compound at 7.58 min; 0.002 ng endosulfan I; 0.01 ng endosulfan II. *Experimental conditions:* HP 5880-A GC with Autosampler; bonded dimethylsilicone WCOT, deactivated with siloxane; 12.5 m $\times$ 0.2 mm i.d. fused silica column; temperature program 62°–250°C at 30°/min; isothermal at 260°C for 3 min; ^{63}Ni EC detector; injection port temp. 200°C; column pressure 13 psi.

ity. The tritium detector has a temperature limit of 225°C. The linear concentration range for both detectors with optimum configuration and operating conditions is about 10^4.

Electron capture detection has proved very useful for the detection of organochlorine pesticides and industrial pollutants due to its high sensitivity for these compounds. However, the high sensitivity also presents the disadvantage of detecting many co-extractives from field and food samples (e.g., PCBs and phthalates). Although having a lower response factor than organochlorine pesticides, these interfering compounds may be present at much higher concentrations and their presence leads to a false interpretation of the chromatograms. The lack of selectivity of EC detectors may be overcome by thorough cleanup and separation of interferences prior to chromatography.

Electrochemical Detectors. These detectors are much more element-specific than the EC detectors but lack their high sensitivity. A pyrolysis step follows the chromatographic separation. In the reductive mode of pyrolysis in the presence of a Ni catalyst, nitrogen compounds are reduced to NH_3. In the reductive or oxidative mode, without catalysts, organohalogen compounds are removed by scrubbing with acid or base. The resultant pyrolysis products are detected in the Coulson conductivity cell by Pt electrodes measuring conductivity of the electrolyte solution by a DC-bridge. The Hall detector is a small-volume conductivity cell and uses AC to overcome polarization problems.

The Coulson microcoulometer cell (Coulson and Cavanagh 1960) detects halogen-, nitrogen-, and sulfur-containing compounds by titrating the pyrolysis products (X^-, NH_3, SO_2) with appropriate ions (Ag^+, H^+, and I_3^-, respectively). The resultant electrical imbalance of the electrolyte solution causes the regeneration of these ions, and the resultant voltage drop produces a recorder output of a usual chromatographic elution pattern. The Coulson microcoulometer may be used as an absolutely quantitative detector, whereas all other GC detectors are relative and require calibration curves from external standards compounds. With the exception of the analysis of *N*-methylcarbamates (e.g., carbofuran), the Coulson microcoulometric cell has been largely replaced by the Hall detector for chlorinated pesticides, the flame photometric detector for P- and S-containing compounds, and the alkali-flame detector for P-compounds.

Flame Photometric Detector. The operation of this detector is based on the specific chemiluminescence produced in a hydrogen-rich flame for sulfur (excited S_2) and phosphorus (HPO). By viewing the emitted light through narrow-band interference filters at two specific wave lengths, a compound containing S and P can be detected simultaneously by the use of photomultipliers. The ratio of P/S can provide valuable clues on the identity of unknown compounds. The sensitivity of this detector is about 10 pg for P and 40 pg for S.

Potassium Chloride Thermionic Detector (TD). This detector is one of the best choices for the analysis of organophosphorus pesticides. It is a hydrogen-flame detector modified with an alkali metal bead (KCl) which produces a highly selective response for P compounds. The principle of this detector is based on the greatly enhanced response by P-containing compounds burning in a hydrogen-rich flame, compared to the conventional flame ionization detector.

Nitrogen/Phosphorus (N/P) Detector. This detector is a variation of the alkali-sensitized detector, described above. This detector incorporates a glass bead containing a rubidium or cesium salt on a platinum wire which can be heated electrically (Holland and Greenhalgh 1981). With a reduced hydrogen flow and consequently lower flame temperature, the response to nitrogen-containing compounds is greatly enhanced and makes it the choice for the determination of N- and P-compounds. Sensitivity under optimum experimental conditions is 0.02 and 0.2 pg/sec for phosphorus- and nitrogen-containing compounds, respectively.

High Performance Liquid Chromatography (HPLC)

The use of column chromatography for cleanup of pesticide residues from fatty food and crop samples was probably one of the earliest applications of this technique and is still being used today. The advent of new types of column packings and bonded phases and the availability of instrumentation, especially reciprocal pumps generating pressure as high as 6000 psi, have opened a completely new field of separation, namely, high performance liquid chromatography (or high-pressure liquid chromatography by its former name). Although the chromatographic resolution and the different types of chromatography approach that of gas–liquid chromatography, HPLC has not become as yet the universal analytical technique due to the lack of selective detectors. At this time, there are three types of HPLC detectors in use for pesticide analysis: UV detectors, fluorescence detectors, and electrochemical detectors. Mass spectrometry has been used as a detection technique but is not used by most food pesticide laboratories, due to high cost and complexicity.

In our experience with HPLC of pesticide residues, a commercial reverse phase column, C_{18}-bonded to silica, will separate many pesticides and can be regenerated innumerable times with pure solvent washes or the use of a guard column with little deterioration or loss of resolving power over many months or even years of continuous use. Others have had the same experience (Moye 1981). An additional advantage of reverse-phase chromatography is that the use of polar solvents such as methanol or acetonitrile, diluted with water, simplifies the cleaning of the delicate pump mechanism and does not cause the salt build-up from buffers such as found with ion-exchange systems. We have found that under such conditions the pumps can be operated 24 hr per day for many weeks without breakdown.

For the laboratory that wishes to have a complete complement of columns, it is recommended that three other columns be kept in reserve for an eventual separation which is not amenable to reverse phase chromatography, a commercial adsorption column (silica) and an anion- and a cation-exchange column. A useful table of the separation of many pesticides by HPLC is found in a recent publication by Lawrence and Turton (1978) and is recommended as a good reference source.

As already mentioned, the detectors for HPLC are limited at this time, but just as progress in GLC of pesticides moved rapidly as selective detectors became available, it is anticipated that similar progress will be forthcoming in this relatively new field of analytical methodology. If a pesticide has an aromatic moiety or other UV-absorbing group, the fixed- or variable wavelength detectors are applicable. In the absence of such UV-absorbing groups within the molecule, pre- and post-column fluorogenic labeling of the pesticide can be performed and has been successfully applied to a number of compounds (Moye and St. John 1980). The chromogenic derivatives are then detected with a commercial fluorometer directly attached to the effluent stream of the HPLC. The reaction apparatus for the formation of fluorogenic derivatives can be built from readily available pumps and glass coils, and heated water baths, and connected to commercial HPLC equipment.

If the pesticide under study has a suitable redox potential, a sensitive electrochemical detector, now commercially available, can be used. The theory and applications of electrochemical detectors have been described by Kissinger *et al.* (1980). Obviously, this type of detector is limited to a few pesticides, e.g. phenols (or hydrolysis products of *N*-methylcarbamates) and aliphatic and aromatic amines (aniline-based herbicides), and nitrophenylthiophosphates (parathion, etc.).

As discussed briefly below (Carbaryl Analysis), HPLC has the advantage of being adaptable to automation of the cleanup and determinative steps. Heat-labile compounds which decompose on GLC columns at elevated temperatures can often be easily separated by HPLC. Polar compounds which do not easily chromatograph on GC without derivatization, often can be chromatographed by HPLC. Thus, it appears that this technique will become more widely used in pesticide residue analysis. Already there is a large number of publications on this subject (Lawrence 1982; Moye and St. John 1980; Hanks and Colvin 1981; Ivie 1980; Follweiler and Sherma 1984).

Thin-Layer Chromatography (TLC)

Thin-layer chromatography has been used as a confirmatory method for the identification of pesticide residues and their metabolites. With the availability of commercial, coated plates, the technique is simple, adequately sensitive, and especially suitable for the separation of organochlorine compounds. This is due to the availability of a universal detection reagent, $AgNO_3$-2-phenoxyethanol and UV-light. Organophosphate insecticides can be detected at nanogram level using as spraying reagent a source of acetylcholinesterase (e.g., animal serum) followed by a suitable substrate like indoxyl acetate or indophenyl acetate (Sherma 1974, 1980).

The Pesticide Analytical Manual (Anon. 1980A) devotes an entire chapter to the semiquantitative determination of pesticide residues by TLC (Chapter 4). The recommended coating is alumina, and ready-made plates (20.3×20.3 cm) with this adsorbent are available. The cleanup recommended for GLC (extraction, partitioning, and Florisil chromatography) is usually sufficient for TLC. High-fat samples may be troublesome and must undergo additional cleanup, like an additional Florisil column. The recommended mobile phase solvents for chlorinated compounds are any of the following: *n*-heptane; 20% acetone in heptane; 2,2,4-trimethylpentane (25% dimethylformamide in ethyl ether as immobile phase transferred to the plate by dipping). For organophosphates, the recommended solvent system is methylcyclohexane (mobile) and 15–20% dimethylformamide (immobile). For the methyl esters of the phenoxyalkanoic acids, *n*-hexane saturated with acetonitrile is the preferred solvent. A good overview of the use of TLC for pesticide residue analysis has been published (Sherma 1974).

Refinements of TLC of pesticides have taken place during the past 10 years by developing quantitative techniques and introducing the concept of "high performance TLC" or HPTLC (Hauck and Amadori 1980). J. Sherma, more than anyone in this field, has promoted the idea of quantitative TLC of pesticides (Sherma 1980). By careful cleanup and preparation of samples and adsorbents, plates, solvents, development, and visualization, and the use of optical densitometers it is possible to achieve an approximate accuracy of 10% under rigorously controlled conditions.

It is questionable, in the opinion of this author, if such stringent cleanup measures are justified to make TLC a quantitative technique. It appears that GLC or HPLC coupled to computerized integrators provide the essential quantitative accuracy necessary for legal or scientific purposes. However, the simplicity of TLC and built-in versatility make it a valuable tool for qualitative identification and confirmation of suspected compounds.

In situ fluorometry, when the parent pesticide or a derivative is fluorescent, can be successfully applied to a limited number of pesticides separated and quantitated by TLC (Mallet 1980). For example, the reaction between a phenol, derived from the hydrolysis of *N*-methylcarbamates, with *N,N*-dimethylaminonaphthylsulfonyl chloride (dansyl) results in a fluorescent derivative which can be chromatographed on thin layers, or alternatively, the dansylation can be carried out directly on the spotted chromatogram before solvent development.

Special HPTLC precoated layers with silica gel 60, RP-18, and microcrystalline cellulose (F 254 s, E. Merck, Darmstadt) have been compared with normal TLC-grade silica gel 60 for the chromatography of pesticides. The HPTLC adsorbents possess a very narrow particle size range and, therefore, are more efficient as chromatographic media. Another interesting development with possible uses in pesticide analysis is the HPTLC-precoated plate with "concentrating zone." This zone is made of porous SiO_2 and is contiguous on the plate with the separation zone, silica gel 60. Using these plates, it is possible to spot a dilute solution of a sample extract which is concentrated at the zones' interphase, and the components of the dilute solution are separated as concentrated and visible bands (Hauck and Amadori 1980).

Mass Spectrometry

Gas Chromatography/Mass Spectrometry. Twenty years ago, the author (Zweig 1964) predicted that once mass spectrometers became practical and within the budget of a well-equipped modern pesticide analytical laboratory, this technique would become an extremely useful tool for the identification of pesticides and metabolic products. A further requirement for successful mass spectrometry was the readily available computer-stored mass spectral data for many pesticides, their impurities, and breakdown and metabolic products. The extensive NIH/EPA Library of mass spectra data is now available to the individual user over commercial telephone lines.

The decision by a pesticide analytical laboratory to purchase this very expensive, although useful, equipment depends on the answers to several questions: What is the workload? Is it cost-effective to purchase the equipment and hire seasoned technical personnel to keep it in operation? If the use of the mass spectrometer is only occasional, it might be more economical to have the MS work farmed out to a contract or university laboratory.

The subject of coupled gas chromatography and mass spectrometry as it pertains to the analysis of pesticide residues has been recently covered (Stan 1981). Since it is outside the scope of this chapter to discuss details of instrumentation, the interested reader is advised to consult the chapter by Stan (1981).

The principle of this elegant technique in the electron impact ionization mode (EI) is the prior purification of a complex mixture by capillary GC and the introduction of the effluent into the ion source of high energy electrons (70 eV) at a high vacuum. The resultant spectra of the effluent stream are stored on computer discs and can be recalled at any time. The EI spectra yield structural information of the molecular fragments which, when computer-matched with reference compounds, can identify the unknown or at least limit the choice to just a few compounds. The MS in the chemical ionization mode (CI) produces fewer ions in the fragmentation pattern but a higher intensity of ions in the molecular ion region.

Another important technique is "selective ion monitoring" in which the compound that is sought is known, and the MS is tuned to one or several known m/e numbers. In this way, the MS serves as a GC detector and remains in the no-scan mode. In summary, Stan (1981) believes that all pesticides for which a GLC separation exists can be analyzed by GC/MS, making this a most versatile and uniform method for confirmational pesticide analyses.

The tremendous pressure drop going from about 760 torr at the outlet of the GC column to 10^{-5} torr at the ion source of the MS had to be compensated by ingenious interfacing devices (e.g. gas diffusion types). However, with the development of capillary GC and high-speed vacuum systems on the ion source housing, it has become much simpler to devise a system for an almost direct interface between the two instruments.

Liquid Chromatography/Mass Spectrometry. As already mentioned, one of the shortcomings of modern HPLC is the lack of specific detectors. If mass spectrometry could be successfully interfaced with HPLC, one would obtain a universal, and yet selective, detection system. McFadden (1980) summarizes the difficulties one must overcome in interfacing these two pieces of equipment: (1) the severe imbalance between the eluent from the liquid chromatograph and the vapor entering the mass spectrometer (as high as 100:1), and (2) the need to vaporize high-melting and heat-labile compounds.

There are three successfully designed interfaces: (1) direct liquid introduction through a pinhole orifice adjacent to the ion source; (2) a moving belt with provisions for preliminary solvent evaporation and direct *in vacuo* passage to a heated chamber where the solute is flash vaporized into the ion chamber; and (3) a thermospray interface in which the LC effluents are directly thermosprayed into the ion source of the quadrupole MS, and the excess vapor pumped away by an added mechanical pump (Blakley and Vestal 1983).

Applications of HPLC/MS are given in the review by McFadden (1980) for the analysis of chloropropham and its animal and plant metabolites; separation and partial resolution of PCBs; the analysis of photoproducts of the herbicide oryzalin; and the analysis of rotenone at the 10-ng level.

For a laboratory with limited funds and space, it would be desirable to be able to use one mass spectrometer for interfacing GC and HPLC, granted, of course, that they would not be run at the same time. However, most commercial equipment is dedicated and probably makes such dual-purpose use impractical.

Fourier Transform Infrared Spectroscopy

The combination of chromatographic separations with real-time infrared spectroscopy has intrigued many chemists. This combination will take advantage of the unique IR-spectral fingerprint of each organic compound and its functional groups. If successful, HPLC, for example, would be provided with another universal, and yet specific, detection system.

The development of commercial rapid-scanning IR Fourier spectrometers using a Michelson interferometer and a mercury–cadmium–telluride IR detector, has made it possible to obtain a useful IR spectrum for 100 ng of anisole injected into a gas chromatograph (Griffiths 1977). Wall and Mantz (1977) obtained spectra of eight components of mirex first separated by GC and then passed through a light pipe into the interferometer.

Lowry and Gray (1980) have explained the principle of this technique as follows: "The Michelson interferometer modulates each wave length in the infrared region at a different frequency in the audio range. The modulated radiation passes through a sample chamber and is measured by the detector. This signal, called an interferogram, is converted by means of the Fourier transform into a single beam spectrum . . . a normal infrared spectrum is obtained by ratioing the single beam spectrum to a reference spectrum."

The LC–IR combination appears to be in the developmental stages, and, although feasible in principle, there are many technical obstacles to overcome. The most serious one is the removal of the mobile solvent before measuring the IR spectrum. Griffiths (1977) has proposed a scheme in which he uses a fraction collector so that the spectra will only be available a short time after the chromatographic run has been completed.

It appears that the time is not propitious to purchase the FTIR equipment for routine analyses for the best-equipped pesticide analytical laboratory. The principle is sound, however, and it is just a matter of time until this method will become practical and help solve many analytical problems that so far have defied solutions.

THE MODERN PESTICIDE ANALYTICAL LABORATORY

A recent visit to a well-equipped modern pesticide analytical laboratory of a major international food processor has revealed that the emphasis is on automation of the analyses and the computerization of the results. The two basic types of analytical equipment were gas chromatographs and high-performance liquid chromatographs. The GLC instruments were equipped with automated samplers, being capable of running 24-hr shifts. Microprocessors were directing the entire programs of temperature control, sample sequence, etc., while a small computer reduced the data to a desired format, e.g., parts per million (ppm) pesticide per sample. The HPLC instruments were equally fitted out with automatic sampling and data reduction systems.

A software program has been written for the screening of "unknown" peaks from a possible hundred pesticides by comparing the experimental elution times on one or several GC columns and the solvent pairs used for Florisil column cleanup and elution. This procedure is very useful in identifying unknown pesticides found in food.

To make the chromatographic analyses more specific, selective detectors for gas chromatography are employed. The Hall detector for Cl^- and nitrogen and the flame photometric detector for P and S are found to be most useful. The electron capture detector, although most sensitive, was found to be generally too nonselective to be useful for samples with many extraneous compounds following the extraction of field samples. The most useful packed columns were OV3, OV17, OV225 silicones or DEGS for more polar compounds. Open tubular capillary columns are finding more widespread use in many laboratories, giving better resolution than packed columns but accommodating smaller sample size, thus reducing sensitivity. The most useful column for HPLC was the C-18 chemical bonded phase for reversed-phase chromatography. Detectors for HPLC are not element-specific and are mainly based on UV absorption or fluorescence emission.

Cleanup of field samples is still being performed manually using disposable Sep-Pak® or Baker Disposable Extraction columns. Using the analytical instrument for automated cleanup ties up a sophisticated apparatus, while manual cleanup can usually accomplish the simultaneous processing of ten samples at one time. It is believed, therefore, that automation for this step would actually slow down the analyses.

Another useful piece of equipment is the Hamilton® automatic diluter which simplifies the dilution of concentrated samples and improves the precision of the analyses. Evaporation of solvent extracts is still being carried out by the "old" and proven Kuderna-Danish method.

This particular pesticide laboratory processes about 10,000 samples per year. Typical samples brought in for analysis are waste samples used for animal feeds: tomato waste; peach waste used for brandy production; miscellaneous crops purchased from a broker or cooperative rather than grown under the direct control of the company, and therefore, having an unknown pesticide spray history; and dried fruit.

Pesticides which are found include the following: carbaryl, which is analyzed by GLC using the nitrogen detector; and chlorinated pesticides, metabolites, and industrial chemicals which are analyzed by the multiresidue methods (discussed below). The most commonly found pesticide residues, usually not above tolerance, are endosulfan (three components); DDT and metabolites in root crops; herbicides such as the triazines, diuron, and treflan; heptachlor and heptachlor epoxide in green chop of pineapples; aldicarb in bananas; azinphosmethyl in peaches; and ethylene dibromide (EDB) in papaya.

MULTIRESIDUE METHODS

The most useful multiresidue methods, capable of analyzing the majority of pesticide residues commonly encountered in foods, have been developed by chemists of the Food and Drug Administration. Details of these methods can be found in several references readily available in most libraries or from the National Technical Information Service (NTIS) (AOAC 1980; Anon. 1980A, B). Another excellent source for detailed chromatographic data and cleanup procedures is the recent chapter by Luke and Masumoto (1982). The discussion below will only highlight the principles of analyzing the three important classes of pesticides in fatty and nonfat foods and crops but will leave the details to the reader who must utilize these methods.

Pesticide Residues in Fatty Foods

The scheme for the cleanup of fatty food samples may be found in Fig. 14.2 and involves basically five steps. The sample is ground and the fat extracted with petroleum ether. The pesticides are partitioned between petroleum ether and acetonitrile. The acetonitrile phase, containing the pesticide residues, is extracted with petroleum ether and chromatographed on a Florisil column and eluted with four solvents: (1) petroleum ether; (2) 6% ethyl ether in petroleum ether; (3) 15% ethyl ether in petroleum ether; and (4) 50% ethyl ether in petroleum ether. The four fractions are concentrated and analyzed by GLC on 5% OV3 on Chromosorb W HP, using a ^{63}Ni-electron capture detector or the electrolytic conductivity cell (Hall detector).

Hexane may be substituted for petroleum ether giving similar results; however, since most of the published literature describes results using petroleum ether, this solvent would appear to be the more useful. The first petroleum ether or hexane fraction contains polychlorinated biphenyls (Reynolds 1969), while the other solvent mixtures elute pesticides of increasing polarity. McMahon and Burke (1978) published a detailed compilation of over 300 pesticides and some industrial chemicals, arranged alphabetically, showing which fraction elutes a particular group of compounds. Some compounds may not clearly separate and might elute in two fractions.

There are about 80 compounds, not necessarily pesticides, which are found in the first two fractions and are early eluters in the normal mode of the GLC operation. The Food and Drug Administration has recommended that these "early eluters" be chromatographed at lower temperatures (130°C) which will aid in the resolution and identification of unknown compounds (Yurawecz 1976). A computer printout of the chromatographic data for many of these compounds is available from the FDA in Washington, DC.

The third and fourth eluates (15 and 50%) may require additional

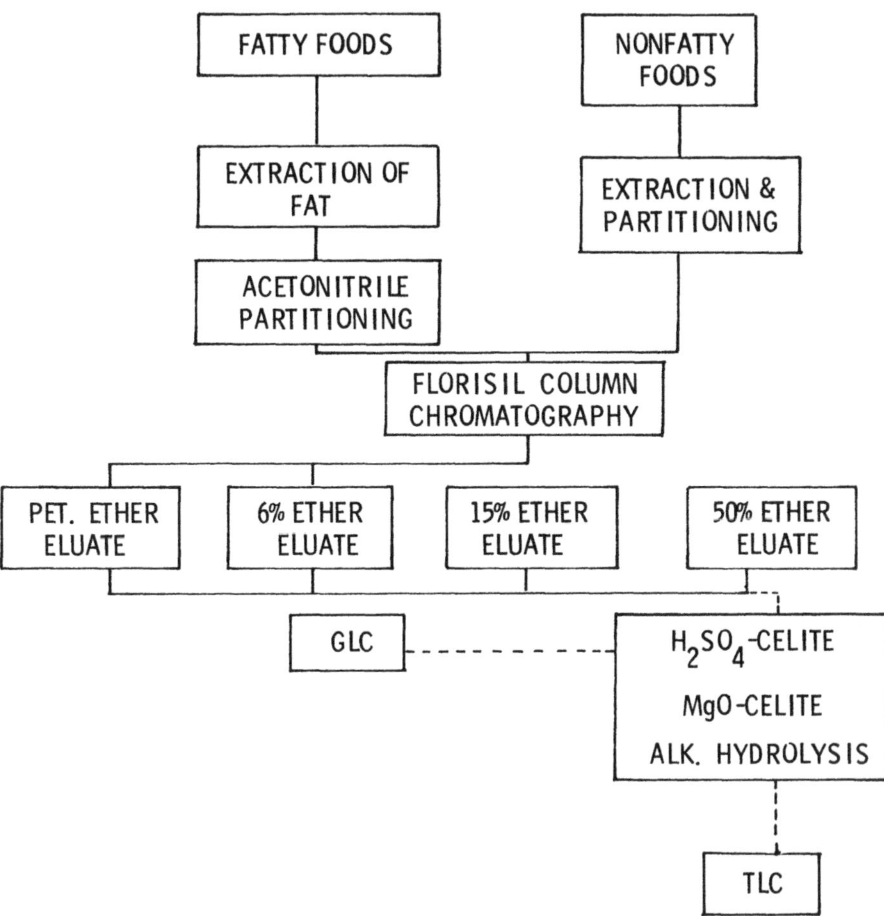

FIG. 14.2. Scheme for multiresidue analysis of nonpolar chlorinated and organophosphorus pesticides: (---) additional cleanup.
From Anon. (1980A).

cleanup prior to gas chromatography or confirmatory tests by TLC. These steps are H_2SO_4–Celite column and MgO–Celite column and alkaline hydrolysis. Table 14.3 shows which pesticides survive these additional cleanup steps.

The multiresidue method is suitable for chlorinated and organophosphorus pesticides, but it is not likely that organophosphorus pesticide residues are found in many fatty samples. Malathion is eluted in the 50% fraction. If organophosphate pesticide residues are sus-

TABLE 14.3. PESTICIDES SURVIVING SUPPLEMENTAL CLEANUP FOR MULTIRESIDUE ANALYSIS

Acid–Celite column	MgO–Celite column	Alk. hydrolysis
Aldrin	Aldrin	Aldrin
BHC	Dieldrin	Dieldrin
Chlordane (techn.)	Endrin	Endrin
o,p'-DDT		PCBs
p,p'-DDT		
o,p'-DDE		
p,p'-DDE		
Heptachlor		
Heptachlor epoxide		
Lindane		
Nitrofen		
o,p'-TDE		
p,p'-TDE		
Toxaphene		

Source: Pesticide Analytical Manual (Anon. 1980A).

pected to be present in the samples, GC detection is carried out with thermionic or flame photometric detection. Additional cleanup for organophosphates is usually not necessary, and in either case, organophosphates do not survive the more drastic cleanup steps following Florisil–chromatography.

Pesticide Residues in Nonfatty Foods

Pesticide residues are extracted from nonfatty foods with acetonitrile or water–acetonitrile (Fig. 14.2). The extract is diluted with water, and the pesticide residues are extracted into petroleum ether. Florisil cleanup of the concentrated extract is performed, as described above for fatty foods, yielding four eluate fractions to be further analyzed by GLC. Chlorinated pesticides are analyzed by electron capture or electrolytic conductivity and organophosphorus compounds by flame photometric or thermionic detectors. As examples of the type of pesticides and crops suitable for this method, the reader is referred to Tables 14.4 and 14.5.

Recently, the multiresidue methods have been modified and simplified. For example Luke *et al.* (1975) recommends the extraction of fruits and vegetables (nonfatty crops) with acetone and partitioning with methylene chloride–petroleum ether to remove water. Without further cleanup, the concentrated extracts are analyzed for organophosphorus and organonitrogen (see below) compounds using a KCl-thermionic detector. For chlorinated pesticides, the usual Florisil chromatographic cleanup technique must be invoked.

TABLE 14.4. ORGANOCHLORINE PESTICIDE RESIDUES AND CROPS WITH OFFICIAL AOAC-APPROVED MULTIRESIDUE METHODS

Pesticides	Crops		
Aldrin	Apples	Cucumbers	Popcorn
BHC	Apricots	Eggplant	Potatoes
o,p'-DDE	Barley	Eggs	Radishes
o,p'-DDT	Beets	Endive	Radish tops
p,p'-DDT	Bell peppers	Grapes	Spinach
Dieldrin	Broccoli	Green beans	Squash
Endrin	Cabbage	Hay	Strawberries
Heptachlor	Cantaloupe	Kale	Sugar beets
Heptachlor epoxide	Cauliflower	Mustard greens	Sweet potatoes
Lindane	Celery	Oats	Tomatoes
Methoxychlor	Collard greens	Peaches	Turnips
Mirex	Corn meal	Pears	Turnip greens
p,p'-TDE	Corn silage	Peas	Wheat
Perthane		Plums	

Source: Pesticide Analytical Manual (Anon. 1980A).

TABLE 14.5. ORGANOPHOSPHATE PESTICIDE RESIDUES AND CROPS SUITABLE FOR MULTIRESIDUE ANALYSIS

Pesticides	Crops		
Carbophenothion	Apples	Grapes	Tomatoes
Diazinon	Barley	Green peppers	Wheat
Ethion	Broccoli	Lettuce	Turnip greens
Malathion	Cabbage	Mustard greens	Carrots
Methyl parathion	Oats	Cauliflower	Potatoes
Parathion	Cucumbers	Squash	

Source: Pesticide Analytical Manual (Anon. 1980A).

Using the Luke's extraction procedure and no additional cleanup, it is possible to obtain excellent resolution for a number of chlorinated and organophosphorus pesticides by GLC with four different columns and the flame photometric and Hall detectors (Anon. 1980A, B). Table 14.6 illustrates the chromatographic systems and the relative retention times of some of the compounds being resolved. More recently, Luke *et al.* (1981) have simplified their procedure even further which makes it possible to eliminate the Florisil cleanup step entirely. Their new modifications consist of two extraction–evaporation steps with petroleum ether, following the initial solvent extraction of the crops. The final concentrate is then directly gas chromatographed using the Hall detector for organic N and Cl compounds and the flame photometric detectors for organic compounds containing phosphorus.

TABLE 14.6. MULTIRESIDUE ANALYSIS OF ORGANOCHLORINE AND ORGANOPHOSPHORUS PESTICIDES WITH FOUR GLC-SYSTEMS[a]

	Organophosphorus				Organochlorine			
	System 1		System 2		System 3		System 4	
Packing:	2% DEGS		10% DC200		2% DEGS + 0.5% H_3PO_4		2% OV-101	
Col. temp.:	180° C		200° C		180° C		200° C	
Detector:	FPD (P)		TD		Hall (Cl^-)		Hall (Cl^-)	

Pesticide	R_P[b]	Pesticide	R_P[b]	Pesticide	R_{HE}[c]	Pesticide	R_A[d]
Mevinphos, α	0.13	Mevinphos	0.13	Aldrin	0.38	BHC, α	0.38
Diazinon	0.14	Demeton, thiono	0.28	BHC, α	0.43	Simazine	0.39
Phorate	0.15	Dimethoate (O)	0.32	BHC, γ	0.63	Dichloran	0.40
Mevinphos, β	0.16	Naled	0.34	Chlorpyrifos	0.82	2,4-D isopropyl	0.41
Fonofos	0.21	Phorate	0.38	*o,p'*-DDE	1.00	Atrazine	0.41
Naled	0.27	Azodrin	0.40	Heptachlor (O)	1.00	BHC, β	0.42
Methamidophos	0.29	Diazinon (O)	0.46	Dacthal	1.02	BHC, γ	0.47
Chlorpyrifos	0.40	Delnav	0.48	Pronamide	1.13	Pronamide	0.5
DEF	0.51	Dimethoate	0.41	Endosulfan I	1.15	Chlorothalonil	0.52
Trichlorfon	0.57	Dyfonate	0.52	Chlorothalonil	1.16	Phosphamidon	0.66
Dicrotophos	0.61	Diazinon	0.53	Dichloran	1.16	Ronnel	0.78
Acephate	0.64	Di-Syston	0.55	Atrazine	1.21	Heptachlor	0.79
Malathion	0.76	Phorate (O)	0.67	*p,p'*-DDE	1.32	Linuron	0.83
Malathion (O)	0.92	Malathion (O)	0.69	Dieldrin	1.48	Chlorpyrifos	0.98
Parathion	1.00	Parathion, methyl	0.70	Simazine	1.50	Dacthal	0.99
Profenofos	1.07	Paraoxon	0.78	Endrin	1.60	Aldrin	1.00
Dimethoate (O)	1.09	Ronnel	0.82	*o,p'*-DDT	1.85	Bromophos	1.09
Fenitrothion	1.10	Malathion	0.90	BHC, δ	1.87	Captan	1.17
Phenthoate	1.17	Parathion	1.00	Perthane	1.97	Folpet	1.21
Parathion (O)	1.23	Dursban	1.02	BHC, β	2.05	Heptachlor (O)	1.24
Phorate, S—O	1.32	Chlorfenvinphos	1.31	*o,p'*-TDE	2.37	Chlorfenvinphos	1.26
Dimethoate	1.36	Supracide	1.43	Linuron	2.48	Chlorbenzide	1.36
Ethion	1.37	Gardona	1.57	Folpet	3.19	Tetrachlorvinphos	1.51
Monocrotophos	1.63	Ethion	2.58	Captan	3.21	Endosulfan I	1.58
Tetrachlorvinphos	1.75	Carbophenothion	2.17	*p,p'*-DDT	3.43	*p,p'*-DDE	1.81
Carbophenothion	1.86			Endosulfan II	4.06	Dieldrin	1.83
Phorate O—S—O	1.89			*p,p'*-TDE	4.16	Nitrofen	1.99
Carbophenothion-methyl	2.25			Nitrofen	4.35	Endrin	2.05
						Endosulfan II	2.08
Mephosfolan	3.49					*p,p'*-TDE	2.33
						Carbophenothion	2.83
Oxydemeton–methyl sulfone	5.42					*p,p'*-DDT	3.03
Phosmet	14.					Dicofol	4.3

[a] Anon. (1980A, B).
[b] R_P = retention time relative to parathion.
[c] R_{HE} = retention time relative to heptachlor epoxide.
[d] R_A = retention time relative to aldrin.

Ionic Organochlorine Pesticide Residues in Fatty and Nonfatty Foods

The pesticides in this category are usually phenoxyalkanoic acids or their esters, widely used herbicides. Dioxin, an impurity of 2,4,5-T, which might be present in extremely small quantities, can be analyzed at the parts-per-trillion (ppt) level by a combination of GLC–mass spectrometry, which will be discussed in the next section.

Figure 14.3 depicts the scheme for the cleanup of polar chlorinated compounds. Chlorophenoxyalkanoic acids are extracted from fatty foods into chloroform, then transferred into alkaline solution, and nonpolar impurities are extracted with an organic solvent and discarded. The acidified aqueous phase is reextracted with chloroform, and the isolated acids are methylated with diazomethane.

Further cleanup and separation of the methyl ether of pentachlorophenol and the methyl esters of the phenoxyalkanoic acids are effected by Florisil column chromatography using two eluting solvents: 20% methylene chloride in petroleum ether (for PCP) and 2.5% di-

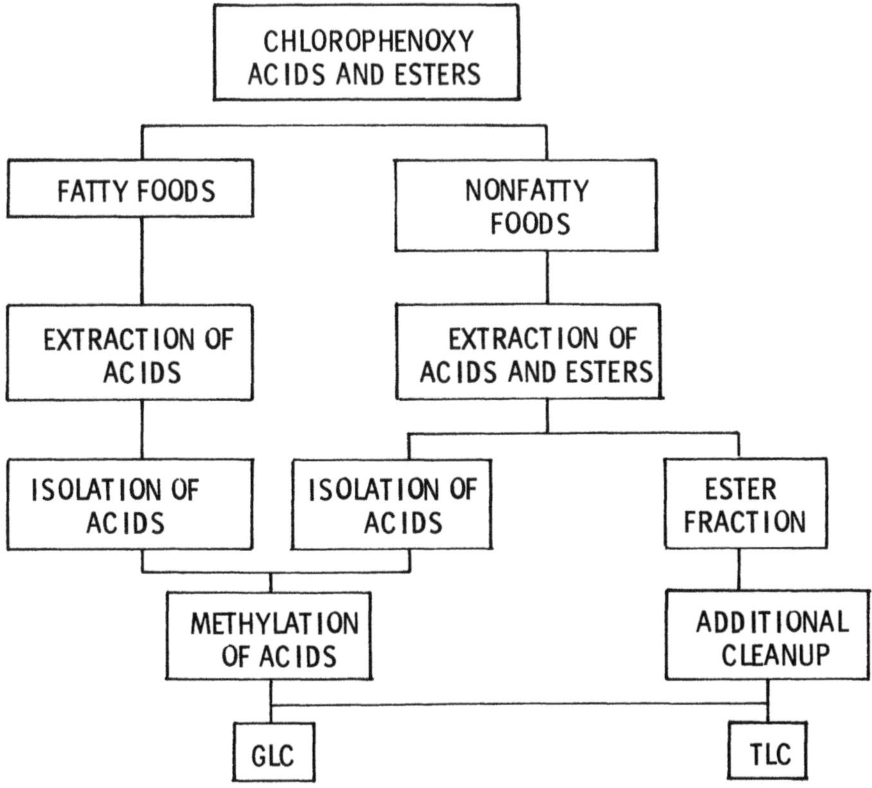

FIG. 14.3. Scheme for multiresidue analysis of polar chlorinated pesticides.
From Anon. (1980A).

ethyl ether in methylene chloride (2,4-D and related acids). The concentrated extracts are analyzed by GLC on a number of silicone-packed columns using electron-capture, microcoulometric, or electrolytic conductivity detectors, described in previous sections.

The ionic chlorinated pesticides are isolated from nonfatty foods by an extraction with acetonitrile or acetonitrile–water (Fig. 14.3). Again, the first extract is made alkaline, washed with organic solvent, and methylated, followed by Florisil-column chromatography, as described above for fatty foods.

Methamidophos, a highly polar, yet nonchlorinated pesticide cannot be analyzed by these generalized methods but is triple extracted with acetone–methylene chloride by the method of Luke *et al.* (1975) and cleaned up on charcoal before the final analysis by GLC with a nitrogen-sensitive thermionic detector.

Gel permeation chromatography (GPC) has been useful for the cleanup of PCP, chlorophenoxyalkanoic acids, chlorinated and organophosphate insecticides (Hopper 1982). The OR PVA 2000 beads used in the early work have now been replaced by Bio-Beads SX-3.

Industrial Chemicals and Pesticide Impurities

Polychlorinated biphenyls (PCBs), a group of chemicals used widely as insulating material for capacitors, are now found to be ubiquitously distributed in the environment. This has resulted in residues in foods, which were in the past sometimes mistaken for chlorinated pesticides, like DDT and similar compounds. There are a number of methods found in FDA's Pesticide Analytical Manual (1980B), but none of these promises to be the ideal solution for all food matrices. A very brief discussion of the cleanup procedures follows, but the reader is advised to consult the literature before attempting to resolve PCBs from nonionic chlorinated pesticides and before interpreting that "unknown peaks" are due to pesticides.

In one method, a hydrated silicic acid column is capable of separating DDT and some of the PCBs by eluting the latter compounds with petroleum ether; DDT and related compounds, as well as some other PCBs are eluted from the column with a mixture of hexane, methylene chloride, and acetonitrile. This method is applicable to the 6% ether–petroleum ether eluate from the Florisil cleanup for the multiresidue method (see above).

Another method takes advantage of the conversion by alkali treatment of DDT and TDE to their corresponding olefins. Oxidation of this solution with Cr_2O_3 converts the alkali products to dichlorophenone but leaves the PCBs intact. Dichlorobenzophenone is separated from

PCBs on a Florisil column and quantitated by GLC using the Hall detector (Cl^-) or similar specific detection systems.

A detailed discussion of the ppt analysis of dioxins is not within the scope of this chapter. If such an analysis is required for certain foods, it is best to contract with one of the few commercial or university laboratories skilled in this highly sophisticated analysis. One of the procedures, developed for the analysis of ppt levels of dioxins in human milk, is probably adaptable to the analysis of dioxins in fatty foods (Langhorst and Shadoff 1980). The procedure involves acid digestion, extraction with an organic solvent, multiple adsorption chromatography on a H_2SO_4–silica gel column, a $AgNO_3$–silica gel column, two reverse-phase HPLC passes, and finally detection by multiple-ion mode mass spectrometry. Using ^{13}C-labeled dioxins as internal standards, 2,3,7,8-tetrachlorodibenzo-*p*-dioxin, the most toxic of the dioxins, and a number of other polychlorinated dioxins can be separately analyzed.

Multiresidue Method for Determination of *N*-Methylcarbamates

The method described by Krause (1980) is suitable for the analysis of seven carbamate insecticides and four related metabolites, aldicarb, bufencarb, carbaryl, carbofuran, methiocarb, methomyl, oxamyl and the metabolites aldicarb sulfoxide and sulfone, 3-hydroxycarbofuran, and methiocarb sulfoxide. The analytical scheme is depicted in Fig. 14.4.

Samples of nonfatty crops are extracted with methanol and plant coextractives removed by acetonitrile partitioning. The pesticide residues are then extracted with petroleum ether, and the concentrated extracts further purified on a silanized Celite-charcoal column using toluene–acetonitrile as eluting solvent. The compounds, thus purified, are next chromatographed by HPLC on an analytical column containing 6-μm Zorbax C-8 or CN spherical particles using an acetonitrile–water gradient elution system. The eluted fractions are hydrolyzed to methyl amine under alkaline conditions and detected with post-column fluorimetry by the reaction with *o*-phthalaldehyde-2-mercaptoethanol. (See also the next section on Residue Analysis of Carbaryl.) Benomyl and carbendazim have been analyzed without derivatization by HPLC (Zweig and Gao 1983).

RESIDUE ANALYSIS OF CARBARYL

The progress which has been achieved in the analysis of pesticide residues can best be illustrated by the example of the residue analysis

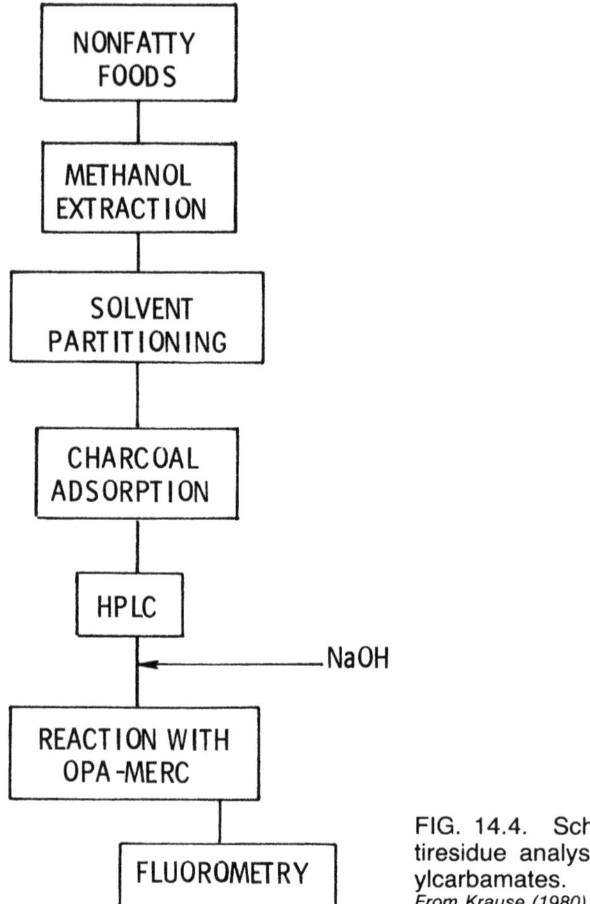

FIG. 14.4. Scheme for multiresidue analysis of *N*-methylcarbamates.
From Krause (1980).

of carbaryl (1-naphthyl *N*-methylcarbamate). When carbaryl (an important carbamate insecticide) became available commercially, about 25 years ago, it was necessary to develop a sensitive analytical method for detection of ppm quantities of residues on crops as required by the government for enforcement of food tolerances of carbaryl. The first analytical methods were colorimetric and involved the alkaline hydrolysis of carbaryl to 1-naphthol, followed by subsequent reaction with aminoantipyrine or *p*-nitrobenzenediazonium fluoborate (Zweig and Archer 1958). Shortly thereafter, two additional and more sensitive methods were developed based on acetylcholinesterase inhibition by carbaryl (Archer and Zweig 1959) and paper chromatography.

With the advent of thin-layer chromatography (TLC), the paper chromatographic method was adapted to TLC on Al_2O_3 with acetone–benzene as solvent (Sherma 1974). The color development of the finished chromatogram takes place in situ by first hydrolyzing the spots with alcoholic KOH, followed by a spray reagent of *p*-nitrobenzenediazonium fluoborate. Both the colorimetric and TLC methods are still listed as official AOAC methods (AOAC 1980). The TLC method is used in the FDA's total diet studies (Anon. 1982).

As GLC became widely used for the analysis of organochlorine and organophosphorus insecticides, the direct chromatography of carbaryl was also tried but proved not to be very successful due to its decomposition on the column at elevated temperatures. However, as the column packings (e.g., 10% DC 200) were deactivated by on-column silanization, and as nitrogen-sensitive and -selective detectors became available, it was possible to analyze carbaryl residues from fruit and vegetables (Seiber 1981). Using OV 101 as the stationary phase and an N/P detector, 60 ng of carbaryl gave 30% full-scale deflection (Anon. 1980A) (Fig. 14.5).

Most recently, carbaryl residues from strawberry leaves have been analyzed by reverse-phase HPLC on a C-18 bonded column with acetonitrile–water (50%) as the mobile phase and detection at 280 nm (Zweig *et al.* 1984). Figure 14.6 shows the response due to 227.5 ng of carbaryl standard and a leaf extract analyzed to contain 455 ng. The determinative step of this fully automated analysis took less than 10 min.

A fully automatic system for the cleanup and analysis of carbaryl

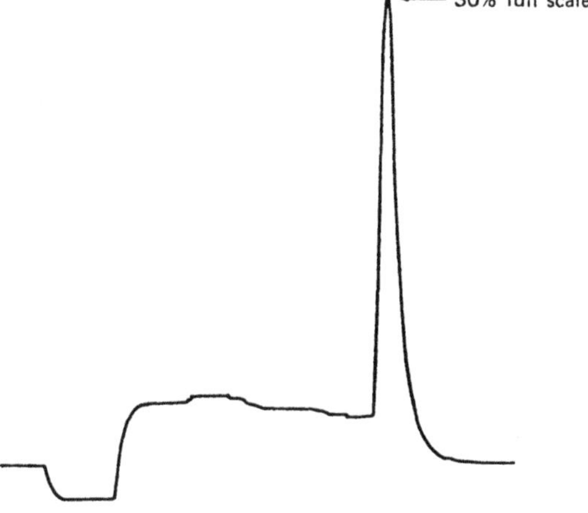

FIG. 14.5. Gas chromatogram of 60-ng carbaryl; 10% OV-101 on Chromosorb column; column temp.: 200°C; N/P detector. Injection point is on left but cannot be indicated because solvent is first vented.

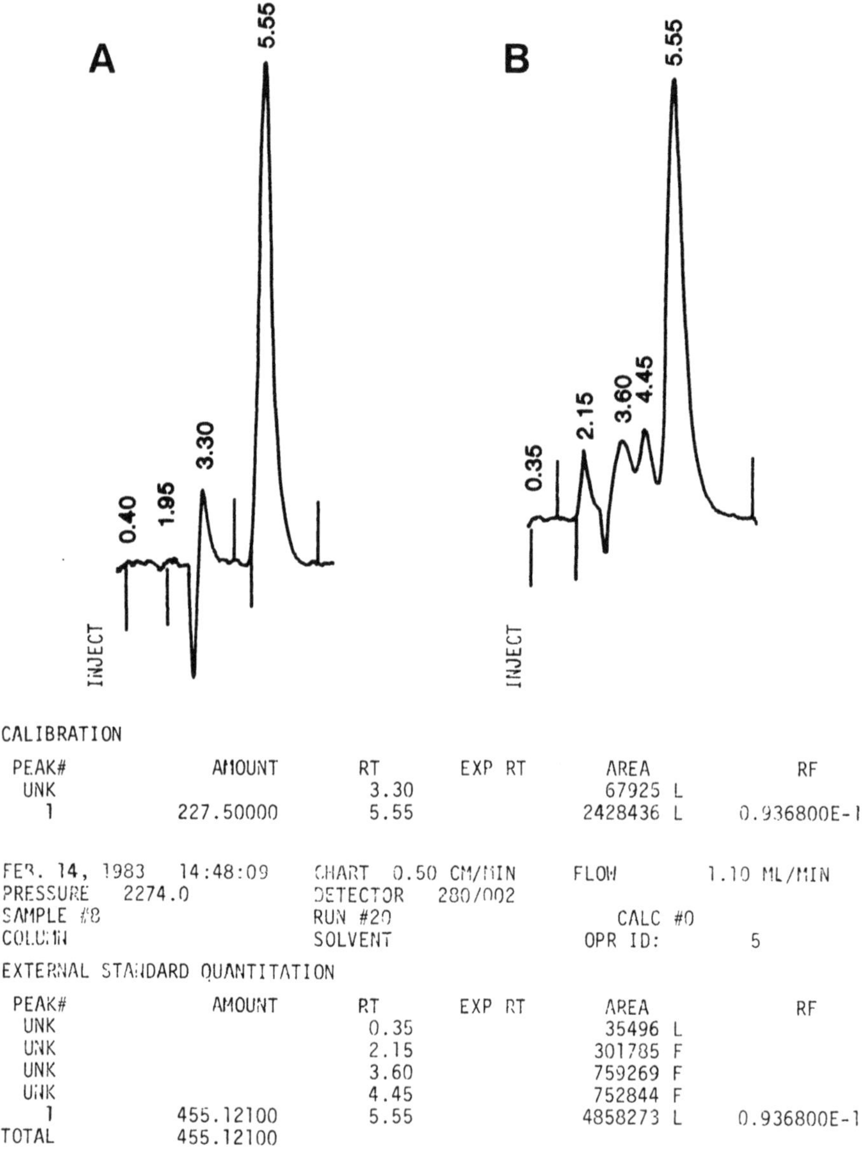

CALIBRATION

PEAK#	AMOUNT	RT	EXP RT	AREA	RF
UNK		3.30		67925 L	
1	227.50000	5.55		2428436 L	0.936800E-1

FEB. 14, 1983 14:48:09	CHART 0.50 CM/MIN	FLOW	1.10 ML/MIN
PRESSURE 2274.0	DETECTOR 280/002		
SAMPLE #8	RUN #20	CALC #0	
COLUMN	SOLVENT	OPR ID:	5

EXTERNAL STANDARD QUANTITATION

PEAK#	AMOUNT	RT	EXP RT	AREA	RF
UNK		0.35		35496 L	
UNK		2.15		301785 F	
UNK		3.60		759269 F	
UNK		4.45		752844 F	
1	455.12100	5.55		4858273 L	0.936800E-1
TOTAL	455.12100				

FIG. 14.6. The HPLC of carbaryl: (A) standard (227 ng); (B) field sample showing 455 ng of carbaryl. *Experimental conditions:* C-18 bonded phase column; mobile phase: acetonitrile:water (1:1); 1.1 mL/min flow; retention time for carbaryl: 5.55 min.
From Zweig et al. (1984).

residues from cabbage by reverse-phase HPLC has been developed (Anon. 1983). In the Waters system, a programmable HPLC pump is capable of injecting a raw acetonitrile extract onto a Bondapak® C_{18}/Porasil® precolumn in order to remove interfering impurities while retaining carbaryl on the column. A programmed solvent gradient consisting of acetonitrile–water increases the organic phase, and moves carbaryl off the precolumn onto the chromatographic column (Bondapak C_{18}) for final separation and analysis. The entire operation takes less than 13 min, and a series of extracts can be run unattended; the only manual operation is the initial solvent extraction of the plant tissue.

PESTICIDE ANALYSIS, THEN AND NOW

What is different about the modern analytical pesticide laboratory in 1983 compared with its operation 20 years ago? Basically, in 1965, the most commonly found pesticides in food were DDT, DDD, methoxychlor, and parathion, and the analyses were performed on a gas chromatograph equipped with a packed silicone column and an electron-capture detector. In 1983, there are over 120 commonly used pesticides which might occur as residues in foods. It is true, however, that most modern pesticides are less persistent than the older chlorinated compounds, and the concentrations of the residues that are found today are much lower. The greatest technical advances in the laboratory have been (1) automated sample injection; (2) automated peak integration and data reduction system; (3) automated temperature-programming of the GC; (4) availability of specific GC detectors, e.g., the Hall detector for N and Cl^-, the Coulson detector for Cl^-, and the flame photometric detector for the simultaneous analysis of P- and S-containing compounds; and (5) the N/P detector, a modification of the thermionic flame detector. Another advance has been the development of HPLC which has greatly expanded the direct analyses of heat-labile compounds or eliminated elaborate cleanup steps. It is unfortunate, however, that specific detectors for HPLC are not available commercially, but there undoubtedly will be progress along these lines during the next 10 years. Probably the greatest advance in the pesticide residue analytical laboratory has been the general availability of GC-MS equipment which has made it possible to identify and quantify traces of unknown or suspected man-made compounds in foods with much greater certainty than was ever imagined 20 years ago.

All of these innovations have speeded up the individual analysis and made it possible to operate the laboratory at greater efficiency and instrument utilization than before.

ACKNOWLEDGMENTS

For generously contributing information and comments to this chapter, acknowledgment and thanks for their time and effort go to Lois Beaver and Ellis Gunderson, FDA, Washington, DC; Dennis D. Manske, FDA, Kansas City Field Office; Milton A. Luke, FDA, Los Angeles District; Robert I. Vetro and Wayne Thornburg, Del Monte Research Laboratories, Walnut Creek, CA.

BIBLIOGRAPHY

AOAC. 1975. Official Methods of Analysis, 12th ed. Assoc. Off. Analyt. Chem., Washington, DC.

AOAC. 1980. Official Methods of Analysis, 13th ed. Methods 29.077–29.085. Assoc. Off. Analyt. Chem., Arlington, VA.

ANON. 1980A. Pesticide Analytical Manual, Vols. I and II. B. McMahon and C.D. Sawyer (Editors). Dept. of Health & Human Services, FDA, Washington, DC.

ANON. 1980B. Modified Luke procedure with limited gas chromatographic determinative step for some organophosphorus and organochlorine pesticides. FDA Laboratory Information Bull. No. 2434 (Los Angeles District Pesticide Team). 13 pp.

ANON. 1982. Compliance Program Report of Findings. FY 79 Total Diet Studies—Infants and Toddlers (7305.002). U.S. Dept. of Health & Human Services, Public Health Service, FDA Bureau of Foods, Washington, DC.

ANON. 1983. Automatically isolate carbaryl from a complex matrix. B65/82549/March, 1983. Waters Assoc., Milford, MA.

ARCHER, T.E., and ZWEIG, G. 1959. Direct colorimetric analysis of cholinesterase-inhibiting insecticides with indophenyl acetate. J. Agric. Food Chem. *7* (3), 178–181.

BLAKLEY, C.R., and VESTAL, M.L. 1983. Thermospray interface for liquid chromatography/mass spectrometry. Anal. Chem. *55* (4), 750–754.

COULSON, D.M., and CAVANAGH, L.A. 1960. Automatic chloride analyzer. Anal. Chem. *32* (10), 1245–1247.

COULSON, D.M., CAVANAGH, L.A., and STUART, J. 1959. Gas chromatography of pesticides. J. Agric. Food Chem. *7* (4), 250–251.

FOLLWEILER, J.M., and SHERMA, J. 1984. Pesticides and related organic chemicals. *In* Handbook of Chromatography. G. Zweig and J. Sherma (Editors). CRC Press, Inc., Boca Raton, FL.

GRIFFITHS, P.R. 1977. Recent applications of Fourier transform infrared spectrometry in chemical and environmental analysis. Applied Spectr. *31* (6), 497–505.

HANKS, A.R., and COLVIN, B.M. 1981. High performance liquid chromatography. *In* Pesticide Analysis. K.G. Das (Editor). Marcel Dekker, New York.

HAUCK, H.E., and AMADORI, E. 1980. Recent developments in high-performance thin-layer chromatography and application to pesticide analysis. *In* Pesticide Analytical Methodology. ACS Symp. Ser. 136. J. Harvey, Jr., and G. Zweig (Editors). Am. Chem. Soc., Washington, DC.

HOLLAND, P.T., and GREENHALGH, R. 1981. Selection of gas chromatographic detectors for pesticide residue analysis. *In* Analysis of Pesticide Residues. H. Anson Moye (Editor). John Wiley & Sons, New York.

HOPPER, M.L. 1982. Automated gel permeation system for rapid separation of industrial chemicals and organophosphate and chlorinated pesticides from fat. J. Agric. Food Chem. *30* (6) 1038–1041.

IVIE, K.F. 1980. High-performance liquid chromatography (HPLC) in pesticide residue analysis. *In* Analytical Methods for Pesticides and Plant Growth Regulators, Vol. XI, Updated General Techniques and Additional Pesticides. G. Zweig and J. Sherma (Editors). Academic Press, New York.

JENNINGS, W. 1980. Gas Chromatography with Glass Capillary Columns. 2nd ed. Academic Press, New York.

KISSINGER, P.T., BRATIN, K., KING, W.P., and RICE, J.R. 1980. Electrochemical detection of picomole amounts of oxidizable and reducible residues separated by liquid chromatography. *In* Pesticide Analytical Methodology. ACS Symp. Ser. 136. J. Harvey, Jr. and G. Zweig (Editors). Am. Chem. Soc., Washington, DC.

KRAUSE, R.T. 1980. Multiresidue method for determination of N-methylcarbamate insecticides in crops using HPLC. J. Assoc. Off. Anal. Chem. *63* (5), 1114–1124.

LANGHORST, M.L., and SHADOFF, L.A. 1980. Determination of parts-per-trillion concentrations of tetra-, hexa-, hepta-, and octachlorodibenzo-*p*-dioxins in human milk samples. Anal. Chem. *52* (13), 2037–2044.

LAWRENCE, J.F., and TURTON, D. 1978. High-performance liquid chromatographic data for 166 pesticides. J. Chromatogr. *159* (2), 207–226.

LAWRENCE, J.F. 1981. Organic Trace Analysis by Liquid Chromatography. Academic Press, New York.

LAWRENCE, J.F. 1982. High Performance Liquid Chromatography of Pesticides, Vol. XII of Analytical Methods for Pesticides and Plant Growth Regulators. G. Zweig and J. Sherma (Editors). Academic Press, New York.

LOWRY, S.R., and GRAY, C.L. 1980. Some applications of Fourier transform infrared spectroscopy to pesticide analysis. *In* Pesticide Analytical Methodology. ACS Symp. Ser. 136. J. Harvey, Jr. and G. Zweig (Editors). Am. Chem. Soc., Washington, DC.

LUKE, M.A., FROBERG, J.E., and MASUMOTO, H.T. 1975. Extraction and cleanup of organochlorine, organophosphate, organonitrogen, and hydrocarbon pesticides in produce for determination by gas–liquid chromatography. J. Assoc. Off. Anal. Chem. *58* (5), 1020–1026.

LUKE, M.A., FROBERG. J.E., DOOSE, G.M., and MASUMOTO, H.T. 1981. Improved multiresidue gas chromatographic determination of organophosphorus, organonitrogen, and organohalogen pesticides in produce, using flame photometric and electrolytic conductivity detectors. J. Assoc. Offic. Anal. Chemists *64* (5), 1187–1195.

LUKE, M.A., and MASUMOTO H.T. 1982. Pesticides and related substances. *In* Handbook of Carcinogens and Hazardous Substances. Chemical and Trace Analysis. M.C. Bowman (Editor). Marcel Dekker, Inc., New York.

MALLET, V.N. 1980. Quantitative thin-layer chromatography of pesticides by in situ fluorometry. *In* Pesticide Analytical Methodology. ACS Symp. Ser. 136. J. Harvey, Jr. and G. Zweig (Editors). Am. Chem. Soc., Washington, DC.

McFADDEN, W.H. 1980. Liquid chromatography/mass spectrometry systems and applications. J. Chromatogr. Sci. *18,* 97–115.

McLEAN, H., and TRUMAN, C. 1983. Unpublished data. Univ. of California, Sanitary Engineering and Environmental Health Res. Lab., Richmond, CA 94804.

McMAHON, B., and BURKE, J.A. 1978. Analytical behavior data for chemicals determined using AOAC multiresidue methodology for pesticide residues in food. J. Assoc. Off. Anal. Chem. *61* (3) 640–652.

MOYE, H.A., and ST. JOHN, P.A. 1980. A critical comparison of pre-column and post-column fluorogenic labeling for the HPLC analysis of pesticide residues. *In* Pesticide Analytical Methodology. J. Harvey, Jr., and G. Zweig (Editors). ACS Symp. Ser. 136. Am. Chem. Soc., Washington, DC.

MOYE, H.A. 1981. High performance liquid chromatographic analysis of pesticide residues. *In* Analysis of Pesticide Residues. H. Anson Moye (Editor). John Wiley & Sons, New York.

REYNOLDS, L.M. 1969. Polychlorobiphenyls (PCB's) and their interference with pesticide residue analysis. Bull. Environ. Contam. Toxicol. *4* (3) 128–143.

SEIBER, J.N. 1981. Carbamate insecticide residue analysis by gas–liquid chromatography. *In* Analysis of Pesticide Residues. H. Anson Moye (Editor). John Wiley & Sons, New York.

SHERMA, J. 1974. Thin layer chromatography: Recent advances. *In* Analytical Methods for Pesticides and Plant Growth Regulators. Vol. VII. G. Zweig and J. Sherma (Editors). Academic Press, New York.

SHERMA, J. 1980. Quantitative thin-layer chromatography (TLC). *In* Analytical Methods for Pesticides and Plant Growth Regulators. Vol. XI. Updated General Techniques and Additional Pesticides. G. Zweig and J. Sherma (Editors). Academic Press, New York.

STAN, H.-J. 1981. Combined gas chromatography—mass spectrometry. *In* Pesticide Analysis. K.G. Das (Editor). Marcel Dekker, Inc., New York and Basel.

THOMPSON, J.F., and WATTS, R.R. 1981. Gas chromatographic columns in pesticide analysis. *In* Analysis of Pesticide Residues. H. Anson Moye (Editor). John Wiley & Sons, New York.

WALL, D.L., and MANTZ, A.W. 1977. High sensitivity infrared spectroscopy of gas chromatographic peaks. Appl. Spectr. *31* (6), 552–560.

YURAWECZ, M.P. 1976. Analytical behavior of early eluting industrial chemicals. GLC data file and retrieval system. FDA Lab. Information Bull. No. *1710 A.*

ZWEIG, G. and ARCHER, T.E. 1958. Residue determination of Sevin (1-napthyl N-methylcarbamate) in wine by cholinesterase inhibition and paper chromatography. J. Agric. Food Chem. *6* (12), 910–913.

ZWEIG, G. (Editor). 1963. Analytical Methods for Pesticides, Plant Growth Regulators, and Food Additives. Vol. 1. Principles, Methods, and General Application. Academic Press, New York.

ZWEIG, G. 1964. Chromatographic methods for pesticide residue analysis. Chromatogr. Rev. *6*, 110–127.

ZWEIG, G., and GAO, R-y. 1983. A rapid analytical method for benomyl of HPLC. Anal. Chem. *55*, 1448–1451.

ZWEIG, G., GAO, R-y, WITT, J.M., POPENDORF, W.J., and BOGEN, K. 1984. Dermal exposure to carbaryl by strawberry harvesters. Abstr. Amer. Chem. Soc. Meet., St. Louis, PEST 52.

15

Flow Injection Analysis: A New Tool for the Automation of the Determination of Food Components

Kent K. Stewart[1]

INTRODUCTION

As one contemplates the future direction of the analytical chemistry of foods, several features predominate in the predictions of what may occur in the coming years. First, the food industry is a highly regulated industry, and it would seem likely that it will remain so. Second, total quality control of food processing and distribution is increasing and it would seem likely that a majority of the industry will implement such an approach in the not too distant future. Third, the concern of food scientists with food toxicology is likely to be a factor for several years to come. Fourth, in the continuing efforts to optimize products, it would appear that there will be increasing numbers of studies on the effects of processing and storage on the nutrient and nonnutrient content of foods.

All of these trends will increase the number of chemical determinations needed in the food analysis laboratory. The develoment of inexpensive, rapid, and precise automated analytical techniques should prove to be extremely useful to the food chemist. There are several types of automation systems available today. One approach is that of

[1]Department of Food Science and Technology, Virginia Polytechnic Institute and State University, Blacksburg, VA 24061

using existing batch operations and replacing the analyst with a machine. Many have taken this approach. The current developments in robotics will make this approach very attractive to many. Such systems permit the easy automation of many existing chemical assays, but the equipment is complex and often quite expensive. Another approach is that of doing sequential analysis in flowing streams. Utilization of flowing streams is attractive since the instrumentation is often less complex and the throughput can be very high. The use of segmented continuously flowing streams was introduced in the 1950s by Skeggs (1957). In the early 1970s nonsegmented continuously flowing stream systems for the sequential analysis of discrete samples were introduced. This process is called flow injection analysis (FIA). Many believe that it has the potential to become the method of choice for many automated and semiautomated assays. A number of reviews (Betteridge 1978; Ruzicka and Hansen 1980; Ranger 1981; Rocks and Riley 1982; Stewart 1983), one textbook (Ruzicka and Hansen 1981), and two views of its early history have appeared (Mottola 1981; Stewart 1981). This analytical process could be a powerful tool for food chemists. The purpose of this chapter is to describe the FIA systems, to suggest some applications of FIA, and to present some speculation on its future. No attempt will be made to provide a comprehensive review of FIA systems since the previously mentioned reviews have discussed this topic.

FLOW INJECTION ANALYSIS

Flow injection analysis has been defined as "an automated or semiautomated analytical process consisting of the insertion of sequential discrete sample solutions into an unsegmented continuously flowing liquid stream with subsequent detection of the analyte" (Stewart 1981). The basic standard FIA systems are shown in Figs. 15.1A and 1B. Figure 15.1A presents the most popular semiautomated FIA system in which a sample is inserted into an unsegmented stream of reagent pumped by a peristaltic pump, the analyte mixed with the reagent by convective and diffusion forces, and the product measured as it passes through the detector (Ruzicka and Hansen 1975). Peak height measurements are normally made. Figure 15.1B is a schematic of an automated FIA system in which the sample is aspirated from a sample cup in a sampler tray into the sample loop of a sample insertion valve, after which the valve is actuated, and the sample is inserted into an unsegmented continuous stream of sample solvent which is then mixed

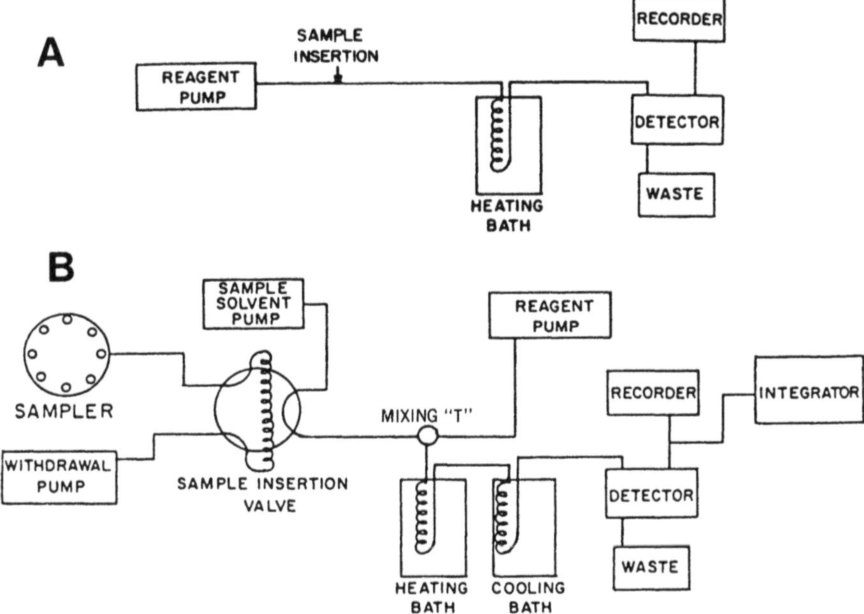

FIG. 15.1. (A) Schematic of a semiautomated FIA system. (B) Schematic of an automated FIA system.
(A) Reprinted with permission from Stewart (1983). (B) Reprinted with permission from Stewart (1980).

with a reagent stream, and the resulting mixture flows on to the detector as before. Depulsed positive displacement pumps are normally used (Stewart 1977). Peak area measurements are usually made.

Either system can perform routine replicate assays at 100+ samples per hour. The results for individual samples are often available within 15 sec after sample injection. Precisions of 0.5–2.0% RSD have been reported for a wide variety of assays including a variety of enzyme determinations. Typical FIA recorder tracings are shown in Fig. 15.2. The use of small bore tubing (typically 0.5-mm i.d.), precise flow rates (typically 3–8 mL/min), and small mixing volumes lead to a minimized, controlled sample dispersion which is an important aspect of FIA. Adherence to these concepts has resulted in the successful development of a large number of FIA analytical systems for a wide variety of analytes. FIA has been used with colorimetric, fluorimetric, flame emission, atomic absorption, inductively coupled plasma, refractive index, chemiluminescence, thermochemical detectors, and a large number of different electrodes as well as a number of electro-

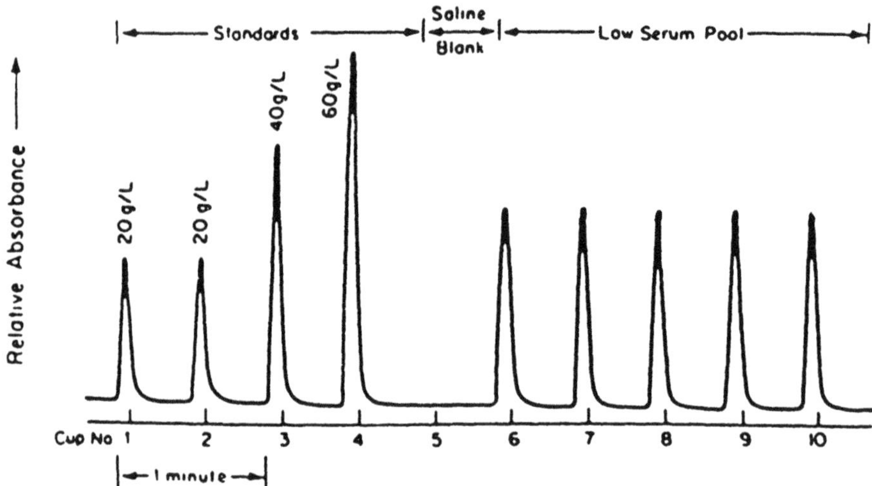

FIG. 15.2. Recorder tracing of a FIA determination of serum albumin with bromocresol green.
Reprinted with permission from Renoe et al. (1980).

chemical detectors including ion selective field effect transistors. Probably any detector that can be used with HPLC systems can be used with FIA systems. Flow injection analysis is appropriate for computer interfacing to provide automatic systems control, data acquisition, calculation of results, etc., and several papers have discussed some aspects of the computerization of FIA systems (Stewart *et al.* 1980; Slanina *et al.* 1980; Brown *et al.* 1981).

Flow injection analysis systems have been described for the automated determination of many analytes of interest to the food chemist and most likely can be used for the automation of most wet chemistries used in the food laboratory. Enzyme assay systems were some of the earliest FIA systems described (Bergmeyer and Hagen 1972; Beecher *et al.* 1975) and the precision of the FIA enzyme determinations make them very attractive for enzymatic food determinations. There are a variety of FIA enzyme assay systems including stopped flow systems, enzyme reactors, and recirculation systems.

Most segmented continuously flowing assay (CFA) systems can probably be readily adapted to FIA systems. Some of the special techniques used in CFA systems such as dialysis, two phase systems, and merging zones have already been adapted for FIA systems. The usual requirements for FIA systems are that the analytes, reagents, and products are soluble in the assay solvent, that sufficient product be developed within less than 60 sec, and that sample dispersion can be rigorously controlled. The shorter reaction times of FIA systems might appear to limit the number of CFA assays which can be adapted for FIA systems; however, some preliminary studies have indicated that while the CFA systems may have reaction times of minutes, the chemically analogous FIA systems often have reaction times of seconds. A probable explanation is that the CFA reaction times were actually longer than necessary for sufficient product formation and/or that there was more limited sample dilution in the FIA systems.

FIA STOPPED FLOW SYSTEMS

The development of FIA stopped flow systems has been described by Ruzicka and his co-workers (1979) and by Malmstadt *et al.* (1980). These special systems permit the automated sequential stopped flow determination of the analyte concentrations of discrete samples with throughput of about 100 samples per hour. In these systems each sample provides its own background correction. Many of the current CFA systems for food analysis are often flawed because of the high backgrounds found in many food extracts. Thus the FIA stopped flow systems have the potential to fill an important gap in the current food assay systems.

FIA DILUTION SYSTEMS

A unique FIA dilution system has been described by Stewart *et al.* (1980). The sample dilution is controlled by the sample loop size, the diluent flow rate, and the time between fractions in the fraction collector (see Fig. 15.3). The authors report that automated dilutions can be performed at 80 samples per hour with precisions of 0.8% RSD.

PSEUDOTITRATIONS AND OTHER EXPONENTIAL DILUTION SYSTEMS

A different type of FIA system is shown in Fig. 15.4. In this mode the sample is manipulated so as to create an exponential dispersion and then mixing in the titrant. The resulting concentration profiles

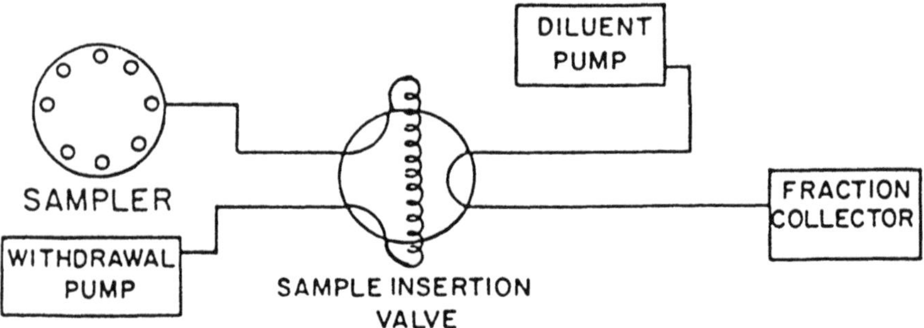

FIG. 15.3. Schematic of an automated FIA-dilution system.
Reprinted with permission from Stewart et al. (1980).

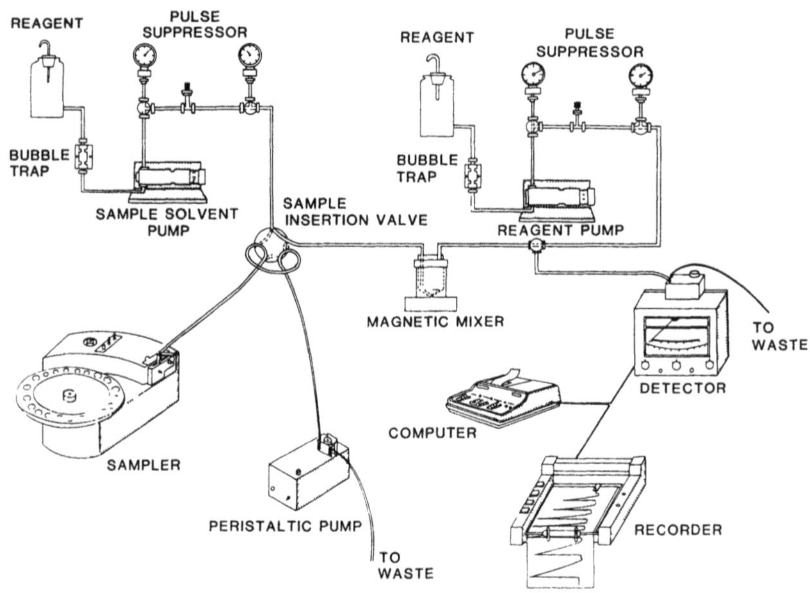

FIG. 15.4. Schematic of an automated FIA system for pseudotitrations.
Reprinted with permission from Stewart (1983).

are shown in Fig. 15.5. Under these conditions Eq. (1) holds where Δt_{eq} is defined in Fig. 15.5, and the other symbols are defined in Table 15.1. This type of system was originally described as an FIA titration system (Ruzicka *et al.* 1977), but due to technical reasons of nomenclature has been more recently called a pseudotitration (Stewart and Rosenfeld 1982). There has been a number of variations upon this ba-

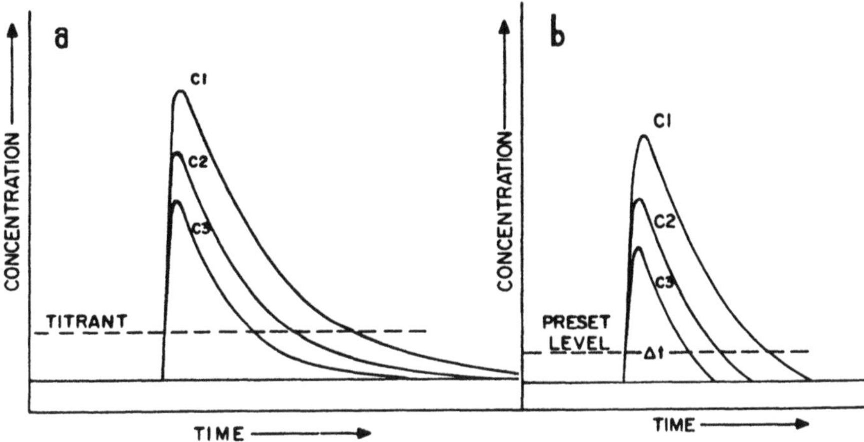

FIG. 15.5. Concentration profiles for FIA pseudotitrations. (a) Concentration profiles for three different samples C1, C2, C3 as the effluent emerges from the mixing chamber if no titrant were present (—). Concentration of the titrant after mixing if no reaction occurs (---). (b) Concentration profiles of the three samples after reacting with the titrant.
Reprinted with permission from Stewart (1983).

sic system (Horvai *et al.* 1976; Nagy *et al.* 1978; Astrom 1979; Stewart and Rosenfeld 1981), and it would seem that these rapid titration-like assays could be adapted for many different determinations of food components.

$$\Delta t_{eq} = K_1 \ln(C^0_{as}/C^0_{reg}) + K_2. \quad (1)$$

Recent work has shown that the time measurements may also be used under much more general conditions and that stirred mixing chambers may also be used in a scale expansion system for many as-

TABLE 15.1. DEFINITION OF SYMBOLS

a	Internal radius of tubing (cm)
C^0_{as}	Original concentration of analyte
C^0_{reg}	Original concentration of titrant
D	Diffusion constant of analyte (cm/sec)
K_1 and K_2	Constants of individual FIA systems
t	Time (sec)
t_a	Time from injection to initial appearance of peak
Δt_b	Time from initial appearance of peak to final appearance of peak
Δt_{eq}	Equivalence time for pseudotitrations and scale expansion systems

say systems (Stewart and Rosenfeld 1982). In this type of system the mixing chamber is placed immediately after the sample insertion valve and the reagent is mixed with the effluent from the mixing chamber. Under these conditions the concentration of the samples can be empirically determined using Eq. (2).

$$t_{eq} = K_1 \ln C^0_{as} + K_2. \tag{2}$$

The general nature of the system was demonstrated by incorporation into FIA systems using colorimetric, fluorimetric, conductometric, and flame emission systems. Scale expansions of two- to thirtyfold were obtained in these cases. Such a scale expansion system can probably be incorporated into any standard FIA system.

THEORY OF FIA SYSTEMS

A simple FIA system can be described as the insertion of a sample bolus into a moving stream inside a segment of narrow bore tubing and the measurement of the analyte at some point downstream. The theoretical description of the time dependent concentration profile of analytes under these simple conditions is contested. There is agreement that the work of Taylor (1953) is the beginning of modern theory, and there is disagreement as to the best current theoretical description. One group prefers the "tanks in series model" (Ruzicka and Hansen 1978) in which the dispersion of an injected bolus is estimated by utilizing a classical model of a series of totally mixed tanks. Others prefer the approach of Vanderslice *et al.* (1981) in which the flow is assumed to be completely laminar and the dispersion of a sample bolus is based on numerical solutions of the diffusion–convection equations. The two key parameters predicted by Vanderslice *et al.* are the time (t_a) from injection to the initial appearance of sample bolus at the detector, and the baseline-to-baseline time (Δt_b) for each sample bolus at the detector (see Fig. 15.5). Equations (3) and (4) show the relationship of the crucial parameters needed to predict these two times. The variables are defined in Table 15.1.

$$t_a = 109a^2D^{0.025}(L/q)^{1.025} \tag{3}$$

$$\Delta t_b = (35.4a^2/D^{0.36})(L/q)^{0.64} \tag{4}$$

THEORY OF PSEUDOTITRATIONS AND EXPONENTIAL DILUTION CHAMBERS

In the FIA pseudotitrations, the sample dispersion is manipulated so as to provide an exponential increase in concentration followed by an exponential decrease in concentration. Pardue and Fields (1981A, B) have developed the general theory for exponential dilution systems including pseudotitrations and the scale expansion system. The general theory developed by these workers encompasses the basis for both the pseudotitrations as well as that of the exponential scale expansion system. Detailed equations were developed for a number of exponential dilution systems. The reader is encouraged to consult their articles for the details.

SOME GENERAL COMMENTS AND OBSERVATIONS

Basically all FIA systems fit into Pardue's classification of a kinetic method of analysis (Pardue 1977), and thus while assays based upon equilibrium theory can be (and frequently are) used in FIA systems, the interpretations of the results must be as those resulting from empirical methods. The FIA pseudotitrations and scale expansion systems are a part of those measurement systems in which the measured *time* is proportional to the concentration of the analyte. This is a special type of measurement system for which there is no unambigious analogy in the traditional equilibrium assay systems. Such time measurement systems could have profound effects on the design of future FIA systems. For example, since the primary concentration measurement is time, the detector is needed only as a trigger and thus the detector response need not be linear with concentration; it only needs to yield reproducible timing points. In a similar fashion, the data acquisition system need be only a relatively simple digital clock coupled with a microprocessor for exponentiation and multiplication. Furthermore, only enough reagent is needed to produce enough product to activate the trigger, and not that needed to react with higher concentrations of the analyte. Thus there is a potential for reagent cost reduction.

It is difficult to visualize where all this will lead. Certainly it suggests that simpler and/or minaturized FIA systems could be developed. Such futuristic systems could prove to be very useful to the food analyst and may well be the basis for food component determinations in the field and on the process line. Such systems could have profound effects on the processing and monitoring of the components of our food supply.

BIBLIOGRAPHY

ASTROM, O. 1979. Single-point titrations. Part 4. Determinations of acids and bases with flow injection analysis. Anal. Chim. Acta *105,* 67–75.

BEECHER, G.R., STEWART, K.K., and HARE, P.E. 1975. Automated high speed analysis of discrete samples: the use of nonsegmented, continuous flow. *In* Protein Nutritional Quality of Foods and Feeds, Part 1. M. Friedman (Editor), pp. 411–421. Am. Chem. Soc., Washington, DC.

BERGMEYER, H.U., and HAGEN, A. 1972. Ein neues prinzip enzymatischer analyse Fresenius Z. Anal. Chem. *261,* 333–336.

BETTERIDGE, D. 1978. Flow injection analysis. Anal Chem. *50,* 832A–846A.

BROWN, J.F., STEWART, K.K., and HIGGS, D. 1981. Microcomputer control and data system for automated multiple flow injection analysis. J. Automatic Chem. *3,* 182–186.

HORVAI, G., TOTH, K., and PUNGOR, E. 1976. A simple continuous method for calibration and measurement with ion-selective electrodes. Anal. Chim. Acta *82,* 45–54.

MALMSTADT, H.V., WALCZAK, K.M., and KOUPPARIS, M.A. 1980. An automated stopped-flow/unsegmented solution storage analyzer. American Laboratory Sept. 27–40.

MOTTOLA, H.A. 1981. Continuous flow analysis revisited. Anal. Chem. *53,* 1312A–1316A.

NAGY, G., FEHER, Z., TOTH, K., and PUNGOR, E. 1978. A novel titration technique for the analysis of streamed samples—the triangle programmed titration technique. Anal. Chim. Acta *100,* 181–191.

PARDUE, H.L. 1977. A comprehensive classification of kinetic methods of analysis used in clinical chemistry. Clin Chem. *23,* 2189–2201.

PARDUE, H.L., and FIELDS, B. 1981A. Kinetic treatment of unsegmented flow systems. Part 1. Subjective and semiquantitative evaluation of flow-injection systems with gradient chamber. Anal. Chim. Acta *124,* 39–63.

PARDUE, H.L., and FIELDS, B. 1981B. Kinetic treatment of unsegmented flow systems. Part 2. Detailed treatment of flow-injection systems with gradient chamber. Anal. Chim. Acta *124,* 65–79.

RANGER, C.B. 1981. Flow injection analysis. Anal. Chem. *53,* 20A–32A.

RENOE, B.W., STEWART, K.K., BEECHER, G.R., WILLS, M.R., and SAVORY, W. 1980. Automated multiple flow-injection analysis in clinical chemistry. Determination of albumin with bromocresol green. J. Clin. Chem. *26,* 331–334.

ROCKS, B., and RILEY, C. 1982. Flow injection analysis: A new approach to quantitative measurements in clinical chemistry. Clin. Chem. *28,* 409–421.

RUZICKA, J., and HANSEN, E. 1975. Flow injection analysis: Part I. A new concept of fast continuous flow analysis. Anal. Chim. Acta *78,* 145–157.

RUZICKA, J., and HANSEN, E. 1978. Flow injection analysis. Part X. Theory, techniques and trends. Anal. Chim. Acta *99,* 37–76.

RUZICKA, J., and HANSEN, E. 1979. Stopped flow and merging zones—a new approach to enzyme assay by flow injection analysis. Anal. Chim. Acta *106,* 207–224.

RUZICKA, J., and HANSEN, E. 1980. Flow injection analysis. Principles, applications, and trends. Anal. Chim. Acta *114,* 14–19.

RUZICKA, J. and HANSEN, E. 1981. Flow Injection Analysis. Wiley, New York.

RUZICKA, J., HANSEN, E., and MOSBAEK, H. 1977. Flow injection analysis. Part IX. A new approach to continuous flow titrations. Anal. Chim. Acta *92,* 235–249.

SLANINA, J., LINGERAK, W.A., and BAKKER, F. 1980. The use of ion-selective electrodes in manual and computer controlled flow injection systems. Anal. Chim. Acta *117,* 91–98.

SKEGGS, L.T. 1957. An automatic method for colorimetric analysis. Am. J. Clin. Path. *28,* 311–322.

STEWART, K.K. 1977. Depulsing system for positive displacement pumps. Anal. Chem. *49,* 2125–2126.

STEWART, K.K. 1981. Flow-injection analysis: A review of its early history. Talanta *28,* 789–797.

STEWART, K.K. 1983. Flow injection analysis: A new tool for old assays—A new approach to analytical measurements. Anal. Chem. *55,* 931A–936A.

STEWART, K.K., and ROSENFELD, A.G. 1981. Automated titrations: The use of automated multiple flow injection analysis for the titration of discrete samples. J. Automatic Chem. *3,* 30–32.

STEWART, K.K., and ROSENFELD, A.G. 1982. Exponential dilution chambers for scale expansion in flow injection analysis. Anal. Chem. *54,* 2368–2372.

STEWART, K.K., BROWN, J.F., and GOLDEN, B.M. 1980. A microprocessor control system for automated multiple flow injection analysis. Anal. Chim. Acta *114,* 119–127.

TAYLOR, G. 1953. Dispersion of soluble matter in solvent flowing slowly through a tube. Proc. Royal Soc. *219,* 186–203.

VANDERSLICE, J.T., STEWART, K.K., ROSENFELD, A.G., and HIGGS, D.J. 1981. Laminar dispersion in flow injection analysis. Talanta *28,* 11–18.

16

Modern Liquid Chromatography: Evolution and Benefits

James R. Kirk[1]

INTRODUCTION

Classical column or open-bed chromatography was highly laborious because it required the preparation of a new column for each separation, development of a skillful technique for sample application, gravity flow for separation of components within the sample, and manual analysis of each fraction collected. Modern liquid chromatography is a closed system; columns are reusable and hundreds of separations can be carried out on a given column. The reusable nature of modern LC columns permit their cost to be prorated over a large number of samples which allows the development of sophisticated, high cost packing materials to ensure high performance and high resolution within the separation. Sample application by injection or sample loop permits precise and proper application of the sample on the column. Solvent pumps provide controlled rapid movement of mobile phase through relatively impermeable, high efficiency columns, resulting in more reproducible column operation and greater accuracy and precision within the analysis. The use of high pressure in the closed chromatographic system has been shown to result in better and faster separation. Thus, modern LC techniques are more convenient, accurate, reproducible and less operator dependent.

[1]Vice President Research and Development, Campbell Soup Company, Camden, NJ 08101

CHROMATOGRAPHIC CONCEPTS

Liquid chromatographic separation is the result of specific interactions between sample molecules, and the stationary and mobile phases. As the sample molecules are moved along the column by the solvent or mobile phase two characteristic features of chromatographic separations occur: differential migration of sample molecules and the separating of the sample molecules in diffuse bands. Differential migration refers to the varying rates of movements due to different equilibrium distribution coefficients of sample molecules between the stationary phase and the mobile phase. As shown in Fig. 16.1, sample component B which has an equilibrium distribution coefficient favoring association with the stationary phase will move through the column more slowly than sample component A which favors association with the mobile phase. As expected the solvent molecules move through the column at the fastest rate possible. The speed with which a compound moves through the column is dependent upon its relative concentration in the mobile phase at any time *(t)* since a sample molecule cannot move through the column while partitioned in the stationary phase.

Because equilibrium distribution characteristics control the migration rate, experimental variables which affect the distribution are as follows: the composition of the mobile phase, composition of the stationary phase, and the separation temperature. The effect of column

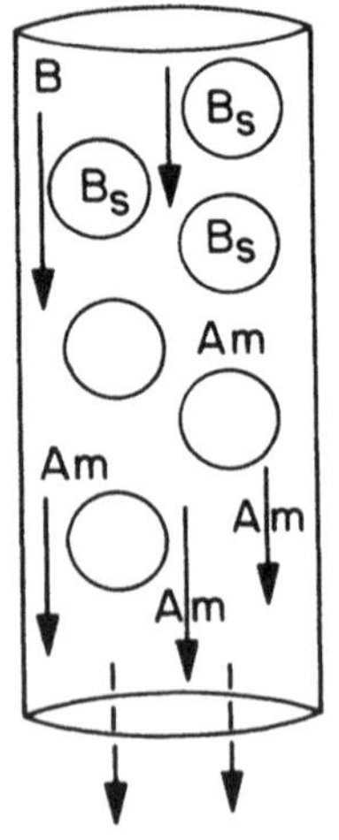

FIG. 16.1. Hypothetical separation of a two-component mixture (A and B) using liquid chromatography. B_s represents component B held by the stationary phase and Am is the movement of component A in the mobile phase.

pressure, which could affect equilibrium distribution and thus migration rate, is negligible at normal column pressures (500–3000 psi).

BAND BROADENING

The second consideration of chromatographic separations is band broadening which is caused by physical or rate processes, leading to diffusion differences between like components within a sample. The physical and rate processes leading to differential migration of identical components within a mixture are as follows: eddy diffusion (multiple flow-pathways), mobile phase mass transfer (differing flow rates within a single pathway), stagnant mobile phase mass transfer (movement of mobile phase into pores in the stationary phase), stationary phase mass transfer (depth of diffusion of component within a stationary phase), and longitudinal diffusion (random diffusion of sample components in all directions). Longitudinal diffusion only has a significant effect at very low flow rates with very small particle column packings. The desired outcome of modern LC separations is the elution of each sample component as a homogeneous bell-shaped band, each band having a characteristic retention time t_R with sufficient time between bands to ensure the integrity of each sample component and minimize band broadening which can affect a decrease in detection sensitivity.

SEPARATION MECHANISMS

Four separate mechanisms can be used to affect the separation of a sample mixture: liquid–liquid (partition), liquid–solid (adsorption), ion-exchange (electrostatic association), and size-exclusion (gel filtration). Two modifications of liquid–liquid chromatography, bonded-phase and ion-pair chromatography, have become very powerful chromatographic techniques. The mechanism of separation associated with bonded-phase chromatography is believed to be the partitioning of the sample molecules between the liquid phases. Ion-pairing is a combination of liquid–liquid and ion-exchange chromatography. The exact mechanism for these later two processes is yet to be confirmed.

Regardless of the mechanism of separation the elution of a component from the chromatography column is related to the velocity of the solvent moving through the column (u_s, cm/sec) and the fraction of molecules of A in the moving phase R; therefore u_a (velocity of A is a function of R) and u and can be mathematically described as: $u_a = quR$. R can also be described from the capacity factor k', a fundamental liquid chromatography parameter, and the migration of compound A through the column is described by:

$$t_R = t_0(1 + k)'), \tag{1}$$

where t_R is the retention time and is expressed as a function of the retention time of the solvent (t_0) and the capacity factor of the column (k') (Snyder and Kirkland 1979). For a given column, mobile phase, temperature and sample component A, k' is normally constant for small samples; thus t_R can be used to tentatively identify compound A within the sample.

The bandwidth t_w in liquid chromatography is expressed by the theoretical plate number:

$$N = 5.16\left(\frac{t_R}{t_w}\right)^2 \quad \text{or} \quad N = 5.54\left(\frac{t_R}{t_{w\frac{1}{2}}}\right)^2. \tag{2}$$

N is approximately constant for different bands in a chromatographic column under constant operating conditions and is a useful measure of column efficiency, the ability of a column to provide narrow bands and thus improve separations. Because N remains constant the previous equation predicts that bandwidth (t_w) will increase proportionately with retention time, except for certain cases such as gradient elution. Therefore, the central goal of liquid chromatography is to attain the maximum number of theoretical plates in a given column, which is proportional to the column length (L). The proportion is expressed as

$$N = \frac{L}{H}, \tag{3}$$

where H is the height equivalent of a theoretical plate. H is smaller for small particles of packing material, low flow rates, less viscous mobile phases, higher separation temperatures, and small sample molecules.

Column efficiency is also affected by size of the sample injected. As the sample size is increased beyond the linear capacity of the column packing H increases and thus N, the theoretical plate number, decreases. This is particularly true for high-efficiency, small-particle columns in use today.

EXTRACOLUMN BAND BROADENING

Extracolumn band broadening takes place to some extent in every LC apparatus but it can be a serious problem, particularly with post column derivatization steps, if the chromatographer is not aware of the factors responsible for this phenomenon. Factors of particular concern are the length and diameter of tubing placed between the injector and column, the column and the detector, the size of the detector cell, and any increase in the volume of the injected sample.

RESOLUTION AND SEPARATION

The goal of modern liquid chromatography is the adequate separation of a given sample mixture. Resolution of two distinct bands is defined as

$$R_s = \frac{t_2 - t_1}{\frac{1}{2}(t_{w_1} + t_{w_2})}. \tag{4}$$

The values t_1 and t_2 correspond to the t_R values of the two peaks and t_{w_1} and t_{w_2} are the peak widths. An $R_s = 1$ represents good separation of the two bands with only a 2% overlap. The larger the R_s value, the better the separation.

To control separation or resolution (R_s) of distinct bands we must know how R_s varies as a function of experimental parameters that can be controlled by the chromatographer.

$$R_s = \tfrac{1}{4}(\alpha - 1)\sqrt{N}\,\frac{k'}{1+k'}, \tag{5}$$

where α is the separation factor (k_2/k_1), N is the theoretical plate number, and k' is the capacity factor (n_s/n_m). The three terms of the equation are essentially independent so we can optimize each term separately. The separation selectivity, as measured by α, is varied by changing the composition of the mobile and/or stationary phase. The second term (N) is a measure of separation efficiency and is varied by changing the length of the column or the velocity of the mobile phase (u). The third term, the capacity factor (k'), is varied by changing solvent strength. If k' for the initial separation falls within the range $0.5 < k' < 2.0$, a decrease in k' results in a decrease in separation, while an increase in k' (weaker solvent) provides an increase in the resolution. As k' increases t_w and t_R also increase. No other change in separation conditions provides as large an increase in R_s for as little effort as a change in k'. However, when k' is near the optimum and resolution is marginal then the best solution is to increase N. Normally with small-particle packing materials, N is increased by increasing the length of the column (L) or decreasing the velocity of the mobile phase (u) rather than changing to another column packing.

Finally, total analysis time is always of concern to the chromatographer and is a function of the t_R for the last band off the column which can be estimated by

$$t \simeq t_n = t_0(1 + k'_n). \tag{6}$$

MODERN LIQUID CHROMATOGRAPHY SYSTEMS

Modern liquid chromatography (LC) is performed on relatively sophisticated equipment as compared to classical liquid chromatography. Modern LC systems should be flexible and meet the needs of the chromatographer with respect to durability, flexibility, convenience, and low maintenance and be able to offer high precision and accuracy and have low operator dependency. The basic components of a modern LC system consist of mobile phase reservoir, pump with pulse suppression, injection valve, column, column temperature control unit, detector, and recorder. It is difficult to construct a single system that will be ideal for all needs and be economical (Hamilton and Sewell 1977).

Mobile phase reservoirs can be highly sophisticated containers or Erlenmeyer flasks plus a magnet stirrer. The essential concepts which must be considered in handling the mobile phase are filtration, degassing, and stirring. Degassing is of particular importance when using aqueous and polar mobile phases especially those used for gradient elution. The use of inert gases after purging to prevent oxygen from redissolving is important if oxygen reacts adversely with the sample, the mobile phase, and/or the stationary phase. Degassing must be carried out judiciously since even minor changes in the concentration of the mobile phase can lead to variation in the k' values.

PUMPING SYSTEMS

The function of the solvent pumping or metering system is to provide a constant, highly reproducible supply of mobile phase through the column. The use of very small particulate packing materials requires the use of a high-pressure pump which is pulse-free or has pulse dampening, delivers mobile phase at a velocity so as not to limit k', delivers mobile phase with reproducibility of $<1\%$, and has a small pump volume for rapid mobile phase changes and gradient elution separations.

Because of their overall performance characteristics, reciprocating pumps with piston displacement volumes of 35–400 μL, multiple chambers and, pulse suppression mechanisms are most widely used. Overall superior performance is obtained with more sophisticated reciprocating pumps which also are the more expensive. An excellent discussion of pump performance and comparison of characteristics of various pumping systems is reported by Snyder and Kirkland (1979).

GRADIENT ELUTION

Separation of a complex mixture having components with widely differing k' values is most often approached through the use of gradient elution. Gradient elution requires two high-pressure pumps which can be microprocessor programmed to change the solvent composition for optimum separation and resolution. This type of gradient former is widely used in commercial instruments with great success. One limitation with the use of high-pressure pumps is that they are less precise at low flow rates; thus the gradients formed at the initial and final stages of the program are imprecise and solute elution characteristics can be erratic. This is averted by extending the gradient both at the low and high side so as not to have solute k' values within the extremes of the gradient program.

SAMPLE INJECTORS

The simplest form of sample introduction onto the column is by the use of a syringe injector through a self-sealing elastomeric system. However, reproducibility is seldom better than 2% and can result in other on-column injection physical problems. Currently, the microsampling injector valve is used almost exclusively with modern LC systems because of the rapid, reproducible, and essentially operator-independent nature. It is important to remember that extracolumn band broadening can occur if the inside diameter of the sample loop is greater than that of the connecting tubing. Automatic sampling devices are available and are highly popular for routine analysis.

DETECTORS

The three most common detectors used for the quantitation of nutrients as food systems are (1) visible-UV photometers/spectrophotometers, (2) fluorometers, and (3) differential refractometers, the latter being used almost exclusively for analysis of carbohydrates. Snyder and Kirkland (1979) indicate that the ideal characteristics of a detector are that it

- shows high sensitivity and predictable response
- responds to all solvents, or has predictable specificity
- shows wide range of linearity
- is unaffected by changes in temperature and mobile phase
- responds independently of mobile phase
- does not contribute to extracolumn band broadening
- is reliable and convenient to use
- shows linear response with increase in solute
- is nondestructive to sample and mobile phase
- has quantitative and fast response time to detected peak

The most widely used detectors in modern LC are fixed wavelength ultraviolet and visible photometers. Light absorption detectors are usually designed to provide an output in absorbance that is linearly proportional to concentration

$$A = \log (I_0/I) = \text{abc}, \tag{7}$$

where I_0 is light intensity, I is intensity of transmitted light, a is the molar absorptivity, b is the cell pathlength (cm), c the concentration of the sample (moles/L), and A is absorbance.

Properly designed UV photometric detectors are relatively insensitive to flow and temperature and have sensitivities of 0.002 absorbance units full scale with $\pm 1\%$ noise. Sensitivity can often be increased through the use of a spectrophotometer which permits operation at the absorption maximum of a solute and may eliminate absorbance from interfering compounds. Modern LC spectrophotometers have stable, low-noise electronics which permit a lower limit of detection and limit of quantitation.

Fluorometric detectors offer unique sensitivity and selectivity for solutes having fluorescent properties or compounds which can be derivatized to a stable fluorophore. Biological compounds are particularly suited for fluorometric detection because they are often symmetrically conjugated and not strongly ionic. In many cases fluorescence detection may be selective for a solute molecule and in most instances it is 100-fold more sensitive than absorbance detection. Fluorescence detection is ideal for use with gradient elution techniques and for trace analysis either because of sample size or low solute concentration within the sample. In all applications of fluorescence detection, concern should be given to the fact that fluorescence detectors can become nonlinear at concentrations far below that of UV and visible absorbance detectors because of internal quenching of fluorescence.

Differential refractometry is a common method of detection with modern LC because it responds to all solutes. However, RI detectors have serious limitations including the following: lack of sensitivity, impracticality due to gradient elution separations, sensitivity to temperature fluctuation, and severe baseline drift with any change in mobile phase character or use of gradient mobile phase system. Despite the sensitivity limitations and impracticality with most mobile phases, differential RI detection is commonly used with size-exclusion chromatography since sensitivity often is not important.

In conclusion, it should be noted that fluorescence detectors can detect less than 1 pg of solute, whereas UV and visible absorption detectors are limited to about 1 ng. Other detectors have a normal limit of detection of approximately 0.1–1 μg under optimum conditions.

COLUMNS

High performance columns that exhibit minimum band broadening of the separated solutes are the heart of the modern LC system. Packing materials can be classified as

- rigid solids, hard gels, or soft gels
- porous, pellicular, or superficially porous particles
- spherical or irregular particles

Rigid solids have a silica base and are most common modern LC packings because they can withstand relatively high pressures (5,000–10,000 psi). Silica particles can be obtained in a variety of sizes, shapes, varying degrees of porosity, and functional groups or polymeric compounds can readily be attached to functional groups on the silica. Hard gels, which are porous particles of polystyrene cross-linked with divinyl benzene, still find some use in ion-exchange and size-exclusion chromatography. However, rigid solids are replacing hard gels as packing materials for these functions (Yost *et al.* 1980).

Liquid chromatography packings are functionally described as pellicular or porous. Pellicular particles are made from solid glass beads which are often coated with a thin layer of stationary phase. Pellicular particles can also be coated with a layer of porous material to give a superficially porous particle which can be coated with a stationary phase or chemically treated to give a bonded-phase packing.

Porous particles are made of silica or diatomaceous earth and are characterized as spherical or irregular. By varying the conditions of aggregation for irregular or spherical particles, the average pore size can be varied. Column packings should be made from particles having a narrow range of particle size to prevent size separation which result in column inefficiency and decreased permeability. Spherical and irregular particles can be packed to give columns of similar efficiency (N) for columns of the same size when similar particle sizes are used for each column, although most studies show a much lower pressure drop associated with the spherical particles and greater physical stability during shipping or rough treatment.

Although no single column packing is best for every separation problem, small porous particles are usually preferred for most modern LC applications. As previously discussed, a decrease in particle size (d_p) in a fixed length column increases the number of plates (N) and the column efficiency but also increases the pressure drop across the column for a given flow rate of the mobile phase.

The maximum sample size for a pellicular packing is much smaller than an irregular packed column of similar efficiency. Pellicular columns have less stationary phase associated with them and the stationary phase present is quickly overloaded by the sample. Estimates are that a porous column can handle five times as much sample as pellicular column of equivalent efficiency. Because of their ability to handle greater sample loads, solute bands on porous columns can be more concentrated and thus are more easily detected. Thus small particle porous columns have the largest N, give the best detection sensitivity, and are preferred for trace analyses. Conversely, large particle porous columns are preferred for preparative modern LC since the bandwidth will be determined by column overloading rather than large H values. Snyder and Kirkland (1979) describe the advantage of pellicular versus small particle column packings and indicate that small particle columns are preferred for quantitation, isolation, and identification of trace constituents such as micronutrients in foods.

Compared to pellicular packings, microparticles offer substantially improved column efficiency, sample capacity, and speed of analysis. Common microparticulate materials used in liquid–solid and liquid–liquid LC columns are silica and alumina and are produced as irregular and spherical particles of various diameters.

The following are common examples of packing material used for modern LC.

Silica (irregular)	*Silica (spherical)*
BioSil A	LiChrosopher Si-100
LiChrosorb Si-60	Nucleosil 50
MicroPak Si	Spherisorb SW
Partisil	Spherosil XOA
μ Porasil	Vydac TP ads.
Sil 60	Zorbax Sil
Alumina (irregular)	*Alumina (spherical)*
ALOX 60 D	Spherisorb AY

The bonded phase LC packings (BPC) used in LLC, RPC, and IPC are prepared by bonding various functional groups to irregular or spherical silica particles and synthetic resins in the case of ion-exchange chromatography (IEC). All of the methods for attaching a bonded phase to a siliceous support rely on the reaction of surface silanol groups. Fully hydrolyzed silica is approximately 8 μmoles Si–OH/m^2 of surface; however, because of steric hindrance only about 4.5 μmoles of silanol groups per square meter are available for bonding. The most widely used method for preparation of bonded phase columns are those made from siloxanes (Si–O–Si–C) because of their stability to hydrolysis through the pH range 2–8.5. Organic bond-phase siloxane coating can be made as a monomolecular layer (e.g., μ BondaPak C_{18} or Zorbax-CN, 8–9% loading) or as a polymerized multilayer coating (e.g., Permaphase or Zorbax ODS, 22% loading) and is available as a pellicular or porous support. Ideally 4 μmoles/m^2 of the silanol groups are reacted with the bonded phase which means that the remaining Si–OH groups on the silica surface are unavailable for chromatographic interactions with most solutes because of steric shielding.

The following is a list of common BPC silica microparticulate reverse phase (RP), normal phase (NP), and ion-pair partition (IPC) packings:

Silica (irregular)	Separation mode	Silica (spherical)	Separation mode
Partisil ODS-1, ODS-2	RP	Spherisorb ODS	RP
μ BondaPak C_{18}, C_8	RP	Vydac RP-TP	RP
LiChrosorb		Nucleosil C-8 and NH_2	RP, NP
RP-18, RP-8, R-2	RP	Spherisorb CN	NP
LiChrosorb DIOL	NP	Zorbax-C_8, CN and NH_2	RP, NP
MicroPak-CN	NP	Hypersil-SAS	IPC
Partisil 10 PAC	NP		

Although the mechanism for retention of solute molecules to BPC packings has not been positively established, the consensus is that in the case of normal-phase BPC chromatography the solute molecules adsorb on to the surface of the defined organic stationary phase and that the solvent molecules compete for the same adsorption sites on the stationary phase. In the case of BPC, it is convenient to envision that the solute molecules are in equilibrium dissolution between the bonded organic stationary phase and the mobile phase as in liquid–liquid chromatography. In either case, selection of a mobile phase parallels that found acceptable for liquid–liquid chromatography and in the case of reverse phase chromatography evaluation of the solvent strength is commonly based on the parameter P' (solvent polarity), which is based on experimental solubility data reported by Rohrschneider (1973) and Snyder (1974, 1978). Experimental adsorption strength ϵ^0 provides a better index of solvent strength for normal phase BPC systems. In either case the mobile phase that gives smaller k' values for a specific sample solute is a solvent that strongly interacts with the solute molecules. In normal phase BPC decreasing the solvent polarity (P') increases k', while in reverse phase BPC as solvent polarity is increased a proportional increase in k' is also observed.

In some cases solvent selectivity is important for the resolution of two overlapping peaks given optimization of k'. Solvent selectivity is controlled by components of the binary or ternary mixture and contributions from donor, acceptor, and dipole characteristics of the mobile phase. Proton donors interact with basic solute constituents while solvents which have a proton acceptor component interact preferentially with hydroxylated molecules. A solvent having a large dipole moment (e.g., methylene chloride) interacts with sample molecules having large dipole moments. Commonly employed mobile phases for normal phase BPC are solvent mixtures of hydrocarbons modified with a polar constituent such as chloroform, ethyl acetate, ethanol, or acetonitrile. Band separation can be accomplished by changing the solvent selectivity or modifying some other variable such as the temperature.

Mobile phases for reverse phase BPC usually contain water as the base solvent. Methanol, acetonitrile, and tetrahydrofuran are the most common modifiers. Selectivity of the mobile phase for reverse phase BPC separations is less easily accomplished because water dominates the mobile phase/stationary phase interactions. Changes in pH can affect the separation selectivity for ionizable solutes because charged molecules would be preferentially distributed into the aqueous phase.

As would be expected variations in pH are not effective in obtaining desired changes in selectivity with solutes that do not ionize.

As previously mentioned, temperature can be used to affect normal and reverse phase BPC separations. Retention generally decreases with an increase in temperature and can be described by the Arrhenius plot, log k' versus $1/T$. Generally a 30°C increase in T leads to a twofold decrease in k', because an increase in temperature causes a decrease in the viscoscity of the mobile phase and an increase in the concentration of solute in the mobile phase, and alters the selectivity of the mobile phase. Separations using normal phase BPC are normally carried out at ambient temperature while reverse phase BPC separations normally employ an aqueous mobile phase heated to temperatures up to 80°C, although 50°–60°C is more common. These temperatures generally result in a doubling of the column plate count.

ION-PAIR CHROMATOGRAPHY

Ion-pair chromatography (IPC) is a special adaptation of normal phase or reverse phase LC through the use of mobile phases which contain an added counterion of an opposite charge to the sample molecule. A comparison of normal phase and reverse phase ion-pairing systems shows that the latter will generally be preferable. Therefore, the discussion will be limited to reverse phase IPC. The stationary phase can be either a silanized silica packing containing an alkyl bonded phase (e.g., C_{18} or C_8) or a silica solid support with a mechanically held water immiscible organic phase, such as 1-pentanol. The mobile phase is similar to that for reverse phase separations except that the aqueous medium is buffered to control the ion species of the sample and also contains an added counterion of the charge opposite that of the sample molecules. In addition, bonded phase separations require a mobile phase which contains an organic co-solvent. In this latter case, the solute and counterions are soluble in the aqueous phase while the former ion pair is soluble only in the organic stationary and mobile phase.

$$R_{aq} + CI^{+}_{aq} \rightleftarrows RCI^{+}_{org}. \tag{8}$$

The extraction constant follows from the mass-action expression as

$$E = \frac{[RCI^{+}]_{org}}{[R]_{aq}[CI^{+}]_{aq}}, \tag{9}$$

where E is the extraction constant for a particular IPC system but is a function of the mobile phase, pH, ionic strength, concentration and organic co-solvent (selectivity), and temperature. The capacity factor (k') is predicted to be proportional to the concentration of the counterion (Snyder and Kirkland 1979). In part, the power of IPC is related to the ability to ionize a specific solute molecule within the sample and to selectively partition these paired ions into the mobile phase.

Ion-pair chromatography is most often applied to polar solutes such as bases, ions, weak acids, and other compounds with multiple ionizable groups. Aqueous samples including physiological fluids and closely related isomers generally have been separated using reverse phase IPC.

PROPERTIES AND CHARACTERISTICS OF SOLVENTS

Successful modern LC separations depend on matching the right mobile phase to a given column and sample. Since many different properties of the solvent must be considered, a rational approach to solvent selection requires a classification of the various solvents according to their properties. The obvious place to begin is by rejecting solvents with physical properties that are incompatible with a particular sample, chromatographic technique, or method of detection. From the remaining solvents we select a solvent or solvents which provide the correct k', α, polarity, and a selectivity characteristics. Snyder and Kirkland (1979) have listed the solvents and their physical properties that are in relatively wide use for modern LC.

Intermolecular reactions between the sample and the mobile phase are critical in obtaining separation and resolution of sample constituents and are dependent on four major interactive forces: dispersion, dipole, hydrogen bonding, and dielectric interaction. The ability of the sample and solute molecules to interact in all four ways is termed the polarity, and polar solvents dissolve polar solutes. Thus, solvent strength increases with polarity in normal phase partition and adsorption chromatography, whereas solvent strength decreases with increasing polarity in reverse phase chromatography. Solvent polarity is best measured by the parameter P', which is based on experimental solubility data reported by Rohrschneider (1973). Because k' is determined by solubility of the sample molecules in the mobile and stationary phases, the P' value will give a good estimate of solvent polarity for partition and adsorption chromatography and for reverse phase systems where ϵ^0 is the dielectric constant.

The P' for a solvent mixture is the arithmetic average of the P' values of the pure solvents weighted according to the volume fraction of each solvent

$$P' = \phi_a P_a + \phi_b P_b + \cdots + \phi_n P_n, \quad (10)$$

where ϕ is the volume fraction of each solvent and P is the polarity of each solvent in the mixture. The final parameter that can be used to effect a change in the resolution of overlapping peaks is to change the selectivity of the mobile phase. This parameter is best described by the interactive forces between solvent and solute molecules. Replacement of a hydrogen donor co-solvent with a co-solvent which has a large dipole moment, dielectric constant, or hydrogen acceptor is likely to cause a change in the positioning of the bands. A systematic approach to solvent selectivity is the use of the selectivity triangle for solvents reported by Snyder (1974, 1978).

SELECTED CHROMATOGRAPHIC TECHNIQUES FOR QUANTITATION OF MICRONUTRIENTS

High performance liquid chromatography (HPLC) is a rapidly expanding technique in food analysis. This method has shown great promise for the quantitation of water- and fat-soluble vitamins in foods including isomeric forms of vitamins. The use of HPLC offers several advantages over conventional chromatographic methods for vitamins in foods including rapid separation, excellent resolution, quantitative recovery, and elimination of interferences by artifacts and degradation products. High performance liquid chromatography can be applied to the separation of vitamins using a wide range of physical and chemical parameters. The most notable advantages of HPLC over other analytical techniques are the ability to separate and quantitate isomeric forms of a nutrient and the ability to quantitate nutrients at very low concentrations (picomolar). Modern HPLC methods for analysis of vitamins in foods center on the use of normal and reserve phase chromatography.

Normal Phase Adsorption HPLC

Various HPLC methods have been reported for the separation and quantitation of fat-soluble vitamins using both reverse phase and normal phase HPLC. However, normal phase adsorption HPLC has become the method of choice for fat-soluble vitamins in foods since silica

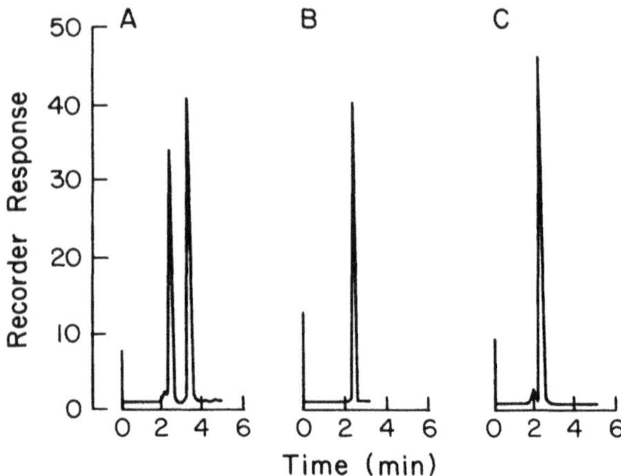

FIG. 16.2. Chromatograms of whole wheat-based commercial cereal extract on μPorasil normal phase column; hexane-$CHCl_3$ (85 + 15) at 1.5 mL/min. (A) Absorbance at 280 nm (first peak, retinyl palmitate absorbance 0.028; second peak, tocopheryl acetate absorbance 0.048). (B) Absorbance at 313 nm (retinyl palmitate absorbance 0.088). (C) Fluorescence of retinyl palmitate (excitation 360 nm, emission 415 nm).
Reprinted from Widicus and Kirk (1979).

columns can withstand the relatively high triglyceride load of the sample extract as compared to reverse phase columns. Widicus and Kirk (1979) reported a rapid method for the simultaneous determination of vitamins A and E in fortified cereal products using normal phase HPLC which permits direct injection of the extracted cereal lipids onto the analytical column without sample cleanup. A μ Porasil column and a hexane:chloroform (85 + 15) isocratic–mobile phase pumped at a flow rate of 1.5 mL/min is used to separate retinyl palmitate from tocopheryl acetate with elution times of 2.46 and 3.40 min, respectively. Tocopheryl acetate is monitored at 280 nm, while retinyl palmitate is monitored at 280 and 313 nm, and by fluorescence (360 nm ex, 415 nm em) as shown by data in Fig. 16.2. No other compounds which correspond to the characteristic retention times of α-tocopheryl acetate or retinyl palmitate are shown to be present or interfere with the quantitation of these two vitamins. The reported average recovery for tocopheryl acetate is 94.9% with a standard deviation of 4.10 and 99.2% for retinyl palmitate with a standard deviation of 4.28.

Recently, Mulry *et al.* (1983) reported a normal phase HPLC procedure for the separation and quantitation of the isomeric forms of vitamin A (all trans, 9-cis, and 13-cis) in foods (Fig. 16.3). The normal

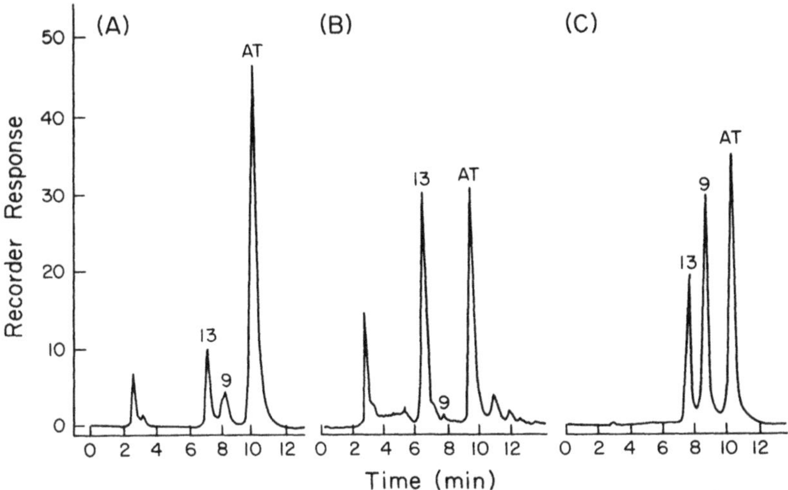

FIG. 16.3. Normal phase HPLC chromatograms from the determination of vitamin A isomers (cis-trans) in (A) ready-to-eat breakfast cereal, (B) UHT-processed milk, and (C) a test mixture of 9-cis, 13-cis and all-trans retinyl palmitate. Column = Supelcosil LS-Si 5 μm; mobile phase = 0.1% methyl tertiary-butyl ether in hexane containing 0.005 trimethylamine; flow rate = 0.9–2.0 mL/min; temperature is ambient.
Reprinted from Mulry et al. (1983).

phase adsorption HPLC procedure is carried out on Supelcosil LC-Si silica column protected by a guard column tap-filled with Pelliguard LC-Si (40 μm) packing. The mobile phase is a binary solvent system containing 0.09% methyl tertiary butyl ether in hexane containing 0.005% trimethylamine, which prevents on-column degradation and/or isomerization of retinyl palmitate due to the reaction between retinyl esters and acidic silica. Fluorescence detection is used to quantitate the retinyl isomers and gives a 100-fold increase in sensitivity over UV absorbance detection. The limit of detection (LOD) and the limit of quantitation (LOQ) for all-trans retinyl palmitate is 0.4 ng (0.7 picomoles) and 1.2 ng (2.3 picomoles), respectively. The LOQ for the 13-cis isomer is 2.4 ng and for the 9-cis isomer is 1.75 ng.

Recoveries of added retinyl palmitate are high for all food products analyzed, exceeding 96 + 2% in all cases. The coefficient of variance of the analysis ranged from 0.3% (ready-to-eat cereal) to 8.1% (margarine).

Reverse Phase HPLC

The principal chromatographic modes of HPLC separation of water soluble micronutrients is reverse phase chromatography for the separation of ionogenic water-soluble compounds and is based on the use of aqueous polar mobile phases with silica columns containing a bonded non polar phase (e.g., octadecylsilica). Detailed studies by Horvath *et al.* (1976, 1977) demonstrated that the retention and separation of water-soluble compounds by the bonded organic phase is the result of their reversible hydrophobic or "solvophobic" association with the nonpolar stationary phase. Retention is described as a function of the ionization and hydrocarbonaceous surface area of the sample components and the mobile phase pH, ionic strength, temperature, and organic modifier content. Shortly after these initial reports, work reported by Gregory and Kirk (1978) showed that the free base B_6 vitamers pyridoxal (PL), pyridoxamine (PM), and pyridoxine (PN) could be separated using a μBondapak C_{18} column and a 0.033 M potassium phosphate (pH 2.2) mobile phase. Under these conditions the pyridine ring nitrogen of the B_6 compounds is protonated, resulting in relatively short retention times and high efficiency. Elevation of mobile phase pH increases the retention time and causes a loss in resolution and efficiency as a result of the ionization of the phenolic hydroxyl group. Data in Fig. 16.4 show the separation of free base B_6 vitamers in a stored dehydrated model food system simulating a ready-to-eat breakfast cereal using UV detection at 280 nm.

A significant advance in the quantitation of micronutrients in foods is the use of pre- or post-column derivatization procedures to form compounds which can be more easily separated and detected. An example of this type of technique is the preinjection derivatization of B_6 vitamers to semicarbazone derivatives and the use of reverse phase chromatography and fluorescence detection reported by Gregory (1980). As shown by data in Fig. 16.5, this methodology increases the retention of pyridoxal and pyridoxal-5-phosphate and enhances the fluorescence response thereby permitting their resolution from interfering compounds and increasing the sensibility of the procedure. This procedure requires a LiChrosorb RP-8 reverse phase BPC column as the stationary phase and a mobile phase of 0.033 M potassium phosphate–2.5% (v/v) acetonitrile, pH 2.2. Mobile phase flow rate is 1.3 mL/min. A modification of the semicarbazone method has been used by Gregory *et al.* (1981) to quantitate B_6 vitamers in rat bioassays; the approximate recovery values for PLP in animal tissues ranged from 75 to 100% with coefficients of variation of approximately 5% (Fig. 16.6). This is a favorable alternative to the enzymatic methods for tissue pyridoxal phosphate (PLP) because of its technical simplicity, specificity and it is a direct physicochemical analysis.

Reverse phase HPLC techniques also have been developed which

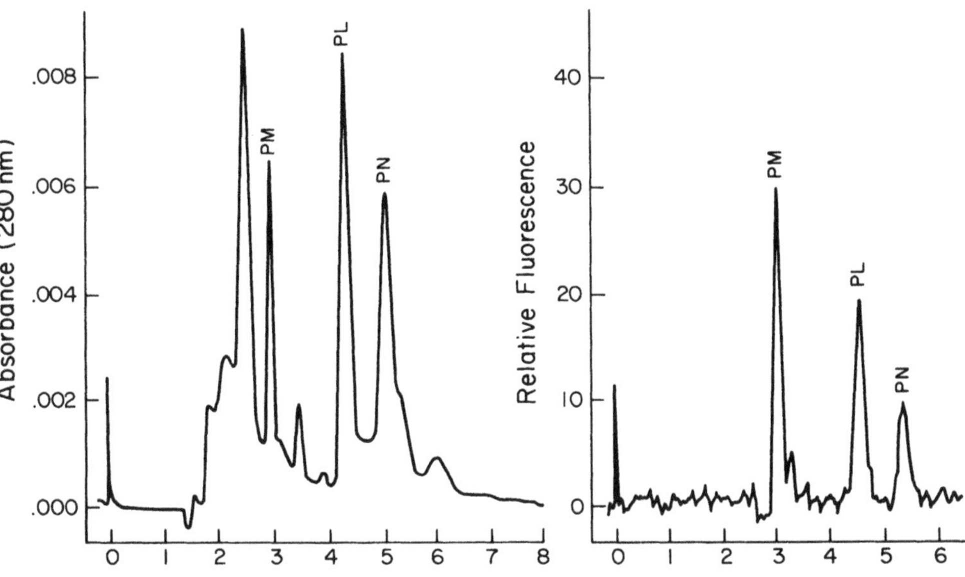

FIG. 16.4. Reverse phase HPLC chromatograms from the determination of vitamin B_6 in a stored dehydrated model system simulating breakfast cereals (storage at 37°C, 97 days, 0.6 water activity). Model system samples were spiked prior to analysis to provide 0.75 μg of each free base vitamer per milliliter of final injection solution. Left chromatogram, 280-nm absorption detection; right chromatogram, fluorometric detection. Column = μBondapak C_{18}; mobile phase = 0.033 *M* potassium phosphate, pH 2.2; flow rate = 2.0 mL/min; inlet pressure = 1800 psi; temperature is ambient.
Reprinted from Gregory and Kirk (1978).

permit the simultaneous determination of multiple water-soluble vitamins. Fellman *et al.* (1982) have reported a C_{18} reverse phase separation for the simultaneous quantitation of low levels of thiamin and riboflavin in selected foods from a single extract following oxidation and sample preconcentration. Following normal acid hydrolysis, treatment with taka-diastase, and pH adjustment, the sample extracts are filtered and the thiamin converted to its fluorescent derivative, thiochrome, with potassium ferricyanide. It is necessary to concentrate the two vitamins in order to meet detection limits, and this is accomplished by passing the oxidized sample extracts through an activated C_{18} Sep-Pak cartridge (Waters Assoc., Milford, MA). Interfering substances are removed by washing with 0.01 *M* phosphate buffer, pH 7.0, and 5% methanol–95% 0.01 *M* phosphate buffer. The vitamins are then eluted from the C_{18} column with 50% aqueous

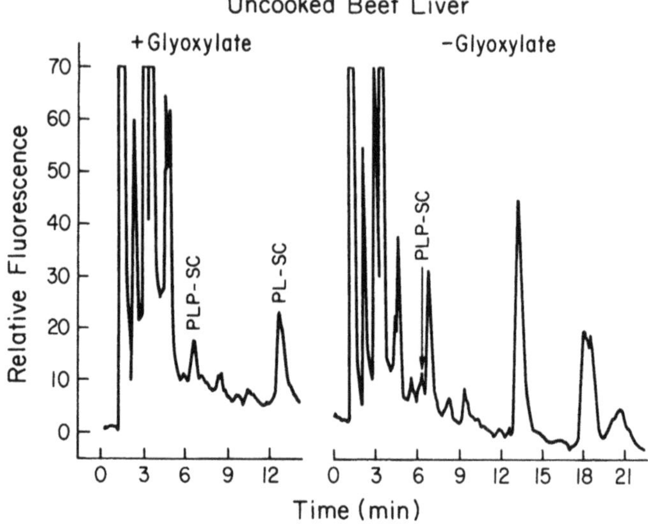

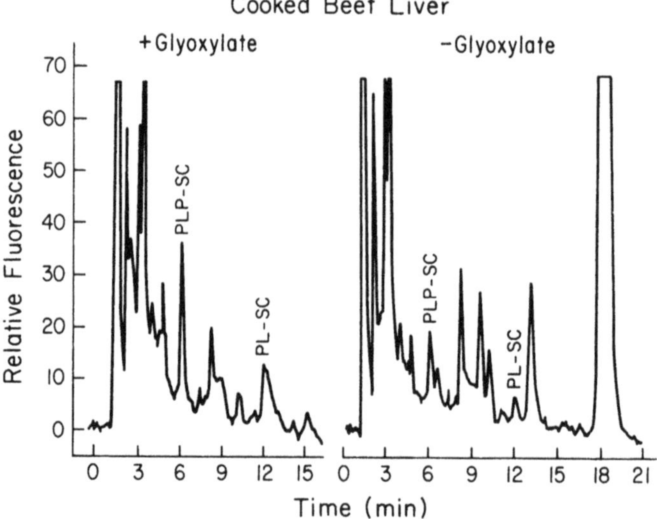

FIG. 16.5. Reverse phase HPLC chromatograms from the determination of semicarbazone derivatives of pyridoxal and pyridoxal phosphate in uncooked and cooked beef liver. Column = Ultrasphere IP 5 μm; mobile phase = 0.033 *M* potassium phosphate, pH 2.2 containing 2.5% (v/v) acetonitrile; flow rate = 2.0 mL/min; temperature is ambient.
Reprinted from Gregory (1980).

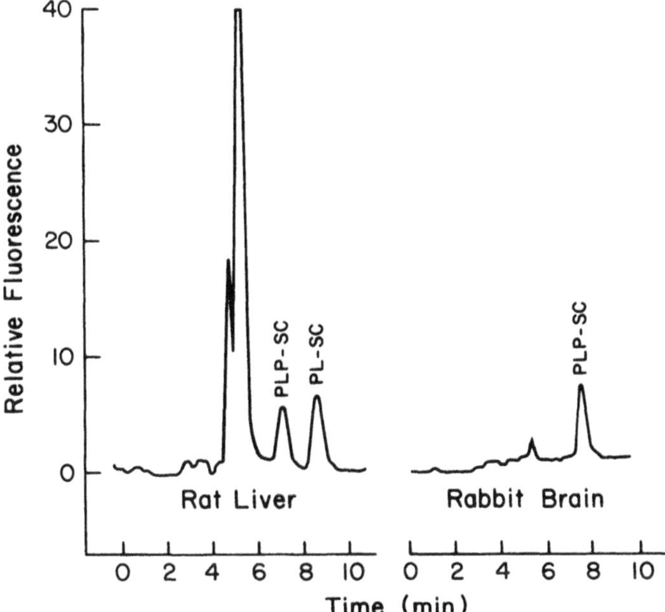

FIG. 16.6. Reverse phase HPLC chromatograms from the determination of PLP and PL in rat liver and rabbit brain as their respective semicarbazone derivatives (PLP-SC and PL-SC). Column = LiChrosorb RP-8; mobile phase = 0.033 *M* potassium phosphate, 2.5% (v/v) acetonitrile, pH 2.2; flow rate = 1.3 mL/min; inlet pressure = 600 psi; temperature is ambient; injection volume = 50 μL; fluorometric detection. Rat liver contained 3.6 μg PLP/g and 3.4 μg PL/g; rabbit brain contained 1.4 μg PLP/g.
Reprinted from Gregory et al. (1981).

methanol. Sample extracts are separated on a Waters 10 cm × 18 mm, 10 μm C_8-Radial PAK column isocratically with 37% methanol/0.01 *M* phosphate buffer, pH 7.0, as shown by data in Fig. 16.7. Detection of thiochrome and riboflavin is accomplished by fluorescence at 360 nm (ex) and 415 nm (em).

A unique feature of this procedure is the use of the C_{18} Sep Pak which provides a rapid selective concentration of thiamin and riboflavin and permits their quantitative determination in foods that have relatively low levels of these vitamins. Second, sample extracts remain stable for approximately 16 hr at ambient temperature if pro-

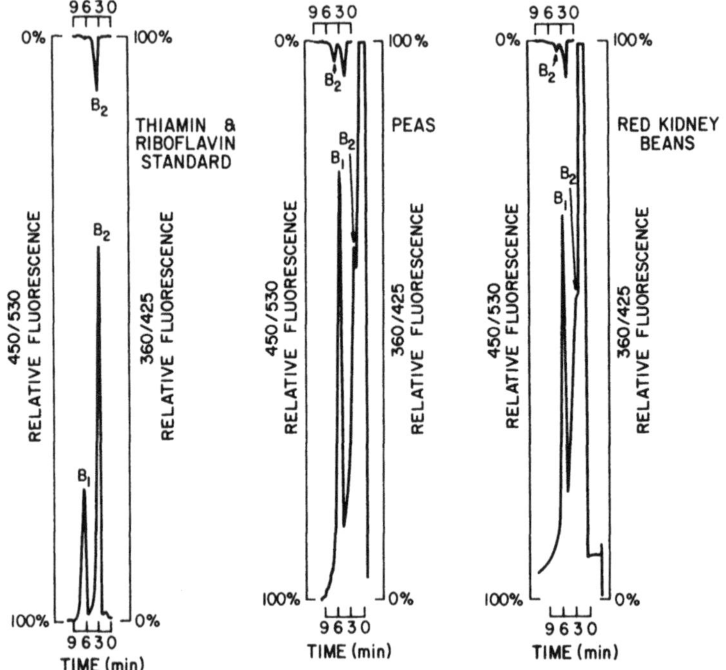

FIG. 16.7. HPLC chromatograms of thiamin and riboflavin in a standard solution and selected food products. Column = C_8 Radial PAK 10 μm; mobile phase = 37% methanol/0.1 *M* phosphate buffer pH 7.0; flow rate 3.0 mL/min at ambient temperature.
Reprinted from Fellman et al. (1982).

tected from light. This allows an automated sample injection system to be used with this assay which greatly increases the number of assays that can be performed per day.

The use of ion-pair reverse phase chromatography is an HPLC method that is becoming popular for the separation and quantitation of many nutrients. Recently, Gregory and Feldstein (1983) reported an ion-pair reverse phase HPLC method employing gradient elution for the determination of B_6 vitamers in biological materials (Fig. 16.8). This method eliminates the need to convert PM and PMP to PL and PLP required for quantitation of B_6 by the semicarbazone technique (Gregory *et al.* 1981). The extraction and preparative cleanup procedures for the sample extract are as reported by Vanderslice *et al.* (1980), except that 4-deoxypyridoxine was used as the internal standard. The B_6 vitamers are extracted from the sample using 5% sulfosalicylic acid

(SSA), which must be removed from the concentrated sample extract using Dowex or Bio Rad AG 2-X8 anion-exchange column because of its fluorescent properties.

The general analytical HPLC method for the ion-pairing and stepwise elution of the B_6 vitamers is adopted from a procedure reported by Tryfiates and Sattsangi (1982). Fifty to 100 μL of sample eluate from the anion-exchange column are injected onto an Ultrasphere IP column (Altex) which has been equilibrated with 0.033 *M* potassium phosphate, pH 2.2, with 4 m*M* C_7 sulfonate and 4 m*M* C_8 sulfonate, 2.5% (v/v) isopropanol (Solvent A). A flow program is initiated with injection (12-min exponential program from 0.5 to 1.0 mL/min), and 3 min after injection the mobile phase is switched to 0.033 *M* potassium phosphate, pH 2.2, with 17.5% isopropanol (Solvent B). This change in buffer polarity is required to affect a change in equilibrium distribution of ion pairs into the mobile phase. Figures 16.8 and 16.9 are representative chromatograms of standards and samples separated and quantitated using this ion-pair reverse phase method.

As with all gradient elution methods reequilibration of the column must be performed before injection of the next sample. Using this pro-

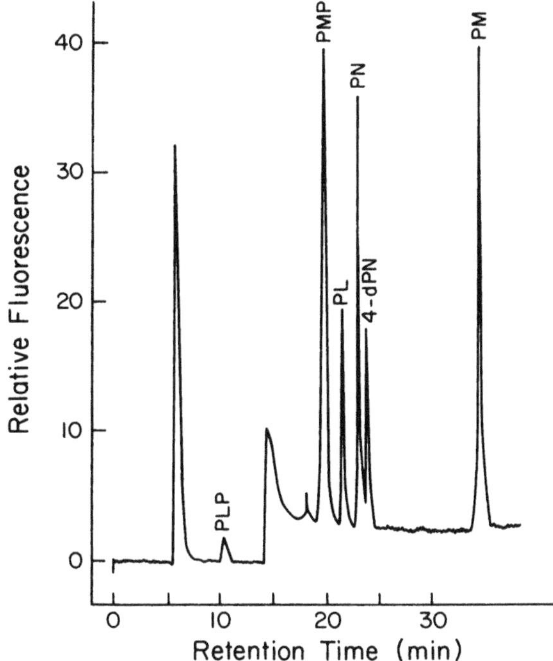

FIG. 16.8. Ion-pair reverse phase HPLC chromatogram showing the separation of B_6 vitamers and their 5-phosphate analogs from a known standard solution (cf. Fig. 16.9 for chromatographic details).
From Gregory and Feldstein (1983).

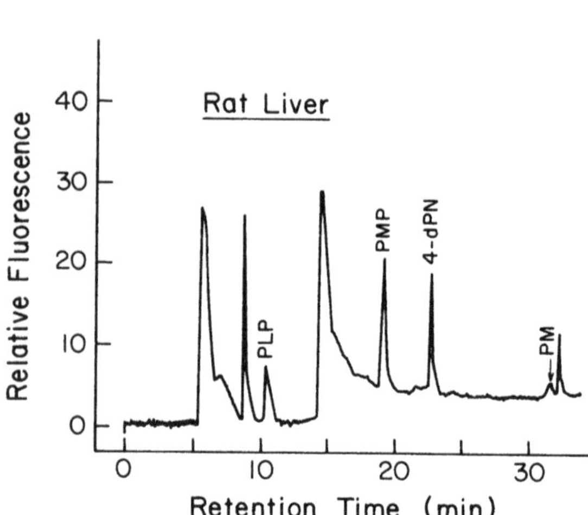

FIG. 16.9. Ion-pair reverse phase HPLC chromatogram showing the separation of B_6 vitamers in rat liver following extraction with 5% sulfosalicylic acid. Column = Ultrasphere IP; mobile phases for Solvent A—0.033 *M* potassium phosphate, pH 2.2, with 4 m*M* C_7 sulfonate and 4 m*M* C_8 sulfonate, 2.5% (v/v) isopropanol; for Solvent B—0.033 *M* potassium phosphate, pH 2.2, 17.5% isopropanol; flow rate = programmed 0.5–1.0 mL/min; temperature is ambient.
From Gregory and Feldstein (1983).

cedure a 10–15-min reequilibration period with solvent A is required, which is one of the drawbacks to using gradient elution techniques.

COMPUTERS AND HPLC

High performance liquid chromatography (HPLC) is very amenable to automation and the interface with computers. Such techniques can greatly simplify and expand the capabilities and use of HPLC as an analytical tool. The use of computers will not only allow the analyst to automate the chromatographic process and program solvent mixing, but to predict the isocratic mobile phase for optimum separation in the minimum time, the resolution between the pair of elution peaks and the retention time for each component, the order of elution of each peak, and the purity of the peaks (Issaq 1983). Numerous systematic approaches to the selection of an isocratic mobile phase for optimum separation have been reported (Issaq *et al.* 1981; Glajch *et al.* 1980; Sachok *et al.* 1980; Belinky 1980; Lindberg *et al.* 1981). These methods also alleviate one of the most difficult and time-consuming aspects of searching for a mobile phase, that is, establishing the identity of the eluted peaks. The computer programs take into consideration peak reversal, peak coalescence, and peak splitting. Thus, with the aid of a computer many complex problems in chromatographic analysis can be easily solved saving both time and money.

BIBLIOGRAPHY

BELINKY, B.R. 1980. Improved resolution in high performance liquid chromatography analysis of polynuclear aromatic hydrocarbons using ternary solvent systems. ACS Symp. Ser. *120,* 149–168.

FELLMAN, J.K., ARTZ, W.E., TASSINARI, P.D., COLE, C.L., and AUGUSTIN, J. 1982. Simultaneous determination of thiamin and riboflavin in selected foods by high performance liquid chromatography. J. Food Sci. *47,* 2048–2050, 2067.

GLAJCH, J.L., KIRKLAND, J.J., SQUIRE, K.M. and MINOR, J.M. 1980. Optimization of solvent strength and selectivity for reversed-phase liquid chromatography using an interactive mixture-design statistical technique. J. Chromatogr. *199,* 57–59.

GREGORY, J.F. 1980. Determination of pyridoxal 5′-phosphate as the semicarbazone derivative using high performance liquid chromatography. Anal. Biochem. *102,* 374–379.

GREGORY, J.F., AND FELDSTEIN, D. 1983. Paired-ion reverse-phase HPLC method for determination of B-6 vitamers in biological materials. Unpublished data.

GREGORY, J.F., and KIRK, J.R. 1978. Assessment of storage effects on Vitamin B_6 stability and bioavailability in dehydrated food systems. J. Food Sci. *43,* 1801–1808, 1815.

GREGORY, J.F., and KIRK, J.R. 1981. Determination of vitamin B_6 compounds by semiautomated continuous-flow and chromatographic methods. *In* Methods in Vitamin B_6 Nutrition, p. 149. Plenum Publishing, New York.

GREGORY, J.F., MANLEY, D.B., and KIRK, J.R. 1981. Determination of vitamin B_6 in animal tissues by reverse-phase high performance liquid chromatography. J. Agric. Food Chem. *29,* 921–927.

HAMILTON, R.J., and SEWELL, P.A. 1977. Introduction to High Performance Liquid Chromatography. John Wiley & Sons, New York.

HORVÁTH, C., MELANDER, W., and MOLNÁR, I. 1976. Solvophobic interactions in liquid chromatography with nonpolar stationary phases. J. Chromatogr. *125,* 129–156.

HORVÁTH, C., MELANDER, W., and MOLNÁR, I. 1977. Liquid chromatography of ionogenic substances with polar stationary phases. Anal. Chem. *49,* 142–154.

ISSAQ, H.J. 1983. Computer-assisted HPLC. Am. Lab. *15,* 41–46.

ISSAQ, H.J., KLOSE, J.R., McNITT, K.L., HAKY, J.E., and MUSCHIK, G.M. 1981. A systematic statistical method of solvent selection for optimal separation in liquid chromatography. J. Liq. Chromatogr. *4,* 2091–2120.

LINDBERG, W., JOHANSSON, E., and JOHANSSON, K. 1981. Application of statistical optimization methods to the separation of morphine, codeine, noscapine and papaverine in reversed-phase ion-pair chromatography. J. Chromatogr. *211,* 201–212.

MULRY, M.C. 1983. Retinyl palmitate isomerization: Analysis and quantitation in fortified foods and model systems. M.S. thesis, Univ. of Florida, Gainesville.

MULRY, M.C., SCHMIDT, R.H., and KIRK, J.R. 1983. Isomerization of retinyl palmitate using conventional lipid extraction solvents (in press).

ROHRSCHNEIDER, L. 1973. Solvent characterization by gas-liquid partition coefficients of selected solutes. Anal. Chem. *45,* 1241–1247.

SACHOK, B., KONG, R.C., and DEMING, S.N. 1980. Multifactor optimization of reversed-phase liquid chromatographic separations. J. Chromatogr. *199,* 317–325.
SNYDER, L.R. 1974. Classification of the solvent properties of common liquids. J. Chromatogr. *92,* 223–230.
SNYDER, L.R. 1978. Classification of the solvent properties of common liquids. J. Chromatogr. Sci. *16,* 223–234.
SYNDER, L.R., and KIRKLAND, J.J. 1979. Introduction to Modern Liquid Chromatography, 2nd ed. Wiley Interscience, New York.
TRYFIATES, G.P. and SATTSANGI, S. 1982. Separation of vitamin B_6 compounds by paired-ion high performance liquid chromatography. J. Chromatogr. *227,* 181–186.
VANDERSLICE, J.T., MAIRE, C.D., DOHERTY, R.F., and BEECHER, G.R. 1980. Sulfosalicylic acid as an extraction agent for vitamin B_6 in food. J. Agric. Food Chem. *28,* 1145–1149.
WIDICUS, W.A., and KIRK, J.R. 1979. High pressure liquid chromatographic determination of vitamins A and E in cereal products. J. Assoc. Off. Anal. Chem. *62,* 637–641.
YOST, R.W., ETTRE, L.S., and CONLON, R.D. 1980. Practical Liquid Chromatography—An Introduction. Perkin-Elmer, Norwalk, CT.

Index

N

T

www.ingramcontent.com/pod-product-compliance
Ingram Content Group UK Ltd.
Pitfield, Milton Keynes, MK11 3LW, UK
UKHW041831200726
13854UKWH00002BA/993
* 9 7 8 9 4 0 1 1 7 3 8 0 3 *